Topics in Combinatorics and Graph Theory

R. Rama

Topics in Combinatorics and Graph Theory

Ane Books Pvt. Ltd.

Springer

R. Rama
Department of Mathematics
Indian Institute of Technology Madras
Chennai, Tamil Nadu, India

ISBN 978-3-031-74254-5 ISBN 978-3-031-74252-1 (eBook)
https://doi.org/10.1007/978-3-031-74252-1

Jointly published with Ane Books Pvt. Ltd.
In addition to this printed edition, there is a local printed edition of this work available via Ane Books in South Asia (India, Pakistan, Sri Lanka, Bangladesh, Nepal and Bhutan) and Africa (all countries in the African subcontinent).
ISBN of the Co-Publisher's edition: 978-81-19662-08-1

This Springer imprint is published by the registered company Springer Nature Switzerland AG
The registered company address is: Gewerbestrasse 11, 6330 Cham, Switzerland

Preface

This book is written to cover some aspects of combinatorics and graph theory, the important topics of Discrete Mathematics. The book covers substantial amount of basic on both the above said topics. Each chapter of combinatorics covers all the introductory aspects with adequate examples. Some chapters also give some expanded topics of combinatorics. First eleven chapters cover all aspects of combinatorics. Chapter 12 covers introduction to discrete probability. Each chapter is providedwith limited number of workable exercise problems. Apart from the worked out examples in each chapter, there is a consolidated number of problems with solutions on this topic 'combinatorics' provided at the end of Chap. 12.

Chapters 13–23 are on topics of graph theory. All the basics required for reading the content are provided. There is a chapter on 'Independent sets, covering and Matchings' which gives the perfect matching theorem and domination. The vertex coloring and edge coloring are given in Chap. 20 highlighting the bounds on vertex coloring, coloring planar graphs, and nice aspects of the 4-color theorem. The general form of Ramsey Number is given with its understanding the usual form of Ramsey Number $R(s, t)$. The bounds computed for some Ramsey Number are stated and proved. There is a chapter on introduction to spectral graph theory. There is a set of worked out examples on graph theory at the end of Chap. 23.

Chennai, India R. Rama

Acknowledgements This book is written on two important topics of Discrete Mathematics. They are combinatorics and graph theory. I wrote the book during my sabbatical (August 2022–February 2023) at IIT Madras. I thank the institute for its support during this period. I also thank the department for the support. I acknowledge the institutes recognition to me as 'Girija Vaidyanathan Chair Professor'. I immensely thank Mr. E. Boopal and my Ph.D. student Mr. Sivasankar for completely helping in editing the book to its present form.

Contents

Part I
Combinatorics

1 Basics of Counting

1.1 Basic Operations on Sets

This chapter introduces the basic tools required for the development of discrete mathematics. The major concepts to be developed are sets, relations and functions. There is no branch of mathematics that can be studied without the above said topics.

Set theory was developed by two key historical mathematicians Georg Cantor (1845–1918) and Richard Dedekind (1831–1916). Cantor realized the significance of 1-1 functions between sets and introduced the notion of cardinality of a set. The German word for a set is Menge, which is the reason Cantor denotes a set by M and its elements by m. Cantor proved that the set of all real numbers is uncountable while the set of all algebraic reals is countable. In 1878 he gave the first formulation of the celebrated continuum hypothesis.

Dedekind is mostly known for his research in algebra and set theory. He was the first to define real numbers by means of cuts of rational numbers. The key concept "ideal" in modern algebra was first introduced by Dedekind. His contributions to set theory as well as the study of natural numbers and modular lattices are remarkable.

Sets: A set M is a collection of objects of any kind. We write $m \in M$ to indicate that m is an element of M. Examples of sets can be found everywhere around us. For example the set of all students in an algebra class, the set of all faculty in an educational institution, the set of all prime numbers.

If a is not an element of a set M then we write $a \notin M$.

Notations: Sets are usually denoted by capital letters $X, Y, Z, \ldots$.

If X is a set and $a, b, c, \ldots$ are its elements then X is written as

$$X = \{a, b, c, \ldots\}.$$

R. Rama, *Topics in Combinatorics and Graph Theory*,
https://doi.org/10.1007/978-3-031-74252-1_1

If X is a set such that it contains elements that satisfy a particular property P then X is written as

$$X = \{x : P(x) \text{ is ture}\}.$$

Examples 1.1 1. $\mathbb{N} = \{1, 2, 3, \ldots\}$ = the set of all natural numbers.
2. $\mathbb{Z} = \{0, \pm 1, \pm 2, \ldots\}$ = the set of all integers.
3. $X = \{x \in \mathbb{N} : x \text{ is prime}\}$ = the set of all prime numbers.

Definitions 1.2 1. Two sets X and Y are said to be equal iff every element of X is an element of Y and vice versa written as $X = Y$.
2. Given two sets X and Y, X is said to be included in Y or X is a subset of Y if every element of X is an element of Y and is written as $X \subseteq Y$.
3. Given two sets X and Y, X is said to be properly included in Y or X is a proper subset of Y if $X \subseteq Y$ and $X \neq Y$.
4. The set which has no element is called an empty set and is denoted by $\emptyset$.
5. For any set X, the set of all subsets of X is called the power set of X and is denoted as $\mathcal{P}(X) = \{Y : Y \subseteq X\}$.

Operations on Sets: We describe methods for generating new sets from existing sets by introducing basic operations on sets.

Union of Sets: Given two sets X and Y, the union of the sets X and Y denoted by $X \cup Y$ is defined as

$$X \cup Y = \{x/x \in X \text{ or } x \in Y\}$$

Intersection of Sets: Given two sets X and Y, the intersection of X and Y denoted by $X \cap Y$ is defined as

$$X \cap Y = \{x/x \in X \text{ and } x \in Y\}$$

Disjoint Sets: Two sets X and Y are said to be disjoint iff $X \cap Y = \emptyset$ (i.e., they have no common elements).

Complement of a Set:

1. Given a set X, the absolute complement of X denoted by $\overline{X}$ is defined as $\overline{X} = \{x/x \notin X\}$.
2. The complement of X with respect to another set A denoted by $A \cap \overline{X}$ is defined as

$$A \cap \overline{X} = \{x/x \in A \text{ and } x \notin X\}.$$

   The set $A \cap \overline{X}$ is also the difference of sets A and X and can also be denoted as $A \backslash X$.

Symmetric Difference: Given any two sets X and Y, the symmetric difference of sets X and Y denoted by $X \oplus Y$ is defined as

$$X \oplus Y = (X \backslash Y) \cup (Y \backslash X)$$

If all sets under consideration in a certain discusion are subsets of a set U, then U is called the universal set.

Examples 1.3 Let $X = \{1, 2, 3\}$, $Y = \{3, 4, a, b\}$. Then

1. $X \cup Y = \{1, 2, 3, 4, a, b\}$
2. $X \cap Y = \{3\}$
3. $\overline{X} = \{n/n \neq 1, 2, 3\}$
4. $X \backslash Y = \{1, 2\}$.

Representation of a Set Pictorially: Sets can be represented pictorially by means of diagrams called as Venn diagrams. The universal set 'U' is usually represented by a rectangle and any subset A of a universal set is represented by a circle inside the universal set (see Fig. 1.1).

Two proper subsets A, B of U such that A and B are disjoint i.e., $A \cap B = \emptyset$ is represented in Fig. 1.2.

Fig. 1.1 .

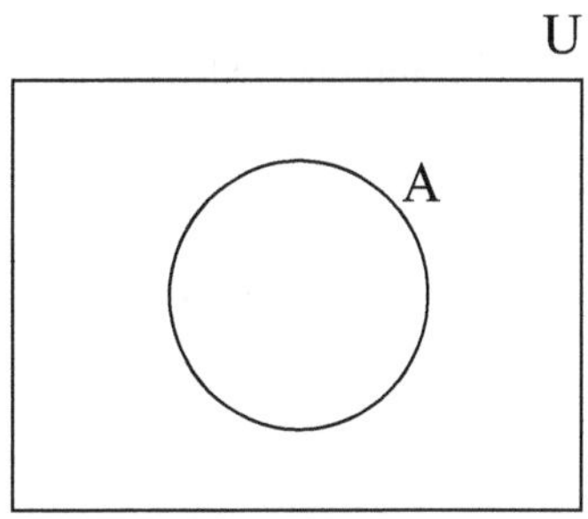

Fig. 1.2 .

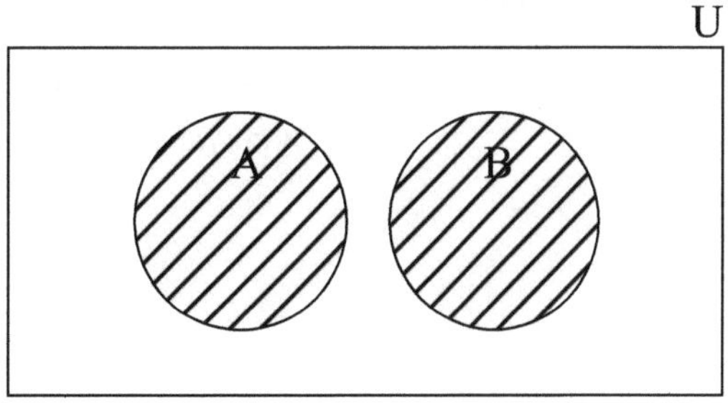

Union of two sets A and B can be represented as

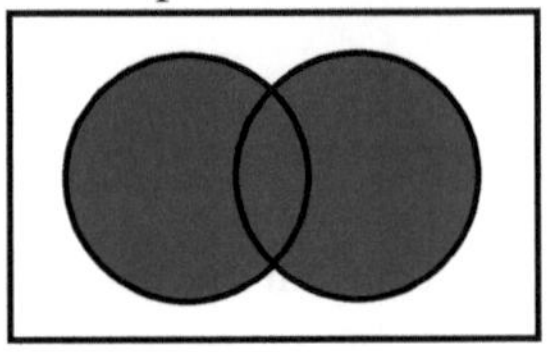

where $A \cap B \neq \emptyset$

Intersection of two sets A and B is represented as

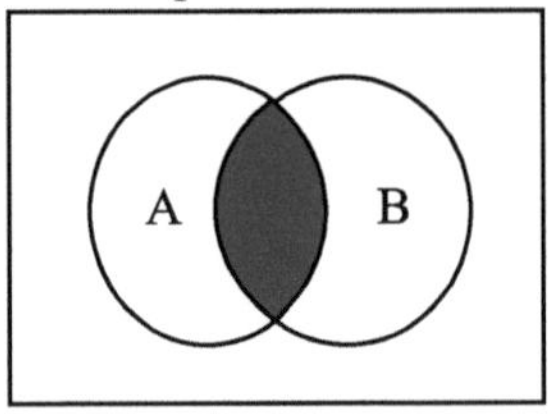

Complement of a set A with respect to the universal set is given as

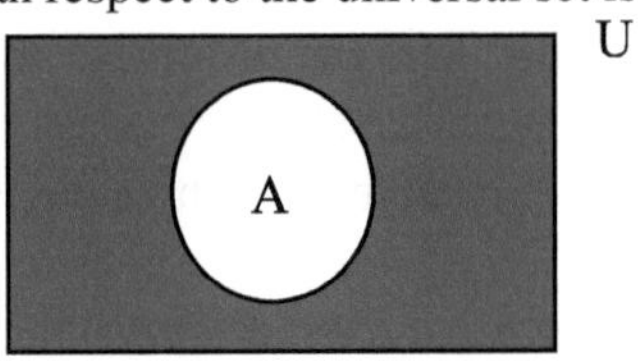

Laws of Set Theory: For any sets A, B, C subsets of a universal set u, we have the following rules:

(i) Commutative Laws:

$$A \cup B = B \cup A, A \cap B = B \cap A.$$

(ii) Associative Laws:

$$(A \cup B) \cup C = A \cup (B \cup C)$$
$$(A \cap B) \cap C = A \cap (B \cap C)$$

(iii) Distributive Laws:

$$A \cup (B \cap C) = (A \cup B) \cap (A \cup C)$$
$$A \cap (B \cup C) = (A \cap B) \cup (A \cap C)$$

(iv) Identity Laws:

$$A \cup \emptyset = A \quad A \cap u = A$$
$$A \cap U = A \quad A \cap \emptyset = \emptyset$$

(v) Inverse Laws:

$$A \cup \overline{A} = u \quad \overline{u} = \emptyset$$
$$A \cap \overline{A} = \emptyset \quad \overline{\emptyset} = u$$

(vi) Double Negation:

$$\overline{\overline{A}} = A.$$

(vii) Idempotent Laws:

$$A \cup A = A$$
$$A \cap A = A$$

(viii) Demorgan's Laws:

$$\overline{A \cup B} = \overline{A} \cap \overline{B}$$
$$\overline{A \cap B} = \overline{A} \cup \overline{B}$$

(ix) Domination Laws:

$$A \cup U = U$$
$$A \cap \emptyset = \emptyset$$

(x) Absorption Laws:

$$A \cup (A \cap B) = A$$
$$A \cap (A \cup B) = A$$

One can easily observe that the power set of the universal set $P(u)$ is closed under union, intersection and complement.

1.2 Relations and Functions

In this section we define the concept of relation within the framework of set theory. A (binary) relation is used in connection with ordered pair of objects. In a set $\{x, y\}$ the order of the elements is irrelevant in general. If the order of the elements is relevant then the idea of an ordered pair is considered and is represented as $\{\{x\}, \{x, y\}\}$ where x is the first element and y follows x in the order. Thus the concept of ordered pair is defined as follows.

An ordered pair of x and y, denoted by (x, y) is the set $(x, y) := \{\{x\}, \{x, y\}\}$. We call x and y the first and second components of the ordered pair (x, y), respectively. In case $x = y$ we have $(x, x) = \{\{x\}\}$.

A relation is a set of ordered pairs.

Cartesian Product of Sets: Let X and Y be sets. The cartesian product of X and Y is defined by

$$X \times Y = \{(a, b)/a \in X, b \in Y\}.$$

i.e., the cartesian product $X \times Y$ is the set of all ordered pairs (a, b) with $a \in X$ and $b \in Y$.

Definitions 1.4 If X and Y are sets and if $\rho \subset X \times Y$, we say that ρ is a relation between X and Y.

If $(x, y) \in \rho$ then we write $x\rho y$.

Examples 1.5 Let X be a subset of real numbers. Define

$$\rho_1 = \{(x, y)/x, y \in X, y = x^2 + 2\}$$
$$\rho_2 = \{(x, y)/x, y \in X, x = y\}$$

Domain and Range of Relations: Let X and Y be two sets and let ρ be a relation between X and Y.

The domain of ρ is defined by

$$D(\rho) = \{x \in X/\exists\, y \in Y, (x, y) \in \rho\}$$

The range of ρ is defined by

$$R(\rho) = \{y \in Y/\exists\, x \in X, (x, y) \in \rho\}$$

A relation 'ρ' is simply a subset of $X \times Y$. For any $(a, b) \in \rho$, one can think of $a \in X$ being assigned to $b \in Y$. One should observe that it is possible to have both $(a, b), (a, c) \in \rho$ for any $a \in X$ and also there can be an $a \in X \ni (a, b) \notin \rho\ \forall\, b \in Y$.

Equivalence Relations: Let X be a set and ρ be a relation in X (i.e., $\rho \subset X \times X$).

1. ρ is called reflexive if $\forall\, x \in X$, $(x, x) \in \rho$.
2. ρ is called symmetric if $\forall\, (x, y) \in \rho$, $(y, x) \in \rho$.
3. ρ is called antisymmetric if $(x, y) \in \rho$ and $(y, x) \in \rho$ implies $x = y$.
4. ρ is called transitive if $(x, y) \in \rho$ and $(y, z) \in \rho$ implies $(x, z) \in \rho$.
5. ρ is an equivalence relation if it is reflexive, symmetric and transitive.
6. ρ is an order relation if it is reflexive, antisymmetric and transitive.

An order relation ρ is said to be a total order relation if for all $(x, y) \in X^2$ either $(x, y) \in \rho$ or $(y, x) \in \rho$. Otherwise ρ is said to be a partial order relation. A totally ordered set is also called a chain.

If ρ is an equivalence relation on the set X, then for each $x \in X$ the set $\{y \in X/(x, y) \in \rho\}$ is called the ρ-equivalence class of the element x.

Theorem 1.6 *Let X be any given non empty set and let ρ be an equivalence relation on X. Then the following are true:*

1. *The collection of distinct ρ-equivalence classes forms a partition of X.*
2. *If P is a partition of X and a relation ρ is defined by $x \rho y$ iff there exists A in P such that $x, y \in A$, then ρ is an equivalence relation.*

Proof(1) Given that ρ is an equivalence relation. Let $[x] = \{y \in X/(x, y) \in \rho\}$. Since ρ is reflexive, $x \in [x]$. Let $x, y \in X$ such that $x \neq y$ and $[x] \cap [y] \neq \emptyset$.
Then $\exists\, z \in [x] \cap [y]$ such that $(x, z) \in \rho$ and $(y, z) \in \rho$ thus $(x, y) \in \rho$ and $(y, z) \in \rho$ since ρ is an equivalence relation, which implies that $[x] = [y]$ which is a contradiction.
Therefore for any distinct $x, y \in X$, $[x] \cap [y] = \emptyset$.
If $z \in X$ then $z \in [z]$ thus $z \in \cup\{[x]/x \in X\} \subset X$. Therefore $X = \cup\{[x]/x \in X\}$.

(2) Given that P is a partition of X and ρ is a relation defined by $x \rho y$ iff there exists $A \in P$ such that $x, y \in A$. Let $z \in X$. By definition of partition P, $\exists\, A \in P$ such that $x \in A$. By definition of ρ, $x \rho x$, thus, ρ is reflexive.
Suppose that $x \rho y$. Then both x and y belong to some element A of P. Since both y and x belong to A, $y \rho x$, thus, ρ is symmetric.
Suppose that $x \rho y$ and $y \rho z$. Then both x and y belong to some element A of P and y and z belong to some element B of P. Since P is a partition of X, $A = B$. Thus $x \rho z$. Therefore ρ is transitive.

Remark 1.7 Let ρ be an equivalence relation on a finite set X. If each equivalence class has r elements, there are $\frac{|X|}{r}$ equivalence classes.

Examples 1.8 Let X = set of all natural numbers union $\{0\}$.Define

$$\rho = \{(x, y)/5x + r \equiv 5y + r\}$$

Then ρ is an equivalence relation. The equivalence classes are:

$$[0], [1], [2], [3], [4].$$

For example [4] = {4, 9, 14, 19, 24, ...}.

Functions: Let X and Y be given sets. By a function $f : X \to Y$ from X into Y we mean a rule that assigns a unique element $y = f(x) \in Y$ to every element $x \in X$. Note that a function is a special case of a relation. Usually X is called as the domain of the function f and the set $\{f(x) : x \in X\}$ is called as the range of f.

Examples 1.9 The relation $\rho = \{(a, 1), (b, 2), (c, 3), (d, 4)\}$ is a function from $X = \{a, b, c, d\}$ to $Y = \{1, 2, 3, 4\}$.

Examples 1.10 Let f be a function defined by the rule $f(x) = 2x^2 + 1$. The domain of the function is the set of all real numbers and the range is the set of all real numbers greater than or equal to '1'. The function can also be represented using (x, y)-co-ordinate system as Fig. 1.3.

Remark 1.11 From the above example it is clear that we can define graph of a function f whose domain and range are subsets of real numbers.

Definitions 1.12 Let $f : X \to Y$ and let $B \subset Y$. Then the inverse image of B denoted by $f^{-1}(B)$ is defined as $f^{-1}(B) = \{x \in X/f(x) \in B\}$.

If B = Range of f, then $f^{-1}(B)$ is called as the inverse of the function f.

Fig. 1.3 .

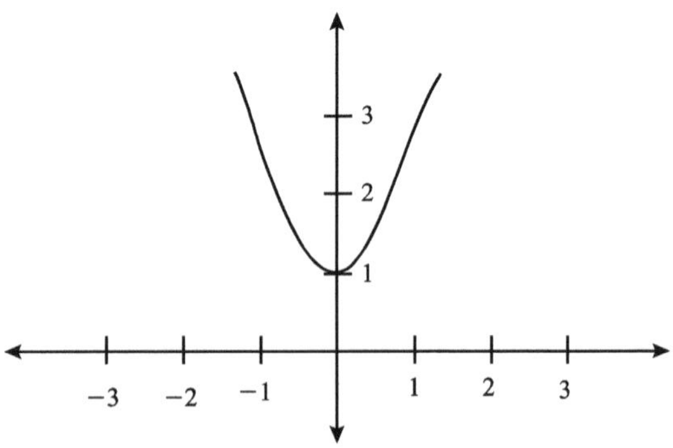

Definitions 1.13 1. If there exists a function $g : A \to Y$ such that $g(x) = f(x)$ where $f : X \to Y$ and $A \subset X$, then g is called as the restriction of f to A and is denoted by $f/A = g$.

2. In particular if B is a singleton set, say $B = \{b\}$, then $f^{-1}(B)$ is called the inverse image of the point b under f denoted by $f^{-1}(b)$.

Proposition 1.14 *For any two subsets A and B of the range Y of a function $f : X \to Y$, we have*

1. $f^{-1}(A \cup B) = f^{-1}(A) \cup f^{-1}(B)$
2. $f^{-1}(A \cap B) = f^{-1}(A) \cap f^{-1}(B)$
3. $f^{-1}(A \backslash B) = f^{-1}(A) \backslash f^{-1}(B)$.

Definitions 1.15 Let X, Y be sets and $f : X \to Y$ be a function. The function f is said to be injective if for all $x, y \in X$, $f(x) = f(y)$ implies $x = y$. f is also said to be one to one.

The function f is said to be surjective if $\forall\, y \in Y$ there exists an $x \in X$ such that $f(x) = y$. i.e., Range of $f = Y$. f is also said to be onto. The function f is said to be bijective if it is both injective and surjective.

Note that if there exists a function $g : Y \to X$ then also f is said to be bijective. Here g is the inverse of f. One can easily prove that f is injective if and only if for each $y \in$ Range of f, the set $f^{-1}(\{y\})$ has exactly one element.

Definitions 1.16 Let $g : X \to Z$ and $f : Z \to Y$ be functions. The composition of f and g denoted by $f \circ g$ is defined as $f \circ g : X \to Y$ such that $f \circ g(x) = f(g(x))$.

Examples 1.17 Let $f, g : \mathbb{R} \to \mathbb{R}$ such that

$$f(x) = x^2,\ g(y) = y + 1$$

Then $f \circ g : \mathbb{R} \to \mathbb{R}$ and $f \circ g(x) = (x + 1)^2$. Note that f is neither injective nor surjective but g is a bijective function.

Examples 1.18 Let $f : \mathbb{N} \to \mathbb{N} \cup \{0\}$ and $g : \mathbb{N} \cup \{0\} \to \mathbb{N}$,

$$f(n) = n - 1,\ g(n) = n + 1$$

then $f \circ g : \mathbb{N} \cup \{0\} \to \mathbb{N} \cup \{0\}$ such that

$$f \circ g(x) = f(g(x)) = n^x.$$

Note that f and g are both one to one and onto. Thus $f \circ g$ is also one to one and onto.

One can show in general that the composition of surjective functions is surjective and the composition of injective functions is injective. As a partial converse to this statement, we have the following result.

Proposition 1.19 *Let $f : Z \to Y$, $g : X \to Z$ and let $f \circ g : X \to Y$, then the following hold true:*

1. *If $f \circ g$ is surjective then so is f.*
2. *If $f \circ g$ is injective then so is g.*

We conclude the section with some examples.

Examples 1.20 Let X be a given set and let A be an arbitrary subset of X.
Define a function $\chi_A : X \to \mathbb{R}$ as

$$\chi_A(x) = \begin{cases} 1 & \text{if } x \in A \\ 0 & \text{otherwise} \end{cases}$$

The function χ_A is called the characteristic function of the subset A in X.

Examples 1.21 Define a function $f : X \to \mathbb{R}$ such that $f(x) = c$ where c is a constant and $c \in \mathbb{R}$. Then f is said to be a constant function.

Finite and Infinite Sets

In this section we define the concept of finite and infinite sets. We observe that given a finite set, the number of elements the set contains is uniquely determined.

Definitions 1.22 A set X is said to be finite if it is either empty or there exists a bijection $f : \{1, 2, \ldots, n\} \to X$ for $n \in \mathbb{N}$. A set X is infinite if it is not finite.

Remark 1.23 Since bijection $f : \{1, 2, \ldots, n\} \to X$ defines an inverse f^{-1}, the finiteness of a set X can also be defined as $g : X \to \{1, 2, \ldots, n\}$.

Proposition 1.24 *A set X is finite if and only if there exists a finite set Y and a bijection $f : X \to Y$.*

Proof If X is finite there exists some finite n and a bijection

$$g : \{1, 2, \ldots, n\} \to X$$

which implies that $f = g^{-1} : X \to \{1, 2, \ldots, n\}$ is well defined. For $Y = \{1, 2, \ldots, n\}$, we have a bijection $f : X \to Y$.

Conversely, suppose there exists a finite set Y and a bijection $f : X \to Y$. Since Y is finite there exists some $n \in \mathbb{N}$ and a bijection $h : Y \to \{1, 2, \ldots, n\}$. Let $g = h \circ f : X \to \{1, 2, \ldots, n\}$. Since f and h are bijections, $h \circ f$ is also a bijection and hence X is finite. □

Remark 1.25 One can also show that for a given set X and for some element $x_0 \in X$, X is finite iff $X \backslash \{x_0\}$ is finite.

Definitions 1.26 1. Given two sets X and Y, we say that X and Y have same cardinality, denoted by $|X| = |Y|$, if there is a bijection $f : X \to Y$.
2. We say that cardinality of X is not greater than cardinality of Y, denoted by $|X| \leq |Y|$ if there exists an injective function $f : X \to Y$.
3. We say that cardinality of X is not less than cardinality of Y, denoted by $|X| \geq |Y|$ if there exists a surjective function $f : X \to Y$ provided $Y \neq \emptyset$.

Note that $|X| < |Y|$ if $|X| \leq |Y|$ and $|X| \neq |Y|$. If $X = \emptyset$, then $|X| = 0$, and if X is finite then $|X|$ = number of elements in X. In the following theorem we see that all finite sets have cardinality less than the set of natural numbers.

Theorem 1.27 *Let X and Y be any two given sets. Then the following statements are equivalent:*

(i) There exists an injective function $f : X \to Y$.
(ii) There exists a surjective function $g : Y \to X$.

Proof The case when $X = \emptyset$ or $Y = \emptyset$ is trivial. If $X \neq \emptyset$ and $Y \neq \emptyset$ then
$(i) \Rightarrow (ii)$ Let $f : X \to Y$ be an injective function and let $x \in X$.

Define $g : Y \to X$ such that

$$g(y) = \begin{cases} f^{-1}(y) & \text{if } y \in R(f) \\ x & \text{if } y \in Y \backslash R(f) \end{cases}$$

It is clear that g is onto.
$(ii) \Rightarrow (i)$ Let $g : Y \rightarrow X$ is surjective, define $f : X \rightarrow Y$ such that $f(x) = y_x$ where $y_x = g^{-1}(x), x \in X$. Clearly f is injective. □

One can see that $|X| \leq |Y|$ iff there exists a surjective function $f : Y \rightarrow X$. In the following we see that two sets X and Y have same cardinality iff there is a injection (surjection) from X to Y and from Y to X.

We use the following lemma without proof.

Lemma 1.28 *Let $\{X_i\}$ and $\{Y_i\}$ are sets such that $f_i : X_i \rightarrow Y_i$ is a bijection $\forall\, i$. If all X_i's are pairwise disjoint and all Y_i's are pairwise disjoint, then there is a bijection $f : \bigcup_i X_i \rightarrow \bigcup_i Y_i$*

1.3 Countable and Uncountable Sets

We have observed so far that a set is either finite or infinite.

$$A = \{1, 2, 3, 4, 5\}$$

is an example of a finite set where as the set of all positive even numbers and the set of all real numbers are examples of infinite sets. We can infact show that the cardinality of the set of all positive even numbers is same as that of the set of all positive natural numbers. This is not the case for real numbers. Hence we further classify infinite sets based on their cardinality as countable or uncountable.

Definitions 1.29 1. A set X is called countably infinite if $|X| = |\mathbb{N}|$.
2. A set X is called countable (countably infinite) if $|X| \leq |\mathbb{N}|$. i.e., there exists a bijection $f : X \rightarrow \mathbb{N}$.
3. A set X is called uncountable if it is not countable i.e., $|X| > |\mathbb{N}|$.

Remark 1.30 Observe that a set X is countable if and only if it is finite or countably infinite.

Examples 1.31 1. $\mathbb{N}$ is countably infinite since $f : \mathbb{N} \rightarrow \mathbb{N}$ such that $f(x) = x$ is a bijection.
2. $\mathbb{N}$ is countably infinite since $f : \mathbb{N} \rightarrow \mathbb{N}$ such that $f(x) = x + 1$ is a bijection.

Properties of Countable sets: Let X, Y be countable sets

(i) For $Z \subseteq X$, Z is countable.
(ii) $X \cup Y$ is countable.
(iii) $X \cap Y$ is countable.

The following result is a generalization of the previous result (ii).

Theorem 1.32 *Countable union of countable sets is countable.*

Proof Exercise. □

Examples 1.33 The set $\mathbb{R}$ of all real numbers is countable.

Solution: Suppose $\mathbb{R}$ is countable, then by properties of countable sets $(0, 1) \subseteq \mathbb{R}$ is countable. Then $\exists$ an bijection $f : \mathbb{N} \to (0, 1)$. Let $0 \cdot a_{i_1} a_{i_2} a_{i_3} a_{i_4} \ldots$ be the decimal representation $f(i)$.

Now we choose an $x \in (0, 1)$ such that the nth digit in the decimal representation of x is different from the nth digit in the decimal representation of $f(x)$.

By the choice of x, $x \neq f(i)\ \forall\ i \in \mathbb{N}^+$ which contradicts our assumption that f is a bijection. Hence $(0, 1)$ is uncountable and so is $\mathbb{R}$. □

Theorem 1.34 (Cantor's Theorem) *The power set of any set X has greater cardinality than the set X i.e., there is no surjection from X to $\mathcal{P}(X)$.*

Proof Suppose f is a surjection from X to $\mathcal{P}(X)$. Let $Y = \{x \in X : x \notin f(x)\}$. Since f is onto, there exists $y \in X$ such that $f(y) = Y$. Note that $Y \in \mathcal{P}(X)$.

If $y \in Y$ then by definition of Y, $y \notin f(y) = Y$.

If $y \notin Y = f(y)$ then by definition of Y, $y \in Y$. We reach a contradiction in both cases, and therefore $|X| \neq |\mathcal{P}(X)|$ holds. □

Corollary 1.35 *There is no injection from $\mathcal{P}(X)$ to X.*

Proof Exercise. □

In the following theorem we state that for any set X there exists a set whose cardinality is same as $\mathcal{P}(X)$.

Theorem 1.36 *Let X be a given set. Then $|\mathcal{P}(X)| = |2^X|$ where 2^X denotes the set of all functions from X into A where $|A| = 2$.*

One can observe from the Cantor's theorem that

$$\mathcal{N}_0 = |\mathbb{N}| \leq |\mathcal{P}(\mathbb{N})| \leq |\mathcal{P}(\mathcal{P}(\mathbb{N}))| \leq |\mathcal{P}(\mathcal{P}(\mathcal{P}(X)))| \leq \dots$$

Thus one can obtain a sequence of sets having greater cardinality than the previous element in the sequence. In the following theorem we show that apart from the first element, every element in the above sequence is uncountable. To be precise we see that the cardinality of $\mathbb{R}$ is $2^{\mathcal{N}_0}$.

Recursive Definition: Recursion is a process in which one or more initial stages of the process are specified and the nth stage of the process is defined using some of the previous stages, using some fixed method. For example method of proof by induction is based on a recursively defined structure. Recursion can be used to define sets, functions and sequences.

Recursive definitions always have two parts:

(a) Base case
(b) Recursive step

Recursive definitions are also called as inductive functions or recurrence relations.

Examples 1.37 The set of all Fibonacci numbers $\{1, 1, 2, 3, 5, 8, \dots\}$ can be defined recursively as:
Base case: $f(1) = 1$, $f(2) = 1$.
Recursive step: $f(n) = f(n-1) + f(n-2), n \geq 3$.

Remark 1.38 Functions such as factorial, Fibonacci, etc., are simpler to define in terms of other functions. Recursion can be used to naturally define such functions.

Definitions 1.39 A function f is recursively defined if at least one value $f(x)$ is defined in terms of another value $f(y)$ where $x \neq y$.

Examples 1.40 Let $X = \{2, 4, 6, 8, \dots\}$. X can be defined recursively as follows: Let $f : \mathbb{N}^+ \to X$ such that
Base case: (a) $f(1) = 2$, $f(2) = 4$.
Recursive step: (b) $f(z) = f(x) + f(y)$ where $x, y < z$.

We now state the general principle of recursive definition which guarantees the existence and uniqueness of a recursively defined function for a given set.

Principle of Recursive Definition: Let X be a set and $x \in X$. If $f : X \to X$, then there is a unique function $g : \mathbb{N} \to X$ such that

$$g(1) = x, g(s(n)) = f(g(n)) \text{ for all } n \in \mathbb{N}$$

where $s(n)$ is defined as

$$s(n) = n + 1 \; s(n + m) = n + S(m).$$

1.4 Number Sequences and Summation

Counting is the fundamental concept of discrete mathematics. A discrete set is a set containing discrete elements as its members. This set may be finite or infinite and may or may not have a property to describe its elements. For example the natural numbers $1, 2, 3, 4, 5, \ldots$ can be said to be a set of natural numbers.

A sequence is a set of elements or numbers in a particular order. One can define a sequence formaly using the definition of a function.

Definitions 1.41 A sequence is a function f from a subset of natural numbers $\mathbb{N}$ to a set S. The values or members of the sequence can be called as terms. Usually a sequence is denoted by $a_0, a_1, a_2, \ldots$ where each a_i is a member of the set S. Note that function f is sometimes referred to as an 'output' function. So we can write $f(n)$ as a_n. The notation $\{a_n\}$ denotes the sequence $a_0, a_1, a_2, \ldots$.

Examples 1.42 (1) 2, 4, 6, 8 is a sequence with 4 terms. Here $a_1 = 2$, $a_2 = 4$, $a_3 = 6$, $a_4 = 8$. It is a finite sequence.

(2) The sequence 1, 2, 3, 5, 8, 13, 21, 34, ... is an inifnite sequence. It is the well known Fibonacci sequence.

(3) The sequence $\{\frac{1}{n}\}$ is a sequence representing $1, \frac{1}{2}, \frac{1}{3}, \frac{1}{4}, \ldots$.

Some Special Sequences

1. ***Arithmetic progression***: It is a sequence with terms $a, a + d, a + 2d, a + 3d, \ldots, a + nd, \ldots$. Here in $\{a + kd\}$, a, d are real numbers $k \in \mathbb{N}$. d is called as common difference.
2. ***Geometric progression***: It is a sequence with terms $a, ar, ar^2, ar^3, \ldots, ar^n, \ldots$. Here the initial term is a and r is called as common ratio. In the sequence $\{ar^n\}$, a, r are real numbers.

Examples 1.43 1. $\{-2 + 3k\}$ is an arithmetic progression with its terms as $-2, 1, 4, 7, 10, 13, \ldots$. Here $a = -2$, $d = 3$.

2. Consider the sequence $\{1, -1, 1, -1, 1, -1, \ldots\}$. It is a geometric progression with $a = 1$, $r = -1$. One can refer $\{(-1)^n\}$ as a formula representing the above sequence.

Remark 1.44 There are sequences that are neither arithmetic progression nor geometric progression.

Examples 1.45 Consider the sequence $1, 5, 13, 25, \ldots$. This sequence is neither an arithmetic progression nor a geometric progression. $\{2n^2 + 2n + 1\}$ is the closed form corresponding to the above sequence.

Summations

Sum of the terms of a sequence is called as summation of the sequence. For example $a_2 + a_3 + a_4 + a_5$ is a summation from the sequence $a_0, a_1, a_2, \ldots$. The summation may range either covering all the terms or a finite number of them.

Definitions 1.46 Let $\{a_n\}$ be a sequence. The sum of the terms $a_t, a_{t+1}, \ldots, a_s$ is written as $a_t + a_{t+1} + \cdots + a_s$ is called a summation where the index is running from t to s.

Notation 1.47 $\sum_{i=t}^{s} a_i$ stands for the summation of $a_t, a_{t+1}, \ldots, a_s$. We say '$i$' is the index of the summation and the index is running from t to s. Also 't' is called the lower index and 's' is called the upper index of the summation.

Examples 1.48 1. Consider the sequence $\{n^2\}$. The value of the summation

$$\begin{aligned}\sum_{i=2}^{4} (i)^2 &= 2^2 + 3^2 + 4^2 \\ &= 29\end{aligned}$$

2. The value of the summation $\sum_{i=3}^{7} (-1)^i = -1$.

More Than One Summation

Most commonly used nested summations are double summations. They occur almost in many parts of mathematics and computer programming. The summations

$$\sum_{i=t}^{s} \sum_{j=\ell}^{m} a_{ij}$$

is called as double summation.

Examples 1.49 Let $f(x) = x + 1$. The value of the summation

$$\begin{aligned}\sum_{x=1}^{4} f(x) &= f(1) + f(2) + f(3) + f(4) \\ &= 14\end{aligned}$$

Here $f(x)$ is describing a sequence $\{n + 1\}$.

Remark 1.50 A summation can be sometimes expressed in a closed form.

Examples 1.51 Closed form of

$$\sum_{i=1}^{n} i^2 = \frac{n(n+1)(2n+1)}{6}.$$

Linearity Property

$$\sum_{m=1}^{n} (ka_m + \ell b_m) = k \sum_{m=1}^{n} a_m + \ell \sum_{m=1}^{n} b_m$$

where k, ℓ are real numbers, $\{a_n\}$, $\{b_m\}$ are sequences.

In particular

$$\sum_{m=1}^{n} (a_m \pm b_m) = \sum_{m=1}^{n} a_m \pm \sum_{m=1}^{n} b_m.$$

Examples 1.52 1. The progression $a, a+d, a+2d, \ldots$ is also sometimes called as arithmetic series. The sum of the first 'n' terms of this series will be

$$\begin{aligned}\sum_{k=0}^{n} (a+kd) &= \sum_{k=0}^{n} a + d \sum_{k=0}^{n} k \\ &= na + \frac{dn(n+1)}{2}.\end{aligned}$$

This is the closed form for the arithmetic series.

2. The progression $a, ar, ar^2, \ldots$ is also sometimes called as geometric series. Its closed form will be $\frac{a(r^{n+1}-1)}{(r-1)}$, $r \neq 1$. Here, $|r|$ may be <1 or >1. It is not difficult to prove that

$$\sum_{k=0}^{\infty} ar^k = \frac{a}{1-r} \text{ when } |r| < 1.$$

1.5 Growth Functions

For many problems in science and engineering, a method of solution is described via algorithms. An algorithm or a sequence of computation will consist of a finite set of instructions leading to a solution for a given problem. The instructions in any algorithm can be in any programming language, English sentences or pseudocode etc. There may be more than one way to solve a problem and hence the sequences of computation will be different. Hence it is important to evaluate an algorithm in terms of number of steps or time taken to complete the assignment. Also such an estimate of time must be independent of the computer or software used. Also such an analysis must be true for any input irrespective of its size. Growth functions are very useful abstraction to capture the running time of any algorithm in a very general way. Consider the following simple example to understand the concept of estimating computation steps or time.

Examples 1.53 Let A be a sequence of n numbers say $A[1], A[2], \ldots, A[n]$. The numbers are indexed as $1, 2, 3, \ldots, n$ for writing down an algorithm for a sequential search in A to check whether a number 't' is present in A or not present in A.

Algorithm M (in pseudocode)

```
i := 1
while (i ≤ n and A[i] ≠ t)
    i ← i + 1
if (i > n), then print 'not present'
return i.
```

Clearly algorithm M will compare t with every number in A in a sequential way. If 't' is present in $A[1]$, then the algorithm will stop so its computation by returning the index '1' to say t is present as the first number in A. If not the instructions are given to compare t with every element in the array A. In the worst case scenario, it may be that t is not present in A. But the algorithm will compare t with all entries in A. This is said to be the worst instance of the problem. In any case, the number of comparisons made with t is given by the index 'i' and it is actually the number of computation steps which is $2n$. Also the index increasing step has to be executed along with the 'while' step; which will be 'n'. Hence one can use '$3n$' steps to solve the problem completely for any input t. Other steps in the algorithm are computed only once. Hence we can say that the running time for the algorithm M is in terms of $3n$ or n. This running time varies as you vary t. Growth functions are very useful for the estimation of the number of computational steps of any algorithm. Growth functions are described to be functions of 'n' where n is the size of the input, which is usually taken to be a large integer r. Complexity of any algorithm is judged only by looking at growth function.

Given any two functions f and g we wish to give systematic method of comparing f and g interms of the common input 'n'. We would like to understand the meaning of f and g grow at the same rate for large values of input 'n' (say). In this process we only concentrate on actual number of computational steps involved for any large size input.

The Big O Notation

Definitions 1.54 Let f and g be any two functions from $\mathbb{N}$ to $\mathbb{R}$ (set of real numbers). We say g asymptotically dominates f or 'f is in big-O of g' if there exist c and k such that

$$|f(x)| \leq c|g(x)| \text{ for all } x \geq k.$$

In notation, $f(x) \in O(g(x))$ or

$$f \in O(g).$$

Note 1.55 It is clear that any f that satisfies the above condition can be said to be in $O(g)$.

We state the following theorem which gives a necessary condition for a function f to be in $O(g)$.

Theorem 1.56 *(i) If* $\lim_{x\to\infty} \frac{|f(x)|}{|g(x)|} = \ell$, *where* $\ell \geq 0$, *then* $f \in O(g)$.
(ii) If $\lim_{x\to\infty} \frac{|f(x)|}{|g(x)|} = \infty$, *then* $f \notin O(g)$.

Remark 1.57 A function $f \in O(g)$ means that f is bounded by a constant multiple of g for large values of x.

Examples 1.58 Show that $n^2 + 7 \in O(n^2)$ using the definition of big-O as well as the above theorem.

Solution: Using the definition of big-O, let $c = 2$, and $k = 3$. Then for $2n^2 = n^2 + n^2 \geq n^2 + 9 \geq n^2 + 7$. Hence $n^2 + 7 \in O(n^2)$.

Solution by the Above Theorem:

$$\lim_{n\to\infty} \frac{n^2 + 7}{n^2} = 1.$$

Here $\ell = 1$. So $n^2 + 7 \in O(n^2)$.

Remark 1.59 A function $f(n) \in O(n^t)$ means that for any value of n, irrespective of value of 'n', the function $f(n)$ is bounded above by n^t, $t \in \mathbb{N}$.

Examples 1.60 Show that $n^4 \notin O(n^3)$.

Solution: Consider

$$\lim_{n\to\infty} \frac{n^4}{n^3} = \infty.$$

Hence by the above theorem, $n^4 \notin O(n^3)$.

Remark 1.61 In complexity measurements, the functions f, g defined above always set to positive real numbers.

The Litte O Notation

Definitions 1.62 If for real valued functions f, g,

$$\lim_{n\to\infty} \frac{f(n)}{g(n)} = 0$$

then we say that f belongs to little-o of g. i.e., $f \in o(g)$.

Remark 1.63 The fact that if $f \in o(g)$, then $g \in O(f)$ can be verified easily. One can see from the definition of big O, $an + b \in O(n^2)$. But this bound is not tight. On the other hand $an + b \in o(n^2)$ but $an^2 \notin o(n^2)$. So little o notation is describing asymptotic upper bound that is not asymptotically tight. For example $an \in o(n^2)$ but $an^2 \notin o(n^2)$.

θ-**notation**: For a given function $g(n)$, $f(n) \in \theta(g(n))$ if there exist positive constants k_1, k_2 and n_0 such that $0 \leq k_1 g(n) \leq f(n) \leq k_2 g(n)$ for all $n \geq n_0$.

Examples 1.64 Consider $f(n) = n^2 - 2n$, $g(n) = n^2$. We can see

$$c_1 n^2 \leq n^2 - 2n \leq c_2 n^2 \ \forall\, n \geq n_0.$$

$$\Rightarrow c_1 \leq 1 - \frac{2}{n} \leq c_2$$

For $c_2, n \geq 3, c_1 \leq \frac{1}{3}$. So choose $c_1 = \frac{1}{3}, c_2 = 1, n \geq 3$ and $f(n) \in \theta(g(n))$.

Ω-Notation (Big-Omega Notation)

For two functions $f(n)$, $g(n)$, if there exist positive constants k and n_0 such that $0 \leq kg(n) \leq f(n)$ for all $n \geq n_0$, then we say $f(n) \in \Omega(g(n))$. Here $f(n)$ grows no faster than $g(n)$. Clearly 'Ω' gives lower bound for functions.

Examples 1.65 (i) Consider $f(n) = 2n + 3$, $g(n) = n$. It is not difficult to see $2n + 3 \in \Omega(n)$.

(ii) Consider $f(n) = \log(n^3) + 2$ and $g(n) = \log(n)$. Clearly

$$\begin{aligned}\log(n^3) + 2 &= 3\log(n) + 2 \\ &\geq \log(n)\ \forall\, n > 1 \text{ and } k = 1\end{aligned}$$

This means $\log(n^3) + 2 \in \Omega(\log(n))$.

o-Notation (Little-'oh' Notation)

For given functions $f(n)$ and $g(n)$, if $0 \leq f(n) \leq kg(n)$ for all $n \geq n_0$, 'k' being any positive constant, then we say $f(n) \in o(g(n))$. Note that little-'oh' is used for denoting an upper bound that is not asymptotically tight one.

Examples 1.66 Consider $f(n) = 3n$ and $g(n) = n^2$. Then clearly $3n \leq kn^2$ for any 'k'.

We have seen briefly all the notations used for indicating the growth functions.

The largest number of computation steps required to solve a given problem counts for time complexity of the algorithm. One may call this measurement as worst-case time complexity. Usually the size of the input is taken to be very large for any given analysis. The constants involved in the counting of steps are usually ignored. We can see that asymptotic notations described above are very useful for describing the worst-case running time $T(n)$ (say). Clearly $T(n)$ is usually defined for integer n.

1.6 Sum and Product Rules

It is equally important to learn enumeration techniques. The enumeration methods are also called as counting methods.

Basic Counting Rules

Many counting problems require the number of distributions of objects or elements in a specified manner. The objects may or may not be distinct.

Examples 1.67 Let n persons arrive at the same time at a railway booking counter, to book tickets. The booking however can be done only sequentially i.e., one at a time. Hence an ordering of these n persons is required so that the tickets are issued to them. Hence we need to know how many different ways are there to order the persons or prioritize the persons so as to issue tickets to them.

Examples 1.68 Three dice are rolled sequentially. The outcome of rolling will be a sequence of numbers between 1 and 6. That is one outcome may be (4, 3, 2). In general outcome will be (k_1, k_2, k_3) where $1 \leq k_i \leq 6$, $1 \leq i \leq 3$. What we call sample space will be

$$A = \{(k_1, k_2, k_3)/k_i = 1, 2, 3, 4, 5, 6, 1 \leq i \leq 3\}.$$

A is the set of all possible outcomes by rolling the dice activity. Suppose the question is: Find the number of elements of A such that $k_1 + k_2 + k_3 = 4$. The exhaustive collection will tell that the number is 3.

In both the above examples the problem is to make an enumeration or number of distributions of objects into holes. Before proceeding let us learn two fundamental principles that are useful for enumeration.

Let Σ be a set, and let $|\Sigma|$ denote the number of elements in Σ.

Addition Rule

Suppose Σ is the union of disjoint nonempty subsets $E_1, E_2, \ldots, E_t$. Then $|\Sigma| = |E_1| + |E_2| + \cdots + |E_t|$. For the disjoint sets E_i, $1 \leq i \leq t$, each element of Σ occurs exactly in one E_i. So, we say that $E_1, E_2, \ldots, E_t$ form a partition of Σ. If E_i's overlap, then more refined method is needed to enumerate Σ.

The addition rule can be understood as an event-activity problem. Suppose E is an event and Σ is the set of ways that E can happen. The question is to count Σ. Suppose $E_1, E_2, \ldots, E_t$ are mutually exclusive events and it is known that E_1 can happen in m_1 ways, E_2 can happen in m_2 ways and so on and E_t can happen in m_t ways. Then the addition principle enumerates the activity as follows:

'The event that E_1 or E_2 or E_3 or … E_t can occur in $m_1 + m_2 + \cdots + m_t$ ways'.

Examples 1.69 Class VIII contains 30 boys and 35 girls. In how many ways can a leader be selected so that the leader is a boy or girl?

Since there are totally 65 students in the class, any one of this 65 can be nominated as a leader. It can be seen that the leader can be one from the 30 boys (or) one from the 35 girls. However the activity will complete only after selecting a leader. Hence the number of ways of selecting a girl leader is 35. Similarly for selecting a boy leader, the number of ways is 30. Hence the total number of ways will be 65.

Examples 1.70 A box contains 10 red balls, 10 blue balls and 10 yellow balls. In how many ways can we pick a blue or yellow ball from the box.

Since there are 10 blue balls and 10 yellow balls, there are 10 + 10 = 20 ways to select a blue or yellow ball.

The above examples are simple and direct as we considered only non-overlapping events. More examples can be seen later.

Multiplication Rule

The cartesian product of two sets X and Y will be

$$X \times Y = \{(x, y)/x \in X, y \in Y\}.$$

If E is an event to be completed in two stages and if there are t_1 outcomes for the first stage E_1 and t_2 outcomes for the second stage E_2. The event E is considered to be complete only when both the stages are complete. Hence the number of ways E to happen will be $t_1 \times t_2$. In general if events $E_1, E_2, \ldots, E_m$ can happen $t_1, t_2, \ldots, t_m$ ways respectively and total event E is considered to be complete only if $E_1, E_2, \ldots, E_m$ in sequence are complete, then E can happen in $t_1 \cdot t_2 \cdot t_3 \cdots t_m$ ways.

Examples 1.71 A social service organization has 10 men and 15 lady members. A person among them has to be nominated as president and a person as treasurer. Find the number of ways such nomination can be done?

Solution: The activity nomination is a pair (x, y) where x is the president,y is the treasurer and x and y are members of the organization. The activity will be complete only when nominations are made for both the posts. Hence one must use multiplication principle. The following are the two cases.

Case (i) A member can be both the president and treasurer at the same time. In this case the number of possible nominations are $25 \times 25 = 625$.
Case (ii) If a president cannot be a treasurer and vice versa, then the number of possible different nominations are $25 \times 24 = 600$.
Case (iii) If members with different gender are to hold the positions, then the number of possible different nominations is

$$10 \times 15 + 15 \times 10 = 300.$$

Here we applied both addition and multiplication principles as the nomination is the union of two disjoint sets.

Examples 1.72 (i) How many 4-digit numbers can be formed using the digits 1–9?
(ii) How many such numbers can be formed so that the last digit is always an even number?
(iii) How many such numbers can be formed so that no digit can be repeated?

Solution:

(i) There are 9^4 such 4-digit numbers as each digit of the number can be selected in 9 ways. The activity of forming a 4-digit number will be complete only when all the 4-digits are filled. Hence by multiplication principle the answer will be 9^4.

(ii) The even numbers are $2, 4, 6, 8$. The last digit should contain only one of these numbers. Hence the last digit can be filled only in 4 ways. Hence the total number of ways in this case will be $9^3 \times 4$.

(iii) The activity is to choose numbers from 1 to 9 for 4 positional digits of the number. Any one position can be filled in 9 ways. Since repetitions are not allowed,the next selection will be only from 8 available numbers. Hence the total number of ways will be $9 \times 8 \times 7 \times 6$.

An enumeration problem many times may require the applications of both addition and multiplication principles. The following example is an illustration of such a situation.

Examples 1.73 Let $\Sigma = \{a_1, a_2, \ldots, a_n\}$, where $|\Sigma| = n$. Show that the number of subsets of Σ is 2^n.

Solution: A subset A of Σ can be formed in n steps as follows. The subset A may contain a_1 or may not, it may contain a_2 or it may not and so on, A may contain a_n or not. Hence each a_i may or may not be put in A. For each pick a_i, there are 2 possible ways in forming A. Hence the number of subsets will be 2^n.

Examples 1.74 How many 5-digit numbers are there that are with exactly one 3 or one 5 in the first digit?

Solution: The activity is to form a 5-digit number using the number 1 to 9. Out of the 5-digits 4-digits can be filled in 9^4 ways except the first one. If the first position contains 3, then there are 9^4 numbers having 3 in the first digit. The situation is identical if the first digit contains 5. Hence the total number of such numbers will be $9^4 + 9^4 = 2 \times 9^4$.

Exercise

1. Prove that
$$\sum_{k=0}^{n} k^2 = \frac{n(n+1)(2n+1)}{6} \; \forall\, k \geq 0$$
2. Find the closed formula for the following sequences

(i) $1, 3, 6, 10, 15, 21, \ldots$
(ii) $1, 1, 2, 3, 5, 8, 13, \ldots$

(iii) $6, 12, 24, 48, \ldots$
(iv) $1, 5, 23, 119, 719, \ldots$

3. Let $A = \{0, 1, 2, 3, 4, 5, 6, 7, 8\}$ and $f(x) = x^2 (\mathrm{mod}\, 9)$. Find the range of f.
4. Sketch the graph of the following relations

(i) $\{(x, y) \in \mathbb{R} \times \mathbb{R} | x \geq y^2\}$
(ii) $\{(x, y) \in \mathbb{R} \times \mathbb{R} | y \geq x^2\}$

5. Let $A = \{0, 1, 2, a, b\}$ and $B = \{a, b, c, 1, 2\}$.
Find the following

(i) $A \cup B$
(ii) $A \cap B$
(iii) $A \backslash B$
(iv) $B \backslash A$

6. Prove the following

(i) $A \cup B = A \cup (\overline{A} \cup B)$
(ii) $\overline{A} \cup \overline{B} = \overline{A \cap B}$
(iii) $A \oplus B = (A \cup B) \backslash (A \cap B)$

7. If $X \oplus A = A$ find X if it exists.
8. Let $A = \{2, 3, 4, 5\}$, $B = \{1, 2, 5, 6\}$, $C = \{1, 3, 4, 6\}$, $U = \mathbb{N}$
Find

(i) $A \cap \overline{B} \cap \overline{C}$
(ii) $A \cap \overline{B} \cap C$
(iii) $\overline{A} \cap B \cap \overline{C}$
(iv) $A \cap \overline{B} \cap \overline{C}$

9. Let X, Y and Z be any given sets. Show that

(i) If $X \subseteq Y$ and $Y \subseteq Z$ then $X \subseteq Z$.
(ii) If $X \subset Y$ and $Y \subset Z$ then $X \subset Z$.
(iii) $X = Y$ iff $X \subseteq Y$ and $Y \subseteq X$.
(iv) $X \in \mathcal{P}(X)$ and $\emptyset \in \mathcal{P}(X)$.
(v) $X \subseteq Y \subseteq Z \subseteq X$ iff $X = Y = Z$.

10. Show that $\{a/a = b^2$ for a real number 'b'$\} = \{a/a \geq 0, a$ is real$\}$.

11. Prove that for all sets X and Y, $\emptyset \subseteq X \cap Y \subseteq X \cup Y$.
12. Given X, Y, Z, prove the following (expand this as laws of set)

    (i) $X \cup Y = Y \cup X$
    (ii) $(X \cup Y) \cup Z = X \cup (Y \cup Z)$
    (iii) $X \cap Y = Y \cap X$
    (iv) $(X \cap Y) \cap Z = X \cap (Y \cap Z)$
    (v) $X \cup (Y \cap Z) = (X \cup Y) \cap (X \cup Z)$
    (vi) $X \cap (Y \cup Z) = (X \cap Y) \cup (X \cap Z)$
    (vii) $\mathcal{P}(X \cap Y) = \mathcal{P}(X) \cap \mathcal{P}(Y)$

13. Determine whether the following are equivalence relations and if so, determine the corresponding equivalence classes.

    (i) $\rho = \{(x, y)/x - y \text{ is divisible by } 7\}$
    (ii) $\rho = \{(x, y)/x, y \in \mathbb{N}, x \leq y\}$
    (iii) $\rho = \{(x, y)/0 \leq x \leq 20, 0 \leq y \leq 20\}$
    (iv) $\rho = \{(x, y)/x, y \in \mathbb{R}, x - y \in Q\}$

14. Determine whether f^{-1} exists. If so find the inverse where $f : \mathbb{R} \to \mathbb{R}$ is defined as

    (i) $f(x) = e^x$
    (ii) $f(x) = \frac{1}{1+x^2}$
    (iii) $f(x) = x^2 + 3x + 1$
    (iv) $f(x) = x$

15. Let f, g, h be any three functions from $X \to X$ such that $X \neq \emptyset$. Show that $f \circ (g \circ h) = (f \circ g) \circ h$.
16. Let $f : X \to Y$ be bijective and $g : Y \to X$ be its inverse. Show that g is unique.
17. Let $f : Z \to Y$ and $g : X \to Z$ such that both f and g are bijective. Show that $f \circ g$ is bijective.
18. Let $f, g : X \to Y$ be two functions. We say that f is equal to g denoted by $f = g$ if $f(x) = g(x)\ \forall\, x \in X$. Define $f, g : \mathbb{Z} \to \mathbb{Z}$ as $f(x) = 2x^2 - x$ and $g(x) = x - 1$. Is $f = g$?
19. Show that the set of all irrational numbers.
20. Let X be an infinite set and let Y be a countable set. Prove

    (i) $|X \cup Y| = |X|$.
    (ii) If $X \backslash Y$ is infinite, then $|X \backslash Y| = |X|$.
    (iii) If Y is finite, then $|X \backslash Y| = |X|$.

21. Show that if X is an infinite set, then there exists $Y \subsetneq X$ such that $|Y| = |X|$.
22. Show that $f(n) \in O(g(n))$ in the following

(i) $f(n) = n^2 \log(n^2 + 1) + 5n^3 + 4$
$g(n) = n^3$
(ii) $f(n) = n^2 + 5n + 10$
$g(n) = 15n^2$

23. Let $f(x) = a_0 + a_1x + a_2x^2 + \cdots + a_tx^t$, with a_t being positive. Prove that

(i) $\theta(f(n)) = \theta(n^k)$
(ii) $O(f(n)) = O(n^k)$.

Induction and Pigeon Hole Principle 2

2.1 The Method of Mathematical Induction and Strong Induction

In this section we discuss the last axiom of Zermelo-Fraenkel set theory known as the axiom of choice which is stated as follows.

Axiom of Choice

Every family of non-empty sets $(A_i)_{i \in I}$ has a choice function

$$f : I \to \bigcup_{i \in I} \{A_i\}$$

such that $f(i) \in A_i$ for each $i \in I$.

The axiom guarantees the existence of a choice function but does not indicate on how to construct such a function. The axiom of choice is indeed equivalent to many other results in different areas of mathematics. Our interest lies in understanding the equivalence of axiom of choice to a well known important proof technique called as the principle of mathematical induction. We list some of the equivalent statements here without proof.

Theorem 2.1 *The following are equivalent:*

1. *Axiom of Choice: Every family of non empty sets has a choice function.*
2. *Zorn's Lemma: If $(X, \leq)$ is a non-empty partially ordered set in which every chain has an upper bound, then, X has a maximal element.*
3. *Well-ordering Principle: For every set X, there is a relation $<$ such that $(X, <)$ is a well-ordering.*

R. Rama, *Topics in Combinatorics and Graph Theory*,
https://doi.org/10.1007/978-3-031-74252-1_2

4. *Bernstein's Theorem: Given any two sets X and Y, there exists either a bijection between X and a subset of Y or a bijection between Y and a subset of X.*
5. *Housdorff Maximal Principle: Every partially ordered set has a maximal chain.*

Since axiom of choice is equivalent to the well ordering principle, it can be shown that the well ordering on $\mathbb{N}$ is equivalent to the principle of mathematical induction which is defined as follows:

Definition 2.2 1. Principle of mathematical Induction (PMI): Weak form: Let $S \subseteq \mathbb{N}$. Suppose $1 \in S$ and $n+1 \in S$ whenever $n \in S$. Then $S = \mathbb{N}$.
2. Principle of mathematical Induction (PMI): Strong form: Let $S \subseteq \mathbb{N}$. Suppose $1 \in S$ and $n+1 \in S$ whenever $1, 2, \ldots, n \in S$. Then $S = \mathbb{N}$.
3. Well ordering Principle for $\mathbb{N}$: (WOP): Every non-empty subset S of $\mathbb{N}$ has a least element.

Theorem 2.3 *The following statements are equivalent.*

(a) PMI weak form
(b) PMI strong form
(c) WOP for $\mathbb{N}$.

Proof $\underline{(c) \Rightarrow (a)}$

Let $P(n)$ be a property that is satisfied for some n.

We assume that

(i) $P(1)$ is true
(ii) $P(K+1)$ is true whenever $P(K)$ is true.

We need to show that $P(n)$ is true for all $n \in \mathbb{N}$. Suppose $\exists\, m \in \mathbb{N}$ such that $P(m)$ is not true.

Let $S = \{m \in \mathbb{N} / P(m) \text{ is false}\}$.

Since $m \in S$, $S \neq \emptyset$ and by well ordering principle there exists a least element $t \in S$.

Since $P(1)$ is true, $t \neq 1$. Therefore $t \geq 2$ and $t - 1 \notin S$ since t is the least element of S. Hence $P(t-1)$ is true. But by hypothesis (ii) $P(t-1) \Rightarrow P(t) \Rightarrow P(t)$ is true, a contradiction as $t \in S$. Therefore our assumption that $\exists\, m \ni P(m)$ is false is not true.

Hence $P(n)$ is true $\forall\, n \in \mathbb{N}$.

$\underline{(a) \Rightarrow (b)}$

PMI weak form $\Rightarrow$ PMI strong form

Assume PMI weak form to be true and let $S \subseteq \mathbb{N}$ such that $1 \in S$ and

$$\{1, 2, \ldots, n\} \subseteq S \Rightarrow n+1 \in S$$

We need to show that $S = \mathbb{N}$

Define $P(n)$ to be the property that $\{1, 2, 3, \ldots, n\} \in S$ for $n \in \mathbb{N}$ and let $A = \{n \in \mathbb{N} : P(n)$ is true$\}$.

$P(1)$ is true since $1 \in S$

Assume $P(K)$ to be true for some K (i.e.,) $K \in A$ (i.e.,) $\{1, 2, \ldots, K\} \subseteq S$.

Then by definition of S, $\{1, 2, \ldots, K\} \subseteq S \Rightarrow K + 1 \in S$.

$\Rightarrow \{1, 2, \ldots, K\} \subseteq S \Rightarrow P(K + 1)$ is true (i.e.,) $K + 1 \in A$.

Thus $P(K) \Rightarrow P(K + 1)$. Hence by PMI weak form $P(n)$ is true $\forall\, n \in \mathbb{N}$.

$\Rightarrow n \in A\, \forall\, n \in \mathbb{N} \Rightarrow A = \mathbb{N}$.

$(b) \Rightarrow (c)$

Let $S \subseteq \mathbb{N} \ni S \neq \phi$ there $\exists\, m \in S$ since $S \neq \emptyset$. We need to show that S has a minimal element. We prove by contradiction. Suppose S has no least element.

Define $P(n)$ to be the property that $n \notin S$.

Observe that $P(1)$ holds as $1 \notin S$

If $1 \in S$ and $S \subseteq \mathbb{N}$ then '1' is the least element of S which is a contradiction since S has no least element.

Let $K \in \mathbb{N}$ such that $\forall\, 1 \leq i \leq K, i \notin S$. Then $P(i)$ holds true $\forall\, 1 \leq i \leq K$.

If $K + 1 \in S$ then $K + 1$ is the least element of S a contradiction. Thus $K + 1 \notin S$ and $P(k + 1)$ holds true. (i.e.,) $\forall\, 1 \leq i \leq K$, $P(i) \Rightarrow P(K + 1)$.

Thus by PMI strong form, $P(n)$ holds true $\forall\, n \in \mathbb{N} \Rightarrow S = \emptyset$, a contradiction to our assumption that S is non empty. Therefore S has a least element. □

It is clear from the above example that the cardinality of $\mathbb{R}$ is greater than that of $\mathbb{N}$. Does there exist a set Y such that $|Y| > |\mathbb{R}|$? The answer is 'yes' and is proved by Cantor. The following results explain that given any set X, $\exists$ a set $\mathcal{P}(X)$ (the power set of X) whose cardinality is greater than that of X.

2.2 Pigeon Hole Principle (PHP)

If $(n + 1)$ pigeons occupy n holes, then there must be atleast two pigeons in a hole. That is if m is the average number of pigeons per hole, then it must be true that some pigeon hole contains at least m pigeons and some must contain at most m pigeons.

Example 2.4 In a group of 85 people show that there must be at least 8 people who were born in the same month.

Solution: Suppose the number of persons born in any month is at most 7. This means the group cannot exceed 84 which is contradiction.

Another form of PHP

Let f be a function from a finite set A to a finite set B, such that $|A| > |B|$. Then there exists an element $b \in B$ and distinct elements $a, a' \in A$ such that $f(a) = f(a') = b$.

Example 2.5 Show that any sequence of 7 distinct integers either has an increasing subsequence of 4 terms or it has a decreasing subsequence of 3 terms.

Solution: Let the sequence be $S = \{x_1, x_2, x_3, x_4, x_5, x_6, x_7\}$. Let a_i be the maximum number of terms in a increasing subsequence of S with x_i as its first term. Similarly b_i be the minimum number of terms in a decreasing subsequence of S with x_i as the last term. If there is some i for which $a_i \geq 4$, then the result is true. Similar situation for b_i.

So assume $a_i \leq 3$ and $b_i \leq 2$ for $i = 1, 2, 3, 4, 5, 6, 7$. Since there are 6 ordered pairs of the form (a, b) where $1 \leq a \leq 3, 1 \leq b \leq 2$, by PHP, there must be a situation that $(a_{i_1}, b_{i_1}) = (a_{i_2}, b_{i_2})$, with $1 \leq i_1 \leq i_2 \leq 7$ as there are 7 elements. That x_{i_1}, x_{i_2} distinct, imply $x_{i_1} < x_{i_2}$ or $x_{i_1} > x_{i_2}$.

- **Case i**: If $x_{i_1} < x_{i_2}$, then any increasing sequence with x_{i_2} can be extended on its left by prefixing x_{i_1} at the beginning. Hence $a_{i_1} > a_{i_2}$.
- **Case ii**: If $x_{i_1} > x_{i_2}$, any decreasing sequence with x_{i_1} at the last can be suffixed with x_{i_2} and hence $b_{i_2} > b_{i_1}$.

Hence the solution.

Example 2.6 205 letters are delivered to 40 apartments. Then show that there must be some apartment that received at most 5 letters.

Solution: Assume that every apartment received only <5 letters. Then the total number of letters delivered will be at most 160. Hence by averaging of PHP, there must be an apartment that received at most 5 letters.

Example 2.7 In a tennis tournament, there are n participants. Any two players play one game against each other. Show that, after a certain period of time, there must be two players who have completed the same number of games.

Solution: The situation is that, a player can finish 0 game, 1 game, 2 games and so on upto $(n - 1)$ games. That is if there is a player who finished $(n - 1)$ games, then there cannot be a player who completed 0 game. Hence the values 0 and $(n - 1)$ both cannot occur in the number of games. Hence the number of possible games played at any given point of time will be at most $(n - 1)$. But the number of possibilities for the number of games finished by any one of them is $n.(0, 1, 2, \ldots, n - 1$ games). Hence by PHP the solution follows.

2.3 Strong Form of PHP

In the above section, in the definition of PHP there is no indication on the capacity of the holes. One can say the activity as below.

There are n pigeons, m pigeon holes, with a restriction that each hole can accomodate only 't' pigeons. Find the minimum number of pigeons to be chosen so that one of the holes is in full capacity.

The answer will be $(t - 1) \cdot m + 1$ number of pigeons. One can make harder restrictions on the capacity of each pigeon hole. We have the following theorem which can be proved easily.

Theorem 2.8 *If there are m pigeon holes each having the capacity as* $t_1, t_2, \ldots, t_m$ *and suppose there are*

$$\sum_{i=1}^{m}(t_i - 1) + 1$$

pigeons to be put in the holes, then there is one hole with its full capacity.

Exercises

1. 161 books are distributed to 40 children. Show that there must be a child who gets 5 books.
2. In a family, there are 12 people.

   (a) Prove that at least two of the members were born on the same day of the week.
   (b) Prove that at least two of the members were born in the same month.
   (c) Suppose the house has 4 bedrooms, show that there must be at least 3 persons sleeping in the same bed room.

3. There are 50 boxes of chocolates. Each box contains no more than 24 chocolates. Show that there are at least 3 boxes containing the same number of chocolates.
4. Show that among 4 numbers one can find 2 numbers so that their difference is divisible by 3.
5. How many names do you need to select to guarantee that there are at least 2 names which have the same first two letters?
6. A group of maths professors hold a annual dinner. No new guests enter the dinner after it has begun and no one leaves the dinner hall until it has ended. Show that at least two professors shook the same number of distinct individual hands.
7. Suppose you randomly select k integers out of the first 2016 positive integers. What is the smallest k that guarantees that at least one pair of the selected integers will sum to 2017?

8. Given 39 positive integers, show that there must be at least 3 of them that have the same remainder when divided by 15.
9. Show that among 11 positive integers less than or equal to 20,there are 2 consecutive integers.
10. Given three permutations of the set $\{1, 2, \ldots, 28\}$, show that two of them have common subsequence of length four.
11. In a town 1500 babies were born in a day. Using PH principle show that two babies were born within a minute of each other.
12. There are 40 boxes containing marbles. Each box contains no more than 12 marbles. Show that there are atleast 4 boxes containing the same number of marbles.
13. Given 13 different 2-digit numbers, show that it is possible to select two numbers so that their difference is a 2-digit number with the first and second digit being the same.
14. A box is filled with balls with three different colors, red, blue and green. Find the least number of balls of each color that must be put in the box so that there are atleast 7 red balls, or atleast 9 blue balls or atleast 8 green balls.
15. If $(t + 1)$ numbers are chosen from the numbers $\{1, 2, 3, \ldots, 2t\}$, then show that there exist two numbers in the choice that will be coprime to each other.
16. A teacher distributes 200 chocolates to a class having 100 students with a condition that every student must get atleast one chocolate and no single student should get more than 100 chocolates. Show that it is possible to find a set of students who together get exactly 100 chocolates.

Binomial Theorem and Binomial Identities 3

3.1 Binomial Coefficients

Earlier, in an example we have seen that $C(n, r)$ stands for the coefficient of $a^r b^{(n-r)}$ in the expansion of $(a + b)^n$. We see a simple combinatorial argument for this as follows.

Theorem 3.1 (Binomial Theorem) $(a + b)^n = \sum_{r=0}^{n} C(n, r)a^r b^{(n-r)}$

Proof

$$(a + b)^n = \underbrace{(a + b)(a + b) \ldots (a + b)}_{n \text{ times}}.$$

It is the product of $(a + b)$, n times. On multiplication, we obtain the sum of terms. The powers of a and b on multiplication will be of the form $a^r b^{(n-r)}$, $0 \le r \le n$. Each such term is obtained by choosing a's from r of $(a + b)$ terms and remaining $(n - r)$ of $(a + b)$ terms for b's so that the product produces a power term of the form $a^r b^{(n-r)}$. Choosing r terms from n of $(a + b)$ terms can be done in $C(n, r)$ ways which is nothing but the coefficient of $a^r b^{(n-r)}$. Hence the theorem. □

Example 3.2 (1) $(1 + x)^n = \sum_{r=0}^{n} C(n, r)x^r$

(2) Setting $x = 2$ in (1), $\sum_{r=0}^{n} C(n, r)2^r = 3^n, n \ge 0$

(3) Setting $x = -1$, $\sum_{r=0}^{n} C(n, r)(-1)^r = 0, n \ge 1$

(4) $C(n, r) = C(n, n - r)$

(5) $\sum_{r=0}^{n} C(2n + 1, r) = 2^{2n}$.

R. Rama, *Topics in Combinatorics and Graph Theory*,
https://doi.org/10.1007/978-3-031-74252-1_3

Example 3.3 A teacher trainee has to work in a school for 5 d in August. The trainee is not allowed to work two consecutive days in the school. Find the number of ways that the trainee can choose the five days to work in the school.

Solution: The trainee can work for example on August 4, 6, 8, 10 and 12th. Without loss of generality let us assume that the trainee works on 5 d a_1, a_2, a_3, a_4 and a_5, each a_i is a day in August. Then the a_i's must satisfy a condition that $1 \leq a_1 < a_2 - 1 < a_3 - 2 < a_4 - 3 < a_5 - 4 \leq 27$, to avoid consecutive days. Instead of choosing $\{a_1, a_2, a_3, a_4, a_5\}$, we may choose the dates $\{a_1, a_2 - 1, a_3 - 2, a_4 - 3, a_5 - 4\}$ out of 27. Hence there are $C(27, 5)$ possible ways for the trainee to complete the work.

Binomial Identities

There are a large number of equations involving binomial coefficients. We state only a few of them.

Identity 3.1: Let m, n, r be positive integers. Then

$$\sum_{i=0}^{r} C(n, r)C(m, r - i) = C(m + n, r).$$

This identity is sometimes called as Vandermonde convolution.

Proof A combinatorial argument looks simple and elegant. Suppose there are n girls and m boys in a class and for an activity one requires r students from this group of $(m + n)$ students. The r students may be formed with i girls and $(r - i)$ boys. The i girls can be selected in $C(n, i)$ ways and $(r - i)$ boys can be selected in $C(m, r - i)$ ways. The selection of r-students then can be done in $C(m, i)C(n, r - i)$ ways. Hence the result. □

Identity 3.2:

$$C(n, r) = C(n - r, r - 1) + C(n - 1, r)$$

Proof Proved later in an example. □

Identity 3.3: (Pascal's identity)

$$C(n + 1, r) = C(n, r - 1) + C(n, r)$$

Proof The activity is to select r students from $(n+1)$ students (say). Let x be a member of the group. The presence of x in the r-subset makes the number of subsets as $C(n, r)$. If a subset does not contain x, such subsets can be counted by the formula $C(n, r-1)$. Hence

$$C(n+1, r) = C(n, r-1) + C(n, r)$$

□

Identity 3.4: Let n and r be non-negative integers with $r \leq n$. Then

$$C(n+1, r+1) = \sum_{k=0}^{n} C(k, r).$$

This is also known as column-sum property.

Proof Proof is by induction on n.

For $n = 0$, $C(1, r+1) = \begin{cases} 1 & \text{if } r = 0 \\ 0 & \text{otherwise} \end{cases}$

Assume $C(n, r+1) = \sum_{k=0}^{n-1} C(k, r), n \geq 1.$

Induction Step: Consider

$$\begin{aligned} \sum_{k=0}^{n} C(k, r) &= \sum_{k=0}^{n-1} C(k, r) + C(n, r) \\ &= C(n, r+1) + C(n, r) \\ &= C(n+1, r+1) \text{ (by Pascal identity).} \end{aligned}$$

□

3.2 Pascal Triangle

Earlier we have seen what are binomial coefficients. We saw some standard binomial identities also.

Pascal's Formula

$$C(n+1, k+1) = C(n, k+1) + C(n, k)$$

is called the Pascal's formula. A simple combinatorial proof was given earlier. The binomial coefficients are in the following triangular manner.

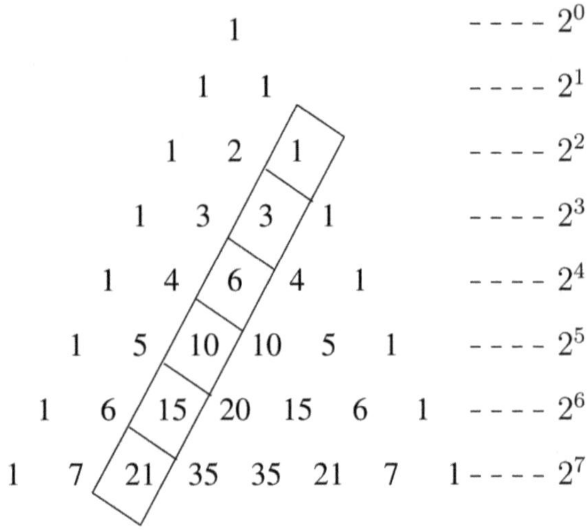

Fig. 3.1 Pascal triangle

$1 = C(0, 0)$
$C(1, 0) = 1$ $1 = C(1, 1)$
$C(2, 0) = 1$ $2 = C(2, 1)$ $1 = C(2, 2)$
1 3 3 1
1 4 6 4 1
1 5 10 10 5 1
1 6 15 20 15 6 1
... ...
$C(n, 0)$ $C(n, 1)$ $C(n, 2)$ ... $C(n, k)$ $C(n, k + 1)$ ... $C(n, n)$

Example 3.4 Consider the following Pascal's Triangle.

Write down the boxed coefficients as $C(n, k)$ and make a comment on the sum based on Pascal's Triangle (Fig. 3.1).

Solution: The sum is nothing but

$$C(2, 2) + C(3, 2) + C(4, 2) + C(5, 2) + C(6, 2) + C(7, 2)$$

The sum is equal to $C(8, 3) = 56$.

Hockey Stick Formula

The above observation can be generalised as

$$C(2, 2) + C(3, 2) + C(4, 2) + \cdots + C(n, 2) = C(n + 1, 3).$$

The validity of the formula can be shown using simple mathematical induction.

Example 3.5 Show that

$$1 \cdot 2 + 2 \cdot 3 + 3 \cdot 4 + \cdots + (n-1) \cdot n = \frac{(n-1)n(n+1)}{3}.$$

Solution: Using the hockey stick formula twice,

$$\begin{aligned} LHS &= 2(C(2,2) + C(3,2) + C(4,2) + \cdots + C(n,2)) \\ &= 2C(n+1,3) \\ &= \frac{(n-1)n(n+1)}{3} \end{aligned}$$

On similar lines, show the following.

$$1 \cdot 2 \cdot 3 + 2 \cdot 3 \cdot 4 + \cdots + (n-2)(n-1)n = \frac{(n-2)(n-1)n(n+1)}{4}$$

The general form of hockey-stick theorem is as follows.

Theorem 3.6

$$C(n,0) + C(n+1,1) + C(n+2,2) + \cdots + C(n+k,k) = C(n+k+1,k).$$

Example 3.7 Find the value of

$$C(47,4) + \sum_{i=1}^{5} C(52-i,3).$$

Solution:

$$\begin{aligned} &C(47,4) + \sum_{i=1}^{5} C(52-i,3) \\ &= C(47,4) + C(51,3) + C(50,3) + C(49,3) + C(48,3) + C(47,3) \\ &= \{C(47,4) + C(47,3)\} + C(51,3) + C(50,3) + C(49,3) + C(48,3) \\ &\quad \text{using } C(n,k) + C(n,k+1) = C(n+1,k) \\ &= \{C(48,4) + C(48,3)\} + C(51,3) + C(50,3) + C(49,3) \\ &= C(52,4). \end{aligned}$$

Example 3.8 If

$$(1+x)^n = a_0 + a_1x + a_2x^2 + \cdots + a_nx^n$$

then show that

$$\frac{a_1}{a_0} + \frac{2a_2}{a_1} + \frac{3a_3}{a_2} + \cdots + \frac{na_n}{a_{n-1}} \text{ is } \frac{n(n+1)}{2}.$$

Solution: From the binomial theorem, we know

$$C(n,k) = \frac{n(n-1)(n-2)\cdots(n-k+1)}{k!}$$

$$\begin{aligned}
&\frac{nC_1}{nC_0} + \frac{2nC_2}{nC_1} + \frac{3nC_3}{nC_2} + \cdots + \frac{nC_n}{nC_{n-1}} \\
&= \frac{n}{1} + \frac{2\cdot\frac{n(n-1)}{2!}}{n} + \frac{\frac{3n(n-1)(n-2)}{3!}}{\frac{n(n-1)}{2!}} + \cdots + \frac{n\cdot 1}{n} \\
&= n + (n-1) + (n-2) + \cdots + 1 = \frac{n(n+1)}{2}.
\end{aligned}$$

3.3 Multinomial Coefficients

A set is a collection of objects or elements where each element occurs exactly once. A multiset is a set that allows its objects to repeat. For example the set $S = \{a, a, a, b\}$ has a's occurring 3 times. However the order is immaterial.

Example 3.9 A bag contains five red balls, three yellow balls and two white balls to be arranged in a row. In how many different ways it can be done?

Solution: A permutation that deals with a multiset is called a multiset permutation.

Suppose each red, yellow, white ball is distinguishable, then the bag contains 10 distinct balls which can be arranged in 10! ways. The arrangement of 5 red balls alone can be done in 5! ways. Similarly yellow and white balls can be arranged in 3! and 2! ways respectively. Then labelling of the 10 balls to distinguish them can be done in $5! \times 3! \times 2!$ ways. If T is the required number of arrangements of the balls in a row, then $T \times 5! \times 3! \times 2! = 10!$ (Multiplication rule). Hence $T = \frac{10!}{5!3!2!}$.

The situation in the above example can be generalized as below. Suppose that $k_1, k_2, \ldots, k_m$ are the occurrence of the objects $a_1, a_2, \ldots, a_m$ where $k_1 + k_2 + \cdots + k_m = n$. Then the number of linear of the n element multiset will be

$$\frac{n!}{k_1!k_2!\ldots k_m!}$$

The following is the theorem which can be proved by an easy method.

Theorem 3.10 (Multinomial Theorem) *Let $n, m, k_1, k_2, \ldots, k_m$ be non-negative integers satisfying $k_1 + k_2 + \cdots + k_m = n$. Consider the multiset of n objects in which a_i, the ith object occurs k_i times for $1 \le i \le m$. Then the number of ways to linearly arrange the n elements will be*

$$\frac{n!}{k_1!k_2!\ldots k_m!}$$

It is called the multinomial coefficient and is denoted by $C(n, k_1, k_2, \ldots, k_m)$.

Example 3.11 Suppose that $k_1, k_2, \ldots, k_m$ are nonnegative integers such that $k_1 + k_2 + \cdots + k_m = n$, then the coefficient of $x_1^{k_1} x_2^{k_2} \ldots x_m^{k_m}$ in the product $(x_1 + x_2 + \cdots + x_m)^n$ is $C(n, k_1, k_2, \ldots, k_m)$.

Solution: $x_1^{k_1}$ is obtained from the product by choosing k_1 terms of

$$\underbrace{(x_1 + x_2 + \cdots + x_m)(x_1 + x_2 + \cdots + x_m) \cdots (x_1 + x_2 + \cdots + x_m)}_{n-times}.$$

This can be done in $C(n, k_1)$ ways. Similarly for the other powers in the product, there are $C(n, k_i)$ ways for $i = 2, \ldots, m$. Hence the coefficient of $x_1^{k_1} \cdots x_m^{k_m}$ is $C(n, k_1)C(n, k_2) \cdots C(n, k_m) = C(n, k_1, k_2, \cdots, k_m)$.

Actually the sum

$$\sum_{\forall k_i \ge 0} C(n, k_1, k_2, \ldots, k_m) = m^n.$$

Some Ball-Box Distribution Examples

Example 3.12 There are 5 balls and 3 boxes. Solve the following two problems under the 4 cases where the balls are either distinct or identical and where the boxes are either distinct or identical.

(a) Find the number of ways to distribute the balls in the boxes.
(b) Find the number of ways to distribute the balls in the boxes so that each box contains atleast one ball.

Solution:

Case 1: Let the 5 balls be distinct and 3 boxes be distinct. Let the boxes be labelled as B_1, B_2, B_3.

(a) Each ball has 3 choices. Hence there are $3^5 = 729$ ways.
(b) Suppose that the box B_1 has i balls. Then the remaining $(5 - i)$ balls are to be put in boxes B_2 and B_3 so that they are nonempty. Hence $(5 - i)$ must be >2. Hence the

number of ways to distribute the remaining $(5-i)$ balls in B_2 and B_3 will be $(2^{5-i}-2)$ as neither B_2 nor B_3 is empty. Hence the total number of ways will be

$$\sum_{i\geq 1, 5-i\geq 2} C(5,i)(2^{(5-i)}-2) = 150.$$

Case 2: Let the 5 balls be distinct and the boxes be identical. The distribution of the balls may be 1, 2, 2 or 1, 1, 3 as each box must contain atleast one ball and the boxes are identical. Let the balls be t_1, t_2, t_3, t_4 and t_5. For (1, 2, 2) choose any of the 5 balls to be put in box B_1 in $C(5, 1)$ ways. The remaining distribution can be done in $C(4, 2)C(2, 2)$ ways. Hence totally there are 30 ways. Since the boxes are identical, the number will be actually $\frac{30}{2} = 15$. The argument will be similar for (1, 1, 3) distribution and it will be 10.

Case 3 and Case 4 are '5 balls are identical 4 boxes are distinct' and '5 balls are identical and 3 boxes are also identical' respectively. Case 3 and 4 are left as exercise.

Example 3.13 (Repeated Counting Example) The problem is to find the number of ways to distribute 5 distinct balls into 3 distinct boxes so that no box is empty.

Solution 1: Let the boxes be named as B_1, B_2, B_3 and the balls be named as b_1, b_2, b_3, b_4 and b_5.

A ball can be put in box B_1 in 5 ways, second ball can be put in B_2 in 4 ways and the 3rd ball can be put in B_3 in 3 ways. Then distribute the remaining 2 balls in the 3 boxes in 3×3 ways. So the answer is $5 \times 4 \times 3 \times 3 \times 3 = 540$.

Solution 1 is incorrect. Suppose balls b_1, b_2 are put in B_1, b_3, b_4 are put in B_2 and b_5 in B_3. The distribution in Solution 1, that is, $(b_1 b_2, b_3 b_4, b_5)$ more than once as the method distinguishes between the order in which the balls are put into the boxes.

Write down the correct solution similar to case 1(b) of above example.

Example 3.14 Find the number of solutions to $x_1 + x_2 + x_3 = 5$, $x_1, x_2, x_3 \geq 0$.

Solution: The integral equation can be understood as distributing 5 identical balls into 3 distinct boxes and finding the number of ways of doing that. Hence it will be

$$\sum_{x_1=0}^{5} \sum_{x_2=0}^{5-x_1} \sum_{0}^{x_3=5-x_1-x_2} 1 = \sum_{x_1=0}^{5} (5 - x_1 + 1) = 21$$

The method of distributing identical balls into distinct boxes is connected to counting multisets, integer solutions of an integral equation and counting number of binary strings over '0 and 1' as follows.

Theorem 3.15 *(a) The number of multisets with k elements chosen from a n-element set (distinct n elements) is equal to the number of integer solutions to* $x_1 + x_2 + \cdots + x_n = k$*,* $x_i \geq 0$*,* $i = 1, 2, \ldots, n$*.*
(b) The number of binary strings of length n with k 0's and (n − 1) 1's is equal to the number of integer solutions to $x_1 + x_2 + \cdots + x_n = k$*,* $x_i \geq 0$*,* $1 \leq i \leq n$*.*
(c) The counting of (a) and (b) will be $C(k + n - 1, k)$*.*

Proof (a) If x_i denotes the number of occurrences of i in the multiset, then it clearly means the number of integer solutions is same as number of multisets with k-objects.
(b) The binary string is constructed with $(n - 1)$ 1's. Place the binary string in a row with 0's placed between two consecutive 1's which includes placing of 0's in the beginning and ending positions of the string. This means each binary string is specified by a tuple $(x_1, x_2, \ldots, x_n)$, $x_i \geq 0$ such that $x_1 + x_2 + \cdots + x_n = k$ where x_i's stands for the ith substring over 0's in the setup.
(c) It is easy to see that the number of binary strings with $(n - 1)$ 1's and r 0's will be $C(n - 1 + r, r)$. □

Remark 3.16 (1) The number of ways of distributing r identical balls into n distinct boxes will be $C(n - 1 + r, r)$.
(2) Number of r-permutations of n objects, each object repeating infinitely will be n^r.

Example 3.17 How many 9 digit binary numbers are there with exactly four 1's?

Solution: Distributing four 1's, there will be 5 positions where the 5 0's can be placed to form substrings of 0's. Hence the answer will be $C(9, 5)$ or $C(9, 4)$.

Example 3.18 Find the number of ways of distributing 10 identical balls into 3 distinct boxes where each box is nonempty.

Solution: Let x_i be the number of balls put in box B_i, $1 \leq i \leq 3$. Then the number of distributions will be the number of integral solutions to $x_1 + x_2 + x_3 = 10$, $x_i \geq 1$, $1 \leq i \leq 3$. Since each box must contain atleast one ball, the problem reduces to finding integer solutions of $x_1' + x_2' + x_3' = 7$, $x_i' \geq 0$. The answer is $C(3 - 1 + 7, 7) = C(9, 2) = 36$.

3.4 Binomial Theorem for Real Exponents

In Sect. 3.1, we saw the expansion of $(a+b)^n$ in a sum of the form

$$(a+b)^n = C(n,0)a^n b^0 + C(n,1)a^{n-1}b + \cdots + C(n,n)a^0 b^n$$

where $n \geq 0$ is an integer. Here $C(n,k)$ are called as binomial coefficients.

Also

$$C(n,r) = \frac{n!}{r!(n-r)!}.$$

Binomial theorem can be applied to expressions like $(a+b)^t$, where t takes negative values. For example

$$\begin{aligned}(a+b)^{-1} &= \frac{1}{a}\left(1+\frac{b}{a}\right)^{-1} \\ &= \frac{1}{a}\left(1-\frac{b}{a}+\frac{b^2}{a^2}-\frac{b^3}{a^3}+\cdots\right), a \neq 0\end{aligned}$$

When $a = 1$,

$$(1+b)^{-1} = 1 - b + b^2 - b^3 + \cdots,\ |b| < 1.$$

So we can see that there is a binomial theorem for a negative integer.

Also we say

$$C(n,k) = \frac{n!}{k!(n-k)!} = \frac{n(n-1)\ldots(n-k+1)}{1.2.3\ldots k}$$

where n, k are integers. However the term

$$\frac{r(r-1)\ldots(r-k+1)}{1.2.3\ldots k}$$

for any real 'r' is a meaningful calculation. For example when $r = \frac{1}{3}$, we have

$$\begin{aligned}&\frac{\frac{1}{3}\left(\frac{1}{3}-1\right)\left(\frac{1}{3}-2\right)\cdots\left(\frac{1}{3}-k+1\right)}{1.2.3\ldots k} \\ &= \frac{\frac{1}{3}\left(\frac{1}{3}-1\right)\left(\frac{1}{3}-2\right)\left(\frac{1}{3}-3\right)}{1.2.3.4} \text{ for } k=4 \\ &= -\frac{10}{243}.\end{aligned}$$

We now state 'Newton's Binomial Theorem' which gives the binomial theorem for any real exponent.

Newton's Binomial Theorem

For any real value r which is not a non negative integer,

$$(1+x)^r = \sum_{j=0}^{\infty} C(r, j)x^j, \text{ where } |x| < 1.$$

The proof for the above theorem comes from the Maclaurin series expansion of $(1+x)^r$ with $|x| < 1$.

Definition 3.19 The expression

$$C(n, r) = \frac{n(n-1)\ldots(n-r+1)}{1.2.3...r}$$

for $r \geq 0$ and n being any real number is called as 'generalised binomial coefficient'.

Example 3.20 Show that

$$C(-n, r) = (-1)^r C(n+r-1, r)$$

for n being positive integer.

Solution:

$$\begin{aligned} C(-n, r) &= \frac{-n(-n-1)\ldots(-n-r+1)}{1.2.3...r} \\ &= \frac{(-1)^r(n)(n+1)(n+2)\ldots(n+r-1)}{1.2.3...r} \\ &= \frac{(-1)^r(n+r-1)!}{r!(n-1)!} = (-1)^n C(n+r-1, r). \end{aligned}$$

Exercise

1. Let n be a positive integer with $n \leq 50$. Show that the sum

   $$C(98, 30) + 2C(97, 30) + 3C(96, 30) + \cdots + 69C(30, 30)$$

   is equal to $C(100, 32)$.
   Hint: Use the binomial identity

   $$C(n, m) + C(n-1, m) + \cdots + C(m, m) = C(n+1, m+1).$$

2. In Pascal triangle where does the sequence $(1, 1, 1, \ldots)$ appear?

3. How many terms are there in the expansion $(x + y)^{10}$? Find the coefficient of the center term whose power will be $x^5 y^5$.
4. The first three terms in the expansion of a binomial are 1, 10, 40. Find the expansion.
5. Find the value of
$$\frac{a_1}{2} + \frac{a_3}{4} + \frac{a_5}{6} + \cdots$$
if
$$(x + y)^n = a_0 + a_1 x + a_2 x^2 + \cdots + a_n x^n$$
Ans: $[\frac{2^n - 1}{n+1}]$.
6. Show that $1 + 3 + 6 + \cdots + C(n, 2) = C(n + 1, 3)$.
7. 20 students enter a contest at school. The school offers a first,second and third prize. How many different combinations of 1st, 2nd and 3rd place winners can there be?
8. A vegetable vendor has four different types of onions,three different types of chillies and 8 different types of greens. How many vegetable combinations can the vendor create if each bag has one type of onion and one type of chilli?
9. In a 12-storey house ten people enter a lift cabin. It is known that they will leave the lift in groups of 2, 3 or 5 people at different storeys. Find the number of ways they can do so if the lift does not stop upto second storey.
10. A committee of 5 is to be formed from 10 women and 7 men. If the committee must contain more women than men, find the number of ways it can be done.
11. A set contains $(2n + 1)$ elements. If the number of subsets of the set containing n elements is 4096, find the value of n.
12. How many integer valued solutions are there to each of the following equations?

    (a) $x_1 + x_2 + x_3 = 14, \forall x_i \geq 0$
    (b) $x_1 + x_2 + x_3 + x_4 = 22, \forall x_i \geq 1$

13. A bag has 450 identical balls. The balls are to be distributed in 65 boxes. One bag marked with '$\star$' will receive atleast 10 balls. 34 bags marked with '$\star\star$' must contain atleast 1 ball in each of them. There is no condition for the remaining 30 bags. Find the number of ways that the balls can be distributed so that there can be some left over balls in the original bag.
14. Determine the coefficient of $x^{10} y^{20} z^{120}$ in $(2x + 3y + z^2)^{100}$.
    Hint: Here z appears in multiple copies.
15. How many coefficients in the polynomial $(1 + x + x^2)^n$ are not divisible by 3?
16. Let $1 \leq r \leq n$. How many integer sequences.
$$1 \leq a_1 < a_2 < a_3 < \cdots < a_r \leq n$$
satisfy $a_i \equiv i \pmod 2$ for all i?
17. How many words can be made from the letters of EAZALA?

18. Write down the binomial expansion of the following functions.

   (i) $f(x) = (2+3x)^{-1}$, $|x| < \frac{2}{3}$
   (ii) $f(x) = (1+5x)^{-\frac{1}{2}}$, $|x| < \frac{1}{5}$
   (iii) $f(x) = (1+x)^{-3}$, $|x| < 1$

19. Show that
$$C(r,k) = C(r-1,k-1) + C(r-1,k)$$
for any real number r.
20. Let n and k be two non-negative integers. Then show that

   (i) $C(n,k) = 2t$ if n is even and k is odd, $t \in \mathbb{N}$.
   (ii)
$$C(n,k) = C\left(\left\lfloor \frac{n}{2} \right\rfloor, \left\lfloor \frac{k}{2} \right\rfloor\right)$$
   when both n and k are even or n and k are being odd.
   (iii)
$$C(n,k) = C\left(\left\lfloor \frac{n-1}{2} \right\rfloor, \left\lfloor \frac{k}{2} \right\rfloor\right)$$
   when n is odd and k being even.

21. In a Pascal triangle of binomial coefficients, show that the number of odd coefficients in row m is 2^t where 't' is the number of 1's in the binary representation of m.
22. Prove that for all integers $m \geq 2$,
$$\sum_{t=2}^{m} t(t-1)C(m,t) = 2^{m-2}(m^2 - m).$$
23. Show that
$$(a+b)^n = \sum_{t=0}^{n} C(n,t)a^{n-t}b^t$$
using mathematical induction.
24. The sequence $a_0, a_1, a_2, \ldots, a_n$ of positive integers is said to be 'log-concave' if $a_t^2 < a_{t-1}a_{t+1}$ for $1 \leq t < (n-1)$. Verify whether the sequence
$$C(n,0), C(n,1), C(n,2), \ldots, C(n,n)$$
is log-concave or not.

4 Partitions

4.1 Set Partitions

Stirling Number of Second Kind

Suppose S is a set having n elements. Then we know that the number of subsets of S is 2^n. That is

$$\sum_{i=0}^{n} C(n, i) = 2^n$$

We have seen the proof of this earlier.

In an identical fashion, one may understand the number of partitions of a set having m-elements into t nonempty subsets. The situation can be understood as, the counting of number of ways of distributing m distinguishable balls into t identical boxes. We give the definition of the Stirling numbers of the second kind as follows.

Definition 4.1 The number of partitions of a set having m elements into t nonempty subsets is denoted by $s'(m, t)$. $s'(m, t)$ are called the Stirling numbers of second kind.

Example 4.2 When $m = 0$, $s'(0, 0) = 1$
When $m \neq 0$, $s'(m, 0) = 0$ and $s'(0, t) = 0$
When $m = t$, $s'(m, m) = 1$, $m \geq 1$ and $s'(m, t) = 0$ when $m < t$

Example 4.3 Find $s'(4, 2)$ and $s'(4, 3)$

Solution: To distribute balls, let b_1, b_2, b_3, b_4 be the balls to be distributed into 2 boxes so that no box is empty. The possible distributions are like,

$(\{b_1\}, \{b_2, b_3, b_4\}) - C(4, 1)$ ways

R. Rama, *Topics in Combinatorics and Graph Theory*,
https://doi.org/10.1007/978-3-031-74252-1_4

$(\{b_1, b_2\}, \{b_3, b_4\}) - 2C(4, 2)$ ways
Total = $C(4, 1) + \frac{1}{2}C(4, 2) = 7$

So $s'(4, 2) = 7$. Similarly, show $s'(4, 3) = 6$.

Remark 4.4 In general it can be shown that the number of ways of distributing m distinguishable balls into t boxes will be

$$\frac{1}{m!}\sum_{i_j>0} c(m, i_1, i_2, \ldots, i_t)$$

where $i_1 + i_2 + \cdots + i_t = m$

The Stirling numbers of the second kind satisfy the following recurrence relation.

Theorem 4.5 *Let m and t be integers greater than zero. Then,*

$$s'(m, t) = s'(m-1, t-1) + t \cdot s'(m-1, t).$$

Proof The proof can be given in terms of ball-box distributing activity. Suppose that we need to distribute m distinct balls into t boxes with no nonempty box. The following are the two cases.

Case (i): Let the box containing the first ball has only that ball in it. Then removing it from the list, we have $(m-1)$ balls and $(t-1)$ boxes. Hence the number of ways will be

$$C(t, 1) \times s'(m-1, t-1).$$

Case (ii): Let the box containing the first ball has atleast 2 balls in it. Then removing the first ball gives the distribution of $(m-1)$ distinct balls into t boxes. Then we need to put back the first ball, for which we have t choices. Hence the total number will be $t \cdot s'(m-1, t)$. Combining Case (i) and Case (ii), we have the total number of distributions $s'(m, t)$. □

The following table gives the value of $s'(m, t)$ for some values of m and t. Hence we assume,

$$s'(m, 0) = s'(0, t) = \begin{cases} 1 & \text{if } m = 0 \\ 0 & \text{if } m \neq 0 \end{cases}$$

Example 4.6 Find the number of surjective functions from a set S containing 6 elements to a set $\overline{S}$ containing 4 elements.

m	0	1	2	3	4	5
$s'(m, 0)$	1	0	0	0	0	0
$s'(m, 1)$		1	1	1	1	1
$s'(m, 2)$			1	3	7	15
$s'(m, 3)$				1	6	25
$s'(m, 4)$					1	10
$s'(m, 5)$						1

Solution: Let g be a surjective function from S to $\overline{S}$. Such a function actually defines a partition of S. That is, for example if $S = \{1, 2, 3, 4, 5, 6\}$, $\overline{S} = \{b_1, b_2, b_3, b_4\}$ and $g(1) = b_1, g(2) = b_2, g(3) = b_3, g(4) = b_4, g(5) = b_1, g(6) = b_2$. Then the set S is partitioned as $\{1, 5\}, \{2, 6\}, \{3\}, \{4\}$. Therefore, as the t-blocks labelled and distinct, we have 6! ways to do this partitioning. Hence it is $6!s'(6, 4) = 6! \times 65$.

Bell Numbers

Suppose the problem is to find all partitions of a set S containing m elements. Then what we get is the following definition.

Definition 4.7 The number of all set partitions of a set S containing m elements into nonempty parts is in general denoted by $\mathscr{B}(m)$ and is called the 'Bell number'.

The following formula can be proved easily using the same technique of ball-box distribution activity.

$$\mathscr{B}(m + 1) = \sum_{i=0}^{m} C(m, i)\mathscr{B}(i).$$

Example 4.8 Find $\mathscr{B}(4)$.

Solution: We assume $s'(4, 1) = s'(4, 4) = 1$. We have earlier computed $s'(4, 3)$ and $s'(4, 2)$. Hence $\mathscr{B}(4) = 1 + 7 + 6 + 1 = 15$.

In terms of ball-box activity, we write the following theorem without proof.

Theorem 4.9 *The number of ways of distributing m distinct balls into t identical boxes, so that no box is empty is $\mathscr{B}(m)$.*

From the above it is clear that the formula for finding the number of ways of distributing m distinct balls into t boxes for $m \le t$ will be $\mathscr{B}(m)$ and for $m > t$, it will be

$$\sum_{i=1}^{t} s'(m,i)$$

We already know what a factorial of a number n is i.e.,

$$n! = \begin{cases} 1 \cdot 2 \cdot 3 \cdots n & \text{if } n \geq 1 \\ 1 & \text{if } n = 0 \end{cases}$$

We now see two new factorials called falling factorial and rising factorial or Pochhammer of a number n.

$$(n)_k = \begin{cases} n(n-1)(n-2)\cdots(n-k+1) & \text{if } k \geq 1 \\ 1 & \text{if } n = 0 \end{cases}$$

will be called as 'falling factorial' of n upto k.

$$[n]_k = \begin{cases} n(n+1)(n+2)\cdots(n+k-1) & \text{if } k \geq 1 \\ 1 & \text{if } n = 0 \end{cases}$$

will be called as 'rising factorial' of n upto k.

Example 4.10 For $n = 4$,

$$\begin{aligned} 4! &= 1 \cdot 2 \cdot 3 \cdot 4 = 24 \\ (4)_3 &= 4(4-1)(4-2) \\ &= 4 \cdot 3 \cdot 2 = 24 \\ [4]_3 &= 4 \cdot 5 \cdot 6 = 120 \\ (4)_4 &= 4! \end{aligned}$$

The following result connects the falling factorial with Stirling numbers.

Theorem 4.11 *Let $m \geq 1$. Let $s'(m,t)$ denote the Stirling numbers of the second kind. Then*

$$x^m = \sum_{t=1}^{m} s'(m,t)(x)_t$$

Proof Clearly

$$(x)_t = x(x-1)\cdots(x-t+1)$$

is a polynomial in x. We can write

$$(x)_{t+1} = (x)_t(x-t).$$

We already know what $s'(m,t)$ is. The proof is by induction on m.
When $m=1$,

$$\begin{aligned} x &= \sum_{t=1}^{1} s'(1,t)(x)_t \\ &= x \end{aligned}$$

Assume the equality for index $(m-1)$. i.e.,

$$x^{m-1} = \sum_{t=1}^{m-1} s'(m-1,t)(x)_t$$

Consider

$$\sum_{t=1}^{m} s'(m\cdot t)(x)_t$$

We know the recursion satisfied by $s'(m,t)$ as

$$\begin{aligned} s'(m,t) &= s'(m-1,t-1) + ts'(m-1,t) \\ \sum_{t=1}^{m} s'(m,t)(x)_t &= \sum_{t=1}^{m-1} [s'(m-1,t-1) + ts'(m-1,t)](x)_t \\ &= \sum_{t=1}^{m-1} s'(m-1,t-1)(x)_t + \sum_{t=1}^{m-1} ts'(m-1,t)(x)_t \\ &= \sum_{t=1}^{m-1} s'(m-1,t-1)(x)_{t+1} + \sum_{t=1}^{m-1} ts'(m-1,t)(x)_t \\ &= \sum_{t=1}^{m-1} s'(m-1,t)(x)_t(x-t) + \sum_{t=1}^{m-1} ts'(m-1,t)(x)_t \\ &= x\sum_{t=1}^{m-1} s'(m-1,t)(x)_t - \sum_{t=1}^{m-1} ts'(m-1,t)(x)_t \\ &\quad + \sum_{t=1}^{m-1} ts'(m-1,t)(x)_t \\ &= x\sum_{t=1}^{m-1} s'(m-1,t)(x)_t \\ &= x\cdot x^{m-1} \text{(by induction assumption)} \\ &= x^m. \end{aligned}$$

□

4.2 Integer Partitions

Integer Compositions

Consider the equation

$$x_1 + x_2 + \cdots + x_t = m \tag{4.1}$$

with each x_i being a non-negative integer, for $1 \leq i \leq t$. For example the equation $x_1 + x_2 + x_3 = 4$ will have the tuple $(1, 1, 2)$ as its solution. That is $x_1 = 1, x_2 = 1, x_3 = 2$ here. In general one can represent the solution of (4.1) as a t-tuple $(a_1, a_2, \ldots, a_t)$ where each a_i is a non negative integer. Such a solution is some times referred to as an integer solution. Such solutions are called weak composition of an integer 'm' into 't' parts. The problem is to find the number of possible weak compositions of n.

Let us consider a modified situation where the composition of solution is $(a_1, a_2, \ldots, a_t)$ where each $a_i > 0$ instead of $a_i \geq 0$, a_i being an integer for $1 \leq i \leq t$. Such a t-tuple is usually referred to as composition.

Consider the activity of distributing n chocolates to 't' students so that each student gets atleast one chocolate. Let the students be arranged in a queue like $s_1, s_2, \ldots, s_t$ being the queue arrangement. Let the 'n' chocolates be arranged linearly on a table, (say). So, student 's_1', the first student in the queue picks some x_1 chocolates by indicating his selection by standing against position 'x_1' in the linear arrangement of the chocolates. It means the student has to select his chocolates from the left most position in the arrangement of chocolates. The activity continues till the student s_t gets his chocolates. One can see that in the linear arrangement of chocolate there are $(n - 1)$ gaps in between the n chocolates. Each student has to choose a position in this $(n - 1)$ gaps to indicate his selection of chocolates. Hence there are$C(n - 1, t - 1)$ possible choices.

We can understand the situation where each $x_i \geq 0$ as an activity where some students do not get anything. We can call this as an unfair distribution of chocolates. So the situation can be understood as some students are not allowed to take any chocolates or may be the distributor says that this student has a chocolate for him already, which is not true. So we can reframe the above activity as distributing $(n + t)$ chocolates to k-students so that each student get atleast one chocolate. This count will be $C(n + t - 1, t - 1)$. Now take away one chocolate from each so that some of these t students may not get any chocolate. Hence the number of weak compositions of n is $C(n + t - 1, t - 1)$.

4.3 Ferrers' Diagram

Integer partitions can be viewed using Ferrers' diagrams or Young's diagram. Suppose $\lambda = (5, 4, 1)$ is a partition. Clearly λ is a partition of 10. We can say λ has 3-parts or λ is a 3-tuple. The diagrammatic understanding of λ is as below:

Every unit is represented by a unit square. To represent the number 5; we draw 5 unit squares linearly with left justification. That is

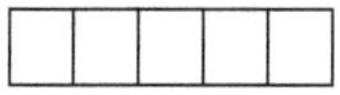

represents 5. We can say this is the first row representation of 5 in λ. For the number 4 in λ, we draw 4 unit squares, left justified, just as second row below the first row. i.e.,

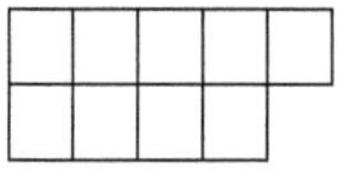

is the diagram for the first 2 components of λ. Proceeding as above, the diagram

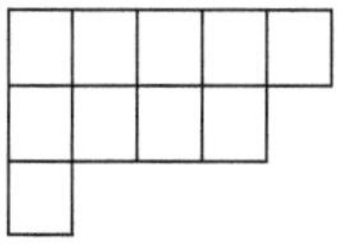

completely represents the partition $\lambda = (5, 4, 1)$ of 10. Clearly area of λ is 10 unit squares. There are more than one partition of 10 and hence different Ferrers' diagram. As the number of partitions grow rapidly with 'n' the number of Ferrers' diagrams increase.

Example 4.12 Let $n = 4$. (2, 1, 1), (3, 1), (1, 1, 1, 1) are partitions of 4. The corresponding Ferrers' diagram are

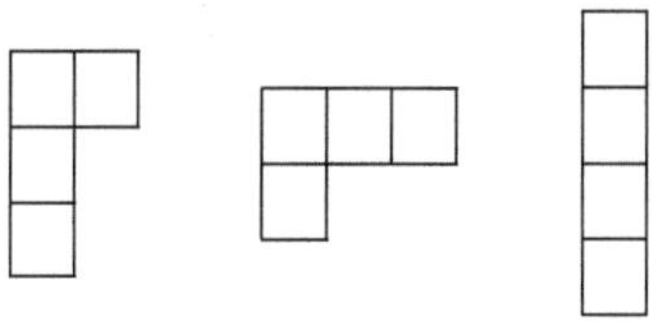

respectively.

Now Ferrers' diagram can be formally defined as below:

$$FD(\lambda) = \{(m, n) \in \mathbb{N} \times \mathbb{N} \mid 1 \leq m \leq k, 1 \leq j \leq \lambda_m\}$$

Note that area of λ is equal to area of $FD(\lambda)$.

Definition 4.13 Conjugate of a given Ferrers' diagram can be obtained by exchanging rows and columns.

Example 4.14 Conjugate of

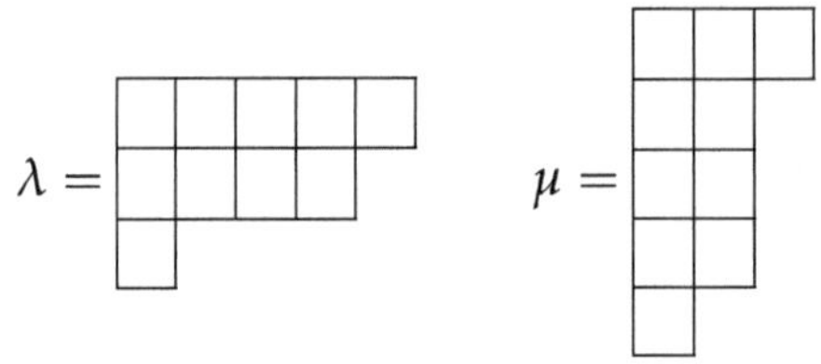

Clearly area of λ is same as area of μ.

Self-conjugate Ferrers' diagram satisfy the property that it is same as its conjugate. For example the Ferrers' diagram

o o o
o o o
o o

is the same as its conjugate. Ferrers' diagrams can be used to prove some nice properties of partitions of numbers. Let $p(n)$ denotes the number of different partitions of an integer n and $p_k(n)$ denotes the number of partitions of n into exactly k parts. Also let $p^*(n)$ denotes the number of self-conjugate partitions of n.

Consider a partition λ of an integer 'n' with its largest partition as 'k' (say). By removing this partition from λ one can get a partition $\bar{\lambda}$ and it will be a partition of $(n-k)$. No part in $\bar{\lambda}$ is larger than k. This preceding process of modifying λ to $\bar{\lambda}$ can be reverted. For example consider the Ferrers' diagram for $\lambda = (5, 2, 2, 1)$.

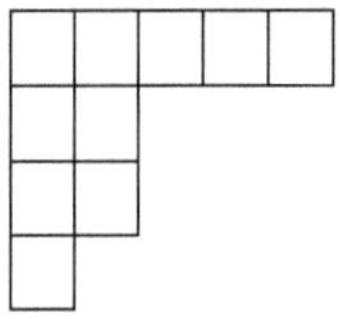

$\bar{\lambda} = (2, 2, 1)$ and its Ferrers' diagram will be

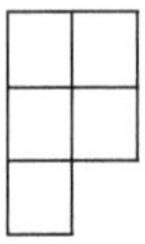

So we can say the following property which identifies a 1-1 relation between the two types of Ferrers' diagrams.

Theorem 4.15 *Let n, k be positive integers and $p_n(k)$ denote the number of partitions of n with its largest part being k. Let $p'_{n-k}(k)$ denote the number of partitions of $(n-k)$ with none of its part being larger than k. Then*

$$p_n(k) = p'_{n-k}(k).$$

Counting $p_n(k)$ is not so easy.

Example 4.16 Consider $n = 6$ and $k = 3$.
The partitions are (3, 3), (3, 2, 1), (3, 1, 1, 1).
So $p_6(3) = 3$.

For any integer n, there is exactly one partition with $k = n$ and $k = 1$. That is (n) and $\underbrace{(1, 1, \ldots, 1)}_{n\text{parts}}$ are the only partitions. Also for $n \geq 2$, $p_n(n-1) = 1$.

The following recursion is not difficult to prove.

Theorem 4.17 *For integers, n and k*

$$p_n(k) = p_{n-1}(k-1) + p_{n-k}(k), 1 < k < n.$$

The following result can be established by direct constructions to show that there exists a 1-1 correspondence between the number of self-conjuge partitions of an integer 'n' with the set of partitions of 'n' with distinct odd number of parts. Let $p^o(n)$ denote the number of partitions of n with distinct odd parts.

Theorem 4.18 *Let n be a positive integer then $p^*(n) = p^o(n)$.*

Example 4.19 Consider the following partition of 21.

$$\lambda = (6, 5, 4, 3, 2, 1).$$

The corresponding Ferrers' diagram will be

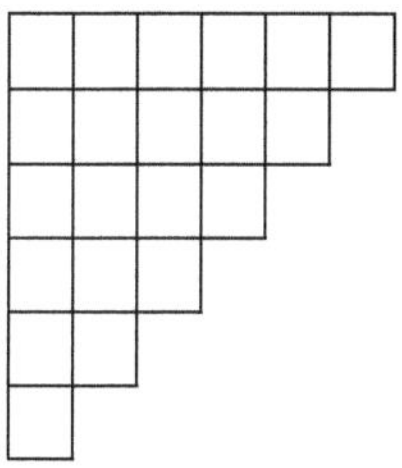

Modify λ in steps so that we get a partition

$$\bar{\lambda} = (11, 7, 3)$$

For any self-conjugate partition, the Ferrers' diagram has symmetry along the diagonal. So we can see that the number of unit squares below the first row lying in the first column is always odd. Now add this many number of unit squares in the first row of $\bar{\lambda}$. Continue the same for second row against second column. We get $\bar{\lambda}$ satisfying the given property of having odd number of distinct parts.

Exercises

1. Find the number of partitions of 19 with odd number of distinct parts.
2. Determine whether the following partitions are self-conjugate or not.

   (i) (5, 4, 3, 2, 1)
   (ii) (9, 6, 3, 2, 1)
   (iii) (14, 7, 2, 1, 1).

3. Show that

   (i) $S'(n, 2) = 2^{n-1} - 1$ for $n \geq 2$
   (ii) $S'(n, n-1) = C(n, 2)$ for $n \geq 1$.

4. Show that the Bell numbers $\mathcal{B}(n)$ satisfy the following recurrence.

$$\mathcal{B}(n) = \sum_{j=1}^{n} C(n-1, j-1)\mathcal{B}(n-j).$$

Permutations, Combinations and Cycles 5

5.1 Permutations on Sets and Multisets

A permutation means simply an arrangement of objects. For example the permutations of the three letters x, y, z are

$$xyz, xzy, yxz, yzx, zxy, zyx.$$

Hence the total number of arrangements of x, y, z will be 6. If it is the arrangement of letters w, x, y, z, then the total number of arrangements will be 24. If it is the arrangement of letters v, w, x, y, z, then the total number of such arrangements will be 120. The number of permutations increases dramatically as the size of the set increases. In the calculations multiplication principle is used for counting.

Example 5.1 Suppose four persons w, x, y, z are in a running race. Determine the possible ways the runners can finish first and second.

Solution: Any one of w, x, y, z can be the winner. The remaining 3 will then be the runners. Also each of the four runners can finish first. Hence applying multiplication principle, there will be $4 \times 3 = 12$ possible ways the runners can finish first and second. The possibilities are described as in Fig. 5.1.

We have the following formal definition of permutation of different sizes.

Definition 5.2 A permutation of any r objects taken from a set of n objects is an arrangement of the r objects. Number of such arrangements will be denoted by $P(n, r)$.

R. Rama, *Topics in Combinatorics and Graph Theory*,
https://doi.org/10.1007/978-3-031-74252-1_5

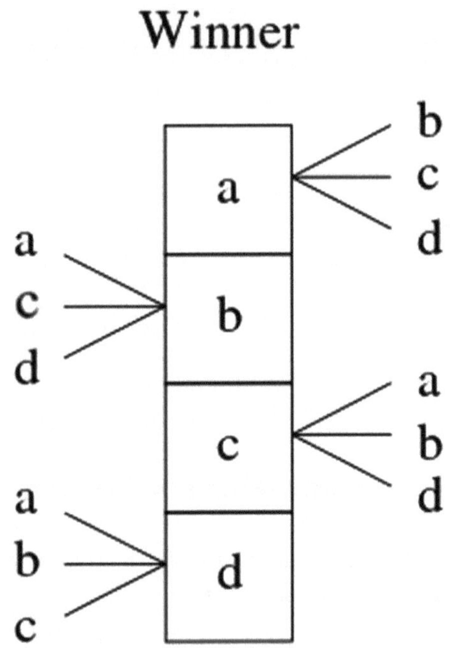

Fig. 5.1 Possibilities

In the following theorem, we actually find $P(n, r)$ for a given n and r using multiplication principle.

Theorem 5.3 *The number of permutations of r objects taken from a set of size n is*

$$P(n, r) = \frac{n!}{(n-r)!} = n(n-1)(n-2)\cdots(n-r+1).$$

Proof The activity is to choose r objects from a set containing n objects. So we have, the first element can be selected in n ways, the second element can be selected in $(n-1)$ ways, the third element can be selected in $(n-2)$ and so on. The rth element can be selected in $(n-r+1)$ ways. Then the arrangement activity of r-elements will be complete in sequence. Hence by multiplication principle we have

$$P(n, r) = n(n-1)(n-2)\cdots(n-r+1)$$

□

Definition 5.4 While counting sets, one often needs to compute the product of the form $n(n-1)(n-2)\cdots(n-r+1)\cdots 2\cdot 1$. This product will be called as 'n factorial' and is denoted by $n!$.

Assuming $0! = 1$, the number of n permutations of n objects is $P(n, n) = n!$.

Example 5.5

$$P(7, 3) = 210$$
$$P(4, 1) = 4$$
$$P(4, 4) = 24$$

Example 5.6 Find the number of words of length 4 formed using the English alphabet set when

(i) no repetition of a symbol is allowed in the thus formed word.
(ii) repetition of a symbol is allowed in the thus formed word.

Solution:

(i) In this case one can see words like 'wood', 'good' cannot be formed. Since we need to use a symbol only once, the number of such words will be 26.25.24.23. That is it is $P(26, 4)$.
(ii) The number of words in this case will be 26^4. Here we allow words of the form 'wood', 'good', 'wool' and so on.

Example 5.7 How many ways one can arrange the seven letters 'ARRANGE'?

Solution: There are two R's and two A's. These are indistinguishable. Let us assume that $A_1 = A_2 = A$ and $R_1 = R_2 = R$. The the set $\{A_1, A_2, R_1, R_2, N, G, E\}$ has 7 elements and the number of arrangements will be 7!. However we see that $A_1R_1R_2A_2NGE$ and $A_2R_1R_2A_1NGE$ are one and the same and such words must be counted only once. The permutation of R_1 and R_2 will be 2! and the permutation of A_1, A_2 will be 2!. Hence we must divide 7! by $2! \times 2!$ to get the correct counting which is $\frac{7!}{2!2!} = 1260$.

Example 5.8 A linear arrangement of balls are to be made. There are 6 red balls and 4 green balls available. Find the number of arrangement of the 10 balls so that

(i) the green balls are always together
(ii) the red balls are always together.

Solution:

(i) Since the green balls must be together always, let the G^3 be the togetherness of the green balls. Then problem is to find the arrangement of 6 red balls and G^3. Hence it will be 7!.
(ii) The situation is identical to (i) and the answer is 5!.

Definition 5.9 (i) Permutation or arrangement of r objects from a n-object set in a row will be called a linear permutation.
(ii) Permutation or arrangement of r objects from a n-object set in a circular fashion is called circular arrangement or circular permutation.

Example 5.10 In how many ways can 7 toys be arranged in a circular fashion?

Solution: In Fig. 5.2 a typical situation will be.

Whether the arrangement is seen clockwise or counter clockwise, actually does not matter. That is they are the same. The enumeration is done by fixing a position in the circular arrangement. That is with respect to T_i the remaining $T_j, j \neq i$ are to be permutated linearly. Hence the number of circular arrangement will be 6!.

A direct simple theorem to count the number of circular arrangement of n objects will be as follows.

Theorem 5.11 *For n different objects, the number of ways they can be arranged in a fixed circle is $(n-1)!$.*

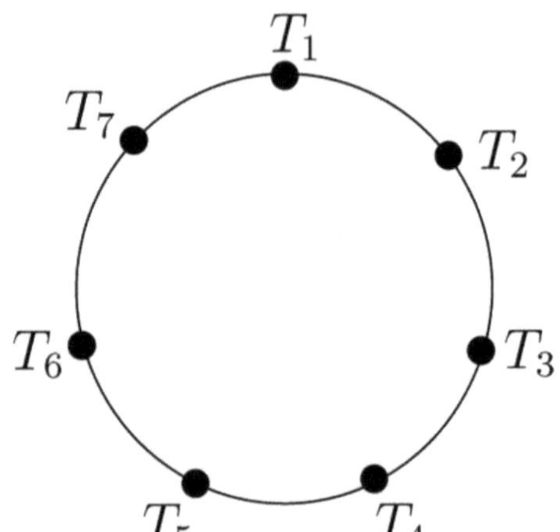

Fig. 5.2 Circular fashion

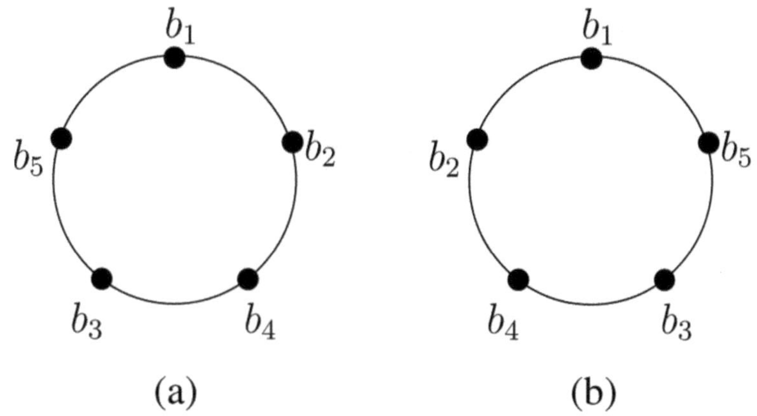

Fig. 5.3 Beads arrangement

The problems on circular permutations get more exciting where restrictions are imposed on the arrangements of the objects.

Example 5.12 Find the number of arrangement of 5 different coloured beads on a necklace.

Solution: Let the beads be numbered as b_1, b_2, b_3, b_4, b_5. Some arrangements will be.

In Fig. 5.3a and b the positions of b_2, b_3, b_4, b_5 with respect to b_1 have flipped. But actually they are same. Hence in this case number of different arrangements will not be $(n-1)! = (5-1)! = 4!$ but it will be $\frac{4!}{2}$.

Remark 5.13 The counting of circular arrangements can be viewed in two ways giving rise to two different counting.

(i) When the positions on the circle are numbered,the arrangement is simply a linear one.
(ii) When the positions are not numbered,the arrangement is not simply a linear one but some repeated situation will occur.

Example 5.14 10 students are made to sit in a round table such that two particular students sit on either side of teacher. Find the total number of such arrangements.

Solution: Three positions on the table are fixed,one by the teacher and two particular students on either side of the teacher. Treating this arrangement as one unit, we need to place remaining 8 students. This can be done in 8! ways. Also the students sitting on the either side of the teacher can be done in 2 ways. Therefore the total number of ways will be $8! \times 2$.

Example 5.15 In how many ways 8 boys and 4 girls can sit around in a circular table so that no two girls sit together?

Solution: 8 boys can sit in a circle in 8! ways. There are 8 spaces in between them. One can place the 4 girls in this 8 spaces in $P(8, 4)$ ways. Hence the total number of ways will be $8! \times P(8, 4) = \frac{8! \times 8!}{4!}$.

Remark 5.16 In general,the number of circular permutations of n objects taken r at a time will be $\frac{P(n.r)}{r}$ if the arrangement along clockwise and counter clockwise are taken as different. Otherwise it will be $\frac{P(n,r)}{2 \times r}$.

Example 5.17 How many ways can 10 people be seated round a table if there are only 7 chairs?

Solution: The activity is first arrangement of 7 out of 10 persons and there is no distinction between the arrangement in clockwise and counter clockwise direction. Hence the answer will be $\frac{P(10,7)}{7}$.

Example 5.18 How many necklaces of 12 pearls each can be made from 30 pearls of different sizes.

Solution: The answer will be $\frac{P(30,12)}{2 \times 12}$ (why?).

5.2 Combinations on Sets and Multisets

Combinations are simply selection of required number of elements without arrangement. For example the combination or subsets of size 2 that can be selected from the 4 letters of the word 'BIRD' are simply

$$\{BI\}, \{IR\}, \{RD\}, \{BR\}, \{BD\}, \{ID\}.$$

In general the number of r-element subsets of a set of n element is of central importance in enumerative combinatorics. Hence we have a formal definition as follows.

Definition 5.19 A combination is a subset of elements of a set. The number of r-element subsets of a n element set is denoted by $C(n, r)$ and is read as 'n choose r'.

Example 5.20 Find the combinations (i.e. subsets) of size 2 and 3 taken from the set $X = \{x, y, z\}$.

Solution: The total number of subsets of X is $2^3 = 8$.

Subsets of size 2:

$$\{x, y\}, \{x, z\}, \{y, z\}$$

Subsets of size 3:

$$\{x, y, z\}$$

Note:
The subset $\{x, z\}$ is same as $\{z, x\}$.

Theorem 5.21 *For all non-negative integers $r \leq n$, the number of combinations of size r which can be selected from a set of size n is $C(n, r)$*

$$C(n, r) = \binom{n}{r} = \frac{n!}{r!(n-r)!}$$

Proof Earlier we have seen what is $P(n, r)$. It is the arrangement of r elements chosen from an n element subset. It is equal to $\frac{n!}{(n-r)!}$. But the choosen r elements can be permutated in $r!$ different ways. But we need to count only one of these $r!$ permutations to obtain the number of combinations. So

$$C(n, r) = \binom{n}{r} = \frac{P(n, r)}{r!} = \frac{n!}{r!(n-r)!}.$$

□

Notation: $C(n, r)$ is alternatively denoted as $\binom{n}{r}$. They are also binomial coefficients because they are the coefficients in the binomial expansion.

Example 5.22 Some binomial expansions are,

$$(a+b)^2 = C(2,0)a^2 + C(2,1)ab + C(2,2)b^2$$
$$= a^2 + 2ab + b^2$$
$$(a+b)^3 = \binom{3}{0}a^3 + \binom{3}{1}a^2b + \binom{3}{2}ab^2 + \binom{3}{3}b^3$$
$$= a^3 + 3a^2b + 3ab^2 + b^3$$
$$(a+b)^4 = \binom{4}{0}a^4 + \binom{4}{1}a^3b + \binom{4}{2}a^2b^2 + \binom{4}{3}ab^3 + \binom{4}{4}b^4$$
$$= a^4 + 4a^3b + 6a^2b^2 + 4ab^3 + b^4$$
$$\vdots$$
$$(a+b)^n = \binom{n}{0}a^n + \binom{n}{1}a^{(n-1)}b + \binom{n}{2}a^{(n-2)}b^2 + \cdots + \binom{n}{n}b^n$$

Example 5.23 How many ways can 11 cricket players be chosen from a team of 14 players?

Solution: The answer is simple and it is $C(14, 11) = \frac{14\cdot 13\cdot 12}{6} = 364$

Example 5.24 Show that $\sum_{r=0}^{n} C(n,r) = 2^n$.

Solution: We have already seen that the number of subsets of an n-element set is 2^n. Each $C(n,r)$ stands for the number of r-element subsets of an n-element set. Hence the solution.

Example 5.25 Show that

$$\binom{n-1}{r-1} + \binom{n-1}{r} = \binom{n}{r}$$

Solution: The r element subset A may or may not contain the tth element a_t of a set $X = \{a_1, a_2, \ldots, a_n\}$. If a_t is in A, the number of such A's will be $\binom{n-1}{r-1}$. Otherwise it will be $\binom{n-1}{r}$.

Example 5.26 Consider strings over 0s and 1s of length n.

(1) There are 2^n different strings of length n since each position of the string will contain either a 0 or 1.
(2) nC_r denote the number of strings having r 0's and $(n-r)$ 1's.
(3) Let us find number of strings with r 0's and $(n-r)$ 1's such that no two 0's are adjacent. Notice that there are $(n-r)$ 1's in each string of length n and these 1's separate the string into $(n-r+1)$ substrings. Each of the substrings must be either empty or must contain exactly one '0'. Thus, each substring is specified by r nonempty substrings out of $(n-r+1)$ substrings. Hence, the number of strings with length n and no adjacent 0's is

$$\sum_{r\geq 0}\binom{n-r+1}{r}=\sum_{r=0}^{\lfloor\frac{n+1}{2}\rfloor}\binom{n-r+1}{r}$$

Let us see some simple direct examples on combinations.

Example 5.27 (1) In how many ways can 5 students be selected from a class of 45 students?
Solution: $C(45, 5)$ ways.
(2) How many ways a committee of 6 or more can be chosen from 8 people?
Solution: $C(8, 6) + C(8, 7) + C(8, 8)$
(3) A group of 9 doctors is composed of 5 cardiologists and 4 dentists. In how many ways can a committee of 5 be formed that includes at least one cardiologist and one dentist?
Solution: $C(5, 1) \times C(4, 1) \times C(7, 3)$
(4) A bag contains 15 distinguishable balls of which 5 are blue, 4 are yellow and 6 are green in colour. Find the number of ways to draw 4 balls so that there is at least one blue ball in the last draw.
Solution: Solution is an exercise.

5.3 Unlimited Repetitions

A set S may be a finite or an infinite set. If we say the set 'S' is an n-element set, it means, it is a finite set having n distinct elements. A subset of S will also be a finite set having some elements of 'S' without repetitions. So one can talk about k-combination of a set S that is used to describe a subset of 'S' having 'k' distinct elements. We know that the number of such subsets will be $2^{|S|}$. But can we not consider subsets of S having repetitions of elements of S in them? Yes, we can form k-combinations of the set S with repetitions.

Example 5.28 Let $S = \{a, b, c\}$, then the 2-combinations of S are

$$\{a, a\}, \{b, b\}, \{c, c\}, \{a, b\}, \{b, c\}, \{c, a\}.$$

Here there are six of them.

Definition 5.29 A multiset is a set with repeated elements.

Example 5.30 $T = \{1, a, a, 2, b, 3, 3, c, c, c\}$ is a multiset.

One can see in T, each element is occurring only a finite number of times. We can call this number to be a frequency of it. That is, frequency of 1 is one, frequency of a is 2, frequency of b is 1, frequency of 3 is 2, frequency of c is 3. Let $freq(\alpha)$ denote the frequency of a symbol 'α'. Clearly

$$\sum_{\alpha \in T} freq(\alpha) = |T|$$

For each multiset, there lies an underlying set S.

One can define sub multiset A of a multiset T to be a multiset by itself and $freq(\alpha)$ in A is less than or equal to $freq(\alpha)$ in T for every $\alpha \in T$. The number of such submultisets will be

$$\prod_{a \in T} (freq(a) + 1).$$

One can easily prove this from our basic understanding of formation of subsets of a set S.

Example 5.31 Find the number of k-element multisubsets of a multiset T.

Solution: Let S be an underlying set for T whose elements are from 'S' in a repeated manner. That is if $freq : S \to \mathbb{N}$ where $freq(a)$ denotes the number of occurrences of α in T. Suppose $S = \{a_1, a_2, \ldots, a_n\}$ and $x_i = freq(a_i)$. Then the integer solution of $x_1 + x_2 + \cdots + x_n = k$ will be a multisubset having k-elements. But we know that number of such solutions will be $C((n + k - 1), k)$.

We can see the above count is also the number of ways of distributing k identical balls into n distinct boxes.

5.4 Sorting a Set

For a set $S = \{a_1, a_2, \ldots, a_n\}$ we know that number of subsets of S is 2^n which includes the empty set. In its counting, one would like to list such subsets of S algorithmically. Let us understand the subset A construction by looking at any element x of S being in A or not. One can construct a binary string over {0, 1} corresponding to every subset of S. For example the binary string $10101\underbrace{000\ldots0}_{n-5}$ corresponds to a subset $A = \{a_1, a_3, a_5\}$ where the positions of the symbols in the string corresponds to the elements of S as they are read from left to right. Sometimes the binary string is also represented as an n-tuple. Also the presence of symbols 'a_i' in the subset is indicated by '1' and '0' by its absence. Also one also can see this string over 0, 1 will correspond to an integer between 0 and $2^n - 1$. For example for $S = \{a_1, a_2, a_3\}$, $n = 3$,

$$\begin{aligned} 011 &= 0 \times 2^0 + 1 \times 2 + 1 \times 2^2 \\ &= 6 \end{aligned}$$

So the subset corresponding to the integer '6' will be $A = \{a_2, a_3\}$. We can now understand the ordering of subsets of a set $S = \{a_1, a_2, \ldots, a_n\}$ based on the number 't' between 0 and $2^n - 1$ that they represent. That is the subset corresponding to 10 appear before the subset corresponding to 11. Subset $\emptyset$ will correspond to 0 and it will appear always first in the list.

Example 5.32 Let $S = \{a_1, a_2, a_3, a_4\}$. Find the subset next to the subset $A = \{a_1, a_3, a_4\}$.

Solution: The binary string corresponding to A is 1011 and the integer corresponding to 1011 will be 11. The next integer is 12 and its binary form is 1100. So the subset $B = \{a_1, a_2\}$ will next to A.

We can write down a simple algorithm for understanding the order of subsets of a given set. It is also said to be a subsets generating algorithm.

Algorithm: Subsets of $\{a_1, a_2, \ldots, a_n\}$

$* \underbrace{(0, 0, \ldots, 0)}_{n-tuple} = \underbrace{(0000\ldots0)}_{n} *$

while $(x_1, x_2, \ldots, x_n) \neq (1, 1, \ldots, 1)$

do

find the largest index i

(between 1 and n) such that $x_i = 0$.

replace x_i with 1, and $x_{i+1}, x_{i+2}, \ldots, x_n$ with 0.

Remark 5.33 The above algorithm is a very simple algorithm which has a natural ordering on the binary strings. Because of such ordering in the corresponding integers in base 2 representation, this is also called as lexicographic ordering of strings over {0, 1} of length n.

We can understand this lexicographic ordering as a squashed ordering of subsets because a complete ordering of subsets is available to us before introducing a new member to the expansion of subsets. We understand more through the following example.

Example 5.34 Let $S = \{a_1, a_2, a_3, a_4\}$. The list will start with $\emptyset$. Then element a_1 is introduced. The new list will be

$$\emptyset, \{a_1\}.$$

Next a_2 is introduced such that the subsets do not contain a_2 come before those containing a_2. The list will be

$$\emptyset, \{a_1\}, \{a_2\}, \{a_1, a_2\}.$$

The list expands as below with a_3

$$\emptyset, \{a_1\}, \{a_2\}, \{a_1, a_2\}, \{a_3\}, \{a_1, a_3\}, \{a_2, a_3\}, \{a_1, a_2, a_3\}.$$

The compute list of subsets in squashed ordering will be

$$\emptyset, \{a_1\}, \{a_2\}, \{a_1, a_2\}, \{a_3\}, \{a_1, a_3\}, \{a_2, a_3\}, \{a_1, a_2, a_3\}, \{a_4\},$$
$$\{a_1, a_4\}, \{a_2, a_4\}, \{a_1, a_2, a_4\}, \{a_3, a_4\}, \{a_1, a_3, a_4\}, \{a_2, a_3, a_4\}, \{a_1, a_2, a_3, a_4\}.$$

Let us briefly see how k-combination of an n-element set can be generated. One direct method is to look into all binary strings of length n and list only those with k 1's. This list will contain all subsets of size k. There are other methods of generating permutations which have some practical improvement in terms of time complexity.

k-Subsets of a Multiset

Let $S = \{a_1, \ldots, a_n\}$ be a set and A be a multiset on S. That A is a set with the elements of S such that each element can occur one or more times. By k-combination of A we mean a multisubset of A with k-elements. That is number of such subsets will be same as number of integer solutions of

$$x_1 + x_2 + \cdots + x_n = k$$

where $0 \leq x_i \leq m_i, 1 \leq i \leq n$. Let $\xi(k; m_1, \ldots, m_n)$ denote such sets.

Example 5.35 Let $S = \{a_1, a_2, a_3\}$ $A = \{3a_1, 3a_2, 3a_3\}$.

$$\xi(3; 3, 3, 3) = \{\{a_1, a_1, a_1\}, \{a_2, a_2, a_2\}, \{a_3, a_3, a_3\}, \{a_1, a_1, a_2\}, \{a_1, a_2, a_2\},$$
$$\{a_1, a_3, a_3\}, \{a_2, a_3, a_3\}, \{a_1, a_1, a_3\}, \{a_2, a_2, a_3\}, \{a_1, a_2, a_3\}\}.$$

Next we briefly look for an algorithm to find such 'k' subsets.

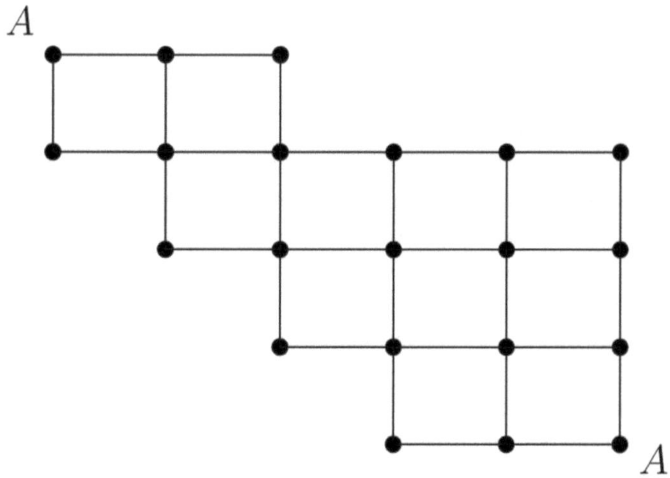

Fig. 5.4 Grid

A simple and direct method to generate all the multisubsets of a multiset will be to completely order the elements of the multiset distinctly and then one can use lexicographic ordering technique as before. For example, the elements of A in the above example can be taken to be $\bar{A} = \{a_1^1, a_1^2, a_1^3, a_2^1, a_2^2, a_2^3, a_3^1, a_3^2, a_3^3\}$. Then the complete set of the subsets of $\bar{A}$ can be generated (Fig. 5.4).

But this method will generate a multisubset more than once. In particular let us assume that we want to find all k-element multisubsets of A. One can assume that the elements of A are ordered. Then a lexicographic ordering of the k-element multisubsets can be written. For example if $A = \{a_1, a_1, a_1, a_2, a_3, a_3\}$. It has 6 elements and the number of subset of A will be 2^6.

Exercises

1. Find the number of natural numbers between 100 to 1000 that have exactly one occurrence of 7 or 9.
2. Three players a, b, c take turns at a game, according to the following conditions. In the beginning a and b play, the loser is replaced by c. That is in the second round, the winner has to play with c. The game continues in this way until a player wins in succession, thus he becomes the winner of the game. We disregard the possibility of a game being ending in a draw. List a few possible winning outcomes.
3. A graduate student has to compulsorily do 4 courses per semester. He must choose a department course from a list of 4 elective courses, and he can choose the remaining courses from the other departments which offer totally 20 courses per semester.

Determine the number of possible ways that a graduate student gets to select his 4 courses.

4. In a country, a unique number board of each car has to be given such that the first position always contains a symbol from the set $\{A, E, F, G, K, L\}$. The board has to contain totally 5 cells. The cells from the 2nd to 5th can contain integers from 1 to 9. Find the number of distinct number boards that can be assigned with.
5. A lady has 4 different hand bags, 5 different hair clips and 6 different pairs of sandals. In how many different ways can this women take a hand bag, wear a hair clip and a sandal while going out to work?
6. Using the digits 1, 2, 3, 4 and 5 how many 3-digit numbers can be formed if

   (a) the last digit must be a 5 and repetition is not allowed?
   (b) first digit is an odd number?
   (c) the number is divisible by 2 and repetition allowed?

7. If a three digit number is formed from the digits 1, 2, 3, 4, 5 and 6, how many numbers are between 101 and 499?
8. Two cards are drawn from a standard deck of 52 cards without replacement. Find the number of ways of drawing a 7 and a king in that order.
9. There are 10 cars in a car race competition. The top 3 winning cars receive money. Find the number of possible money winning orders for the car race.
10. How many lines can be drawn using 3 non-collinear points A, B and C in a plane?
11. A team of 5 students are to be arranged in circular manner from a group of 12 students. Find the number of ways it can be done.
12. In how many different ways can the letters of the word 'YELLOW' be arranged in such a way that the vowels always occur together?
13. How many 6 letter words can be formed from an English alphabet that contain 2 or 3 vowels?
14. Find the number of ways of selecting a 2-member committee for a farmer society which consists of 20 members.
15. A cricket association consists of 9 teams. How many games will be played over the course of the year if each team play every other team exactly 10 times?
16. How many 2-element sets are there in the set $\{x \in \mathbb{N}/1 \leq x \leq 50\}$?
17. In how many ways a 10 digit positive integer can be formed if the digits in odd positions are always odd and the digits in the even positions are always even?
18. Anne has bought 15 chocolates to her 4 children Bob, Andrew, Mary and Kelab. In how many ways can she distribute the chocolates such that each child gets atleast one chocolate?
19. How many ways can I distribute 90 note books to a class of 15 boys and 25 girls if it is the condition that each boy gets one notebook and each girl gets at least 2 notebooks?

20. How many 4-digit even numbers greater than 1500 can be formed using the digits 7, 1, 4, 2 and 9?
21. How many 3 letter words can be formed using the letters in the word 'M A H A B A L L I U R A M' that contains atmost one vowel?
22. (i) Find the number of ways of placing 10 red balls and 4 white balls in a row.
    (ii) Find the number of ways of placing 10 red balls and 4 white balls in a row so that red balls are together and the white balls are together.
    (iii) Find the number of ways of placing 10 red balls and 4 white balls in a row so that the arrangement always starts with a white ball and ends with a red ball.
23. Find the number of ways of arranging 12 students in a class room in to 6 disks to perform a group project in a competition.
24. Find the number of ways of arranging the letters in the words 'W O D E P E A L S' and 'S A U C I L E T A'.
25. Consider the following grid with a marked source 'A' and target 'B'. A robot needs to move along the lines starting from source 'A'. The only allowed moves are right and down ways. Find the number of paths from A to B.
26. A postman has to visit 6 cities each 7 times. In how many different ways can the postman make his visits so that, always his starting city is different from ending city?
27. Find the number of ways of selecting a 11-member cricket team and a 7-member kabaddi team from a group of 4 people if

    (i) a person cannot be in both teams.
    (ii) a person can be in both.
    (iii) exactly one person can be in both in teams.

28. On a festival day a teacher in a class distributes 65 chocolates to 10 students. How many different distributions are there so that each student gets atleast one chocolate?

Generating Functions 6

In this chapter, we will define what are recurrence relations and we discuss the methods to solve them. Such relations can be used for solving counting problems.

A power series of the function $f(x)$ will be of the form $f(x) = \sum_{n=0}^{\infty} a_n x^n$. Such representation of functions have numerous applications. However in this chapter, we are going to see how these power series representations can be used for counting purpose. These power series when summed up will be called generating functions. We ultimately use them to solve recurrence relations.

6.1 Power Series

Sequence of the form $(1, 1, 2, 3, 5, 8, 13, 21, \ldots)$ is called the Fibonacci sequence of Fibonacci numbers. A sequence of the form $(a, a, a, a, \ldots)$ is called a constant sequence. Any general sequence can be written as

$$(a_0, a_1, a_2, a_3, a_4, \ldots).$$

A sequence is a string of objects that follow a particular pattern.

Consider a power series with a_i as the coefficient of x^i and write it as $a_0 + a_1 x + a_2 x^2 + a_3 x^3 + \cdots$. That is $\sum_{n=0}^{\infty} a_n x^n = a_0 + a_1 x + a_2 x^2 + a_3 x^3 + \cdots$.

This power series is called as the generating series or generating function of $(a_0, a_1, a_2, a_3, a_4, \ldots)$.

R. Rama, *Topics in Combinatorics and Graph Theory*,
https://doi.org/10.1007/978-3-031-74252-1_6

Examples 6.1 1. Consider the sequence $(1, 1, 1, 1, \ldots)$. The generating function for this sequence will be $1 + x + x^2 + x^3 + \cdots$.

$$i.e., \sum_{n=0}^{\infty} x^n = \frac{1}{1-x}.$$

2. Consider the sequence $(2^0, 2, 2^2, 2^3, 2^4, \ldots)$. The generating function for this sequence will be

$$\sum_{n=0}^{\infty} 2^n x^n = \frac{1}{1-2x}.$$

3. Let the generating function $f(x) = \frac{1}{1+x}$. The power series expansion for $f(x) = \dfrac{1}{1+x} = \sum_{n=0}^{\infty}(-1)^n x^n$. The sequence will be $(1, -1, 1, -1, 1, \ldots)$.

Remark 6.2 We only take the representations of sequences as power series. We are not going to test any convergence conditions of such series. It is only a representation through which a view of the nth term or more terms can be got.

Operations on Power Series

The basic operations like addition, subtraction, multiplication and division can be defined on power series which represent sequences.

Addition

If $\sum_{n=0}^{\infty} a_n x^n$ and $\sum_{n=0}^{\infty} b_n x^n$ are two power series, then $\left(\sum_{n=0}^{\infty} a_n x^n\right) + \left(\sum_{n=0}^{\infty} b_n x^n\right) = \sum_{n=0}^{\infty}(a_n + b_n)x^n$.

Multiplication

The multiplication of $\sum_{n=0}^{\infty} a_n x^n$ and $\sum_{n=0}^{\infty} b_n x^n$ will be $\left(\sum_{n=0}^{\infty} a_n x^n\right)\left(\sum_{n=0}^{\infty} b_n x^n\right) = \sum_{n=0}^{\infty} c_n x^n$ where

$$\begin{aligned} c_n &= a_0 b_n + a_1 b_{n-1} + a_2 b_{n-2} + \cdots + a_n b_0 \\ &= \sum_{k=0}^{n} a_k b_{n-k}. \end{aligned}$$

Subtraction

If $\sum_{n=0}^{\infty} a_n x^n$ and $\sum_{n=0}^{\infty} b_n x^n$ are two power series, then $\left(\sum_{n=0}^{\infty} a_n x^n\right) - \left(\sum_{n=0}^{\infty} b_n x^n\right) = \sum_{n=0}^{\infty} (a_n - b_n) x^n$.

Division

Let $\sum_{n=0}^{\infty} a_n x^n$ be a power series with coefficients $a_i \neq 0$, $i \geq 0$. If a_t is the first nonzero coefficient, then

$$\sum_{n=0}^{\infty} a_n x^n = \sum_{n=t}^{\infty} a_n x^n = x^t \sum_{n=0}^{\infty} a_{n+t} x^n.$$

Let $\sum_{n=0}^{\infty} a'_n x^n$ be the multiplicative inverse of $\sum_{n=0}^{\infty} a_n x^n$ where $a_n \neq 0$, then $\left(\sum_{n=0}^{\infty} a_n x^n\right) / \left(\sum_{n=0}^{\infty} a'_n x^n\right)$ can be defined.

Example 6.3 Let $f(x) = \sum_{n=0}^{\infty} x^n$. Consider $g(x) = (1 - x)$. Then

$$\begin{aligned} f(x).g(x) &= (1 + x + x^2 + \cdots)(1 - x) \\ &= 1. \end{aligned}$$

$g(x)$ is the multiplicative inverse of $f(x)$.

Hereafter we will not call the series as 'power series'. We will call it as generating function. We represent sum by $F(x)$, i.e.,

$$F(x) = \sum_{n=0}^{\infty} a_n x^n$$

will be the generating function, for future reference.

Example 6.4 The following are some generating functions.

$$F_1(x) = \sum_{n=0}^{\infty} 3^n x^n$$

$$F_2(x) = \sum_{n=4}^{\infty} 2^n x^n$$

$$F_3(x) = \sum_{n=0}^{\infty} n^2 x^n$$

Definition 6.5 Two generating functions $F_1(x) = \sum_{n=0}^{\infty} a_n x^n$ and $F_2(x) = \sum_{n=0}^{\infty} b_n x^n$ are said to be equal iff $a_n = b_n$ for $n \geq 0$.

So far we have seen the power series representations of infinite sequences. Now we write power series representations of a few sequences that have finite number of terms.

Examples 6.6 1. The generating function of $(2, 4, 6, 8, -4)$ will be $2 + 4x + 6x^2 + 8x^3 - 4x^4$.
2. The sequence associated with the generating function $1 + x + x^3 + x^5 + x^7$ is $(1, 1, 0, 1, 0, 1, 0, 1)$.

Definition 6.7 Let $F(x) = \sum_{n=0}^{\infty} a_n x^n$ and $G(x) = \sum_{n=0}^{\infty} b_n x^n$. Suppose $a_0 = 0$. Then, the generating function $F(G(x))$ is obtained by substituting $G(x)$ for each x in $F(x)$.

$$F(G(x)) = \sum_{n=0}^{\infty} a_n (G(x))^n,$$

where $(G(x))^n$ is obtained by multiplication method.

Remark 6.8 In any multiplication, we must work with finitely many multiplications and additions. If $a_0 \neq 0$, then the constant term in $F(G(x))$ will be $a_0 + a_1 b_0 + a_2 b_0^2 + a_3 b_0^3 + \cdots$ which is an infinite sum of numbers. In this case we need to know its sum or its convergence, which is undesirable.

Definition 6.9 Let $F(x) = \sum_{n=0}^{\infty} a_n x^n$. Then the term-by-term differentiation of $F(x)$ will be

$$\frac{d}{dx}(F(x)) = \sum_{m=0}^{\infty}(m+1)a_{m+1}x^m.$$

Example 6.10 Let $F(x) = (1+x)^m$ and $G(x) = (1+x)^n$. Then

$$F(x)G(x) = (1+x)^{m+n}.$$

Using binomial theorem,

$$\left(\sum_{t=0}^{m} C(m,t)x^t\right)\left(\sum_{r=0}^{n} C(n,r)x^r\right) = \sum_{l=0}^{m+n} C(m+n,l)x^l.$$

By multiplication rule of power series,

$$\sum_{i=1}^{k} C(m,i)C(n,k-i) = C(m+n,k).$$

This is called Vandermonde Convolution.

Example 6.11 Let a sequence of numbers satisfy the following:

$$a_0 = 1,$$
$$a_n = 2a_{n-1}.$$

Find the generating function of the sequence.

Solution: It is clear that the terms of the sequence are powers of 2. The generating function

$$\begin{aligned}
F(x) &= \sum_{n=0}^{\infty} a_n x^n \\
&= a_0 + \sum_{n=1}^{\infty} a_n x^n \\
&= 1 + \sum_{n=1}^{\infty} 2a_{n-1}x^n \\
&= 1 + 2x\sum_{m=0}^{\infty} a_m x^m \\
&= 1 + 2xF(x) \\
\Rightarrow F(x) &= \frac{1}{1-2x} = \sum_{n=0}^{\infty} 2^n x^n
\end{aligned}$$

The other power series one can use as generating functions are as follows:

1. Exponential Functions:

$$e^x = \sum_{n=0}^{\infty} \frac{x^n}{n!}$$

2. Logarithmic Functions:

$$log(1+x) = \sum_{n=1}^{\infty} \frac{(-1)^{n-1}x^n}{n}$$

Some uses of generating functions

Some of the combinatorial problems that we solved earlier can be modeled and solved by generating functions.

Example 6.12 Suppose the situation is to find the number of ways of distributing 20 chocolates to 5 children so that each child gets atleast one chocolate. The answer is $C(19, 4)$. There can be more restrictions imposed on the distribution, like 'a child can not receive more than 5 chocolates'. If the restriction is on more than one child, PIE can be used. For the problem of distributing n chocolates to 1 child, the number of ways is

1. The sequence that enumerates for all $n \geq 1$ be (a_n) and $a_n = 1$. Hence the generating function will be $x + x^2 + x^3 + x^4 + \cdots = \frac{x}{1-x}$.

Suppose the problem is to distribute n apples to 5 children. Then the product $(x + x^2 + x^3 + x^4 + \cdots)^5$ represents and enumerates the number of distributions via its coefficients. The coefficient of x^n is (say) of the form $x^{k_1}x^{k_2}x^{k_3}x^{k_4}x^{k_5}$ where each x^{k_i} is taken from the ith factor of the product. That is $k_1 + k_2 + k_3 + k_4 + k_5 = n$. This means child 1 gets k_1 chocolates, child 2 gets k_2 chocolates and so on. Hence, the number of integral solutions of $k_1 + k_2 + k_3 + k_4 + k_5 = n$, which is the coefficient of x^n will be the number of distributions of n chocolates to 5 children.

Example 6.13 If a_n = the number of integral solutions of $k_1 + k_2 + k_3 + k_4 + k_5 = n$ where $0 \leq k_1 \leq 3$, $0 \leq k_2 \leq 3$, $1 \leq k_3 \leq 5$, $2 \leq k_4 \leq 7$. Find the generating function for (a_n).

Solution: The generating function for the condition $0 \leq k_1 \leq 3$ will be $F_1(x) = 1 + x + x^2 + x^3$.

Similarly, the other generating functions are:

$0 \leq k_2 \leq 3 : F_2(x) = 1 + x + x^2 + x^3$.

$1 \leq k_3 \leq 5 : F_3(x) = x + x^2 + x^3 + x^4 + x^5$.

$2 \leq k_4 \leq 7 : F_4(x) = x^2 + x^3 + x^4 + x^5 + x^6 + x^7$.

Then the generating function that we require will be
$F_1(x)F_2(x)F_3(x)F_4(x)$
$= (1+x+x^2+x^3)^2(x+x^2+x^3+x^4+x^5)(x^2+x^3+x^4+x^5+x^6+x^7)$

Example 6.14 Find the coefficient of x^{54} in $(1+x^2+x^3+x^4)^{10}$

Solution: The answer will be $\frac{10!}{3!6!}$.

Example 6.15 Find the number of ways to split n distinct cookies in 3 parts. The first part can have 0 or more cookies. The second part can have only odd number of cookies. The third part can contain only even number of cookies.

Solution: Let $F_1(x) = \sum_{n\geq 0} 2^n x^n = \dfrac{1}{1-2x}$, which is the generating function for the first part. It can seen that the number of odd subsets of an n-element set is 2^{n-1}. Hence the generating function for the second part will be $F_2(x) = \sum_{n=1}^{\infty} 2^{n-1}x^n = \frac{x}{1-2x}$. Also, the number of subsets having even number of elements will be 2^{n-1} for $n \geq 1$ and is 1 if $n = 0$. Hence the generating function for the third part will be $F_3(x) = 1 + \frac{x}{1-2x} = \frac{1-x}{1-2x}$. Let $G(x)$ be the generating function product

$$\begin{aligned}F_1(x).F_2(X).F_3(x) &= \frac{x(1-x)}{(1-2x)^3} \\ &= \frac{-1}{4}\frac{1}{1-2x} + \frac{1}{4}.\frac{1}{(1-2x)^3}.\end{aligned}$$

By binomial theorem,

$$\begin{aligned}(1-2x)^{-3} &= \sum_{n=0}^{\infty} C(-3,n)(-2x)^n \\ &= \sum_{n=0}^{\infty} C(n+2,2)2^n x^n\end{aligned}$$

Therefore, coefficient of x^n in $G(x)$ will be

$$\frac{C(n+2,2)2^n - 2^n}{4} = 2^{n-3}n(n+3), \quad n \geq 0$$

6.2 Exponential Generating Functions

Some counting problems cannot be solved by using the ordinary generating functions introduced in the previous section. Those problems where the counting is tied to a specific k-element subsets for each k, require a more structured generating functions. Then, in this case, for each sequence (a_n) we will associate a generating function of the form $\sum_{n=0}^{\infty} a_n \frac{x^n}{n!}$.

We formally define it as follows.

Definition 6.16 Let (a_n), $n \geq 0$ be a sequence of real numbers. Then the formal power series $F(x) = \sum_{n=0}^{\infty} a_n \frac{x^n}{n!}$ is called the exponential generating function for $(a_n)_0^{\infty}$

Examples 6.17 1. Consider the sequence $\{1, 1, 1, \ldots\}$. The exponential generating trivially will be $F(x) = \sum_{n=0}^{\infty} 1 \frac{x^n}{n!}$ which is nothing but e^x.

2. Let (a_n) be a sequence such that $a_n = 1$ if $n = 1$ and $a_n = 0$ otherwise. The exponential generating function will be

$$F(x) = \sum_{n=0}^{\infty} a_n \frac{x^n}{n!} = x.$$

3. Let X be an empty set. Then the exponential generating function will be $F(x) = \sum_{n=0}^{\infty} a_n \frac{x^n}{n!} = 1$.
4. Let X be a non empty set. Then the exponential generating function will be $F(x) = e^x - 1$.
5. Let X be a set having even size. Hence the generating function will involve $x^2, x^4, x^6, \ldots$ terms. Hence,

$$F(x) = \frac{e^x + e^{-x}}{2}$$

Remark 6.18 Addition and multiplication operations can be defined similarly for exponential generating functions.

Example 6.19 Find the exponential generating function for

$$\{a_n \mid a_n \text{ is the number of subsets of an } n\text{element set}\}$$

Solution: Let A_1 be a subset of an n-element set X. For each $x \in X$, there are two possibilities. They are, either x belongs to A_1 or x belongs to $X \setminus A_1$. Let $F(x)$ be the exponential generating function e^x for both the number of subsets for A_1 and $X \setminus A_1$. Hence, the exponential generating function will $e^x.e^x = e^{2x} = \sum_{n=0}^{\infty} 2^n \frac{x^n}{n!}$. We already know that the number of subsets of an n element set is 2^n.

Example 6.20 Let $\Sigma = \{0, 1, 2\}$ and ternary strings be formed over Σ such that the strings contain only even number of zeros, any number of 1's and 2's. Find the number of such numbers of length n, $n \geq 0$.

Solution: As there are no restrictions on 1's and 2's, we can assume the exponential generating function e^x for each of them. Already we have seen the generating function for even parity will be

$$\begin{aligned} 1 + \frac{x^2}{2!} + \frac{x^4}{4!} + \cdots &= \sum_{n=0}^{\infty} \frac{x^{2n}}{(2n)!} \\ &= \frac{e^x + e^{-x}}{2} \end{aligned}$$

In order that the condition on strings over $\{0, 1, 2\}$ to satisfy, we need to use the multiplication principle so that all the conditions are satisfied. Hence, the exponential generating function for the situation will be

$$\begin{aligned} \left(\frac{e^x + e^{-x}}{2}\right)(e^x)(e^x) &= \frac{e^{3x} + e^x}{2} \\ &= \frac{1}{2}\left\{\sum_{n=0}^{\infty} 3^n \left(\frac{x^n}{n!}\right) + \sum_{n=0}^{\infty} \frac{x^n}{n!}\right\} \end{aligned}$$

The number strings of length n will be $\left(\frac{3^n+1}{2}\right)$.

Some combinatorial aspects of generating functions are discussed below. We can see the approach of generating function method is useful for giving good interpretations for standard combinatorial problems.

1. **Ball-Box Distributions**

   (a) The number of ways of distributing n indistinguishable balls into m distinguishable boxes is the coefficient of x^n in

$$(1 + x + x^2 + x^3 + \cdots)^m = \frac{1}{(1 - x)^m}.$$

Interpretation: Suppose there are m distinguishable boxes and x^k represents k indistinguishable boxes. Each box, may contain '0' ball represented by x^0, or 1 ball represented by x^1, or two balls represented by x^2 and so on. That is each box containment can be represented by

$$(1 + x + x^2 + x^3 + \cdots).$$

Here assume unlimited number of balls being distributed. Hence, the distribution in the m boxes can be represented by

$$(1 + x + x^2 + x^3 + \cdots)^m. \tag{6.1}$$

Now the problem can be modified as distribution of n indistinguishable balls in m balls and the corresponding representation will fetch the coefficient of x^n satisfying,

$$x^{n_1} x^{n_2} \ldots x^{n_m} = x^n.$$

That is the number ways will be the number of non negative integral solutions of the equation $n_1 + n_2 + n_3 + \cdots + n_m = n$. The answer will be $C(m + n - 1, n)$.
(b) Suppose the problem is to find the number of ways of distributing n indistinguishable balls into m distinguishable boxes with atmost n_k balls in the kth box. In this case we need to find the coefficient of x^n in the product

$$\prod_{k=1}^{m} (1 + x + x^2 + \cdots + x^{n_k}).$$

(c) Suppose the problem is to find the number of ways of distributing n indistinguishable balls into m distinguishable boxes with atleast t_k balls in the kth box. Then we need to find the coefficient of x^n in the product

$$\prod_{k=1}^{m} x^{t_k} (1 + x^2 + \cdots) = \frac{x^{t_1 + t_2 + \cdots + t_m}}{(1 - x)^m}$$

2. **Ordered Object Selection**
The number of ways of selecting an ordered n objects where the objects may repeat from a set containing m objects is the coefficient of $\frac{x^n}{n!}$ in the product

$$\left(1 + \frac{x}{1!} + \frac{x^2}{2!} + \frac{x^3}{3!} + \cdots\right)^m = e^{mx}.$$

Example 6.21 Find the 3-permutations of the set $A = \{x, x, y, y, z\}$.

Solution: By generating functions, the answer will be the coefficient of $\frac{x^3}{3!}$ in the product

$$\left(1+\frac{x}{1!}+\frac{x^2}{2!}\right)\left(1+\frac{x}{1!}+\frac{x^2}{2!}\right)\left(1+\frac{x}{1!}\right)$$

and it will be

$$\frac{3!}{1!}+\frac{3!}{1!2!}+\frac{3!}{2!}+\frac{3!}{1!2!}+\frac{3!}{2!1!}=18$$

Exercise

1. Find the coefficient of x^{24} in $(1+x^4+x^8)^{10}$.
2. Find a generating function for a_n which is the number of ways of distributing n similar balls in numbered baskets where the first, third, fifth, seventh and ninth baskets are non empty.
3. Find the generating function for a_n, the number of integral solutions to the equation $k_1+k_2+k_3+k_4+k_5=n$ if
   (i) $0 \le k_i \le 2$ for each i.
   (ii) $k_i > 0$ for i = 1,2.
   (iii)$1 \le k_1 \le 2$ and $k_2, k_3, k_4 \ge 0$.
4. Let a basket contain 2 apples,one orange and 5 bananas. Fruits of the same type are identical. Then find the number of ways of choosing k fruits.
5. Find the number of integral solutions to $x+y+z=4$, $0 \le x \le 2$, $0 \le y \le 1$, $0 \le z \le 5$.
6. Show that $f(x)=\frac{1}{1-4x}$ is the generating of the sequence $\{C(2n,n)|n \ge 0\}$
7. How many binary strings with each number of zero's can be formed?
8. Find the exponential generating function for each of the following sequences.
   (i) $a_n = 3^n$
   (ii) $a_n = (-1)^n 3^n$
   (iii) $a_n = n!$
   (iv) $a_n = 7^n$
9. For each exponential generating functions given below, find the sequence $\{a_n : n \ge 0\}$ they represent.
   (i) e^{3x}
   (ii) $x^3 e^{2x}$
   (iii) $\frac{1}{1-x}$
10. Find the coefficient of $\frac{x^5}{5!}$ in $\frac{e^x-e^{-x}}{2}$.
11. Expand the generating function $F(x)=\frac{1}{(1-2x)(1+4x)^2}$ in powers of x. Find the coefficient of x^{102}.
12. A country floats coins in denominations of 2, 3, 5 and 7. How many different ways are there to spend an amount of 27 exactly?

13. How many strings of length n can be made using the symbols $\{a, b, c, d, e, f\}$ if atleast one a and two c's are present in each of the string?
14. Find the number of ways of placing n balls into 3 boxes so that the first and second boxes contain atmost 10 balls and the third box can hold any number of balls. Use the generating function method.
15. Let $\{a_1, a_2, a_3, \ldots\}$ and $\{b_1, b_2, b_3, \ldots\}$ be two sequences and let $b_m = \sum_{t=1}^{m} a_t$. If $F(x)$ is the generating function of $\{a_1, a_2, a_3, \ldots\}$ and $G(x)$ is the generating function of $\{b_1, b_2, b_3, \ldots\}$. Find the relationship between $F(x)$ and $G(x)$.

7 Recurrence Relations

7.1 Generating Function Method

Consider the sequence of numbers,

$$0, 1, 1, 2, 3, 5, 8, 13, \ldots$$

This sequence is called the Fibonacci sequence of numbers. It is easy to see that each term is the sum of the previous two terms. So how do we define it mathematically. Let $\{F_n\}$ denote the terms of the Fibonacci sequence. Then,

$$\begin{aligned}
F_0 &= 0 \\
F_1 &= 1 \\
F_2 &= F_1 + F_0 \\
F_3 &= F_2 + F_1 \\
&\vdots \\
F_n &= F_{n-1} + F_{n-2} \quad \text{for } n > 1
\end{aligned}$$

This relation '$F_n = F_{n-1} + F_{n-2}$' is known as the recurrence relation between nth, $(n-1)$th and $(n-2)$th terms.

Definition 7.1 For any sequence $\{a_n\}$, $n \geq 0$, if there is a single formula that will allow us to compute a_n is known as the closed form formula for $\{a_n\}$, $n \geq 0$.

If for any sequence $\{a_n\}$, $n \geq 0$, we have a closed form formula, then we need not use the fact that, how each term is formed / related from the previous terms. Hence we can

R. Rama, *Topics in Combinatorics and Graph Theory*,
https://doi.org/10.1007/978-3-031-74252-1_7

use generating functions to derive such closed form solution. The steps involved in such a procedure will be:

(1) Find the generating function from the recurrence relation.
(2) Manipulate the generating function to a closed form.
(3) Invert the function in step 2 to obtain the numbers a_n.

Example 7.2 Consider the Fibonacci recurrence relation given above that

$$F_n = F_{n-1} + F_{n-2}, \quad n \geq 2 \tag{7.1}$$

Shift the indices so that we can avoid initial conditions F_0 and F_1. So, write the relation (7.1) as

$$F_{n+2} = F_{n+1} + F_n. \tag{7.2}$$

Multiply, (7.2) throughout by x^{n+2} to get

$$F_{n+2}x^{n+2} = F_{n+1}x^{n+2} + F_n x^{n+2}$$

And sum over $n = 0$ to ∞ throughout.

$$\sum_{n=0}^{\infty} F_{n+2}x^{n+2} = \sum_{n=0}^{\infty} F_{n+1}x^{n+2} + \sum_{n=0}^{\infty} F_n x^{n+2}. \tag{7.3}$$

Let $f(x) = \sum_{n=0}^{\infty} F_n x^n$. Then (7.3) becomes

$$f(x) - F_1 x - F_0 = f(x)x - xF_0 + x^2 f(x)$$
$$\Rightarrow f(x) = \frac{F_0 + F_1 x - F_0 x}{1 - x - x^2}.$$

Put $F_0 = 0$ and $F_1 = 1$.

$$f(x) = \frac{x}{1 - x - x^2}$$

The form

$$f(x) = \frac{x}{1 - x - x^2}$$

is known as the closed form formula. $f(x)$ is the generating function of the Fibonacci sequence.

Next we need to invert $f(x)$ obtained above to find 'a_n' which is F_n here. Consider

$$f(x) = \frac{x}{1 - x - x^2} = \frac{x}{1 - \phi x - \hat{\phi} x^2}$$

where $\phi = \frac{1+\sqrt{5}}{2}$ is the famous golden ratio and $\hat{\phi} = \frac{1-\sqrt{5}}{2}$.

Using partial fraction method,

$$f(x) = \frac{x}{(1 - \phi x)(1 - \hat{\phi} x)} = \frac{k_1}{1 - \phi x} + \frac{k_2}{1 - \hat{\phi} x}$$

where, $k_1 = \frac{1}{\phi - \hat{\phi}}$ and $k_2 = \frac{1}{\hat{\phi} - \phi}$. Therefore,

$$\begin{aligned} F(x) &= k_1(1 - \phi x)^{-1} + k_2(1 - \hat{\phi} x)^{-1} \\ &= k_1 \sum_{n=0}^{\infty} \phi^n x^n + k_2 \sum_{n=0}^{\infty} (-1)^n \hat{\phi}^n x^n \\ \therefore F_n &= \frac{1}{\phi - \hat{\phi}} (\phi^n - \hat{\phi}^n) \end{aligned}$$

Definition 7.3 A recurrence of the form

$$f(n) = a_1 F_{n-1} + a_2 F_{n-2} + \cdots + a_k F_{n-k}$$

is said to be a linear recurrence relation with constant coefficients where $a_1, a_2, \ldots, a_k$ are called the coefficients and $a_1, a_k \neq 0$. If $f(n) \equiv 0$, the recurrence relation is said to be homogeneous.

Example 7.4 The following are the linear homogeneous recurrence relations with constant coefficients

(i) $F_n - F_{n-1} - F_{n-2} = 0$, $F_1 = 0$, $F_2 = 1$
(ii) $F_n - F_{n-1} - F_{n-2} = 0$, where $F_1 = 1$, $F_2 = 3$
(iii) $F_n - F_{n-2} - F_{n-3} = 0$, $F_1 = F_2 = F_3 = 1$
(iv) $F_n - 2F_{n-1} - F_{n-2} = 0$, $F_1 = 0$, $F_2 = 1$

We have seen above, how to find the solution to (i) in Example 7.4 which is the famous Fibonacci sequence. By similar way, we can find the solutions to (ii), (iii) and (iv) in Example 7.4 using generating function method. We have used a linear generating function associated with the solution of Fibonacci recurrence relation.

Example 7.5 Solve the following homogeneous recurrence relation by generating function method.

$$F_n - 2F_{n-1} + F_{n-2} = 0, \quad n \geq 2 \text{ with}$$
$$F_0 = 0, \; F_1 = 1$$

Solution: Consider,

$$F_n - 2F_{n-1} + F_{n-2} = 0 \tag{7.4}$$

Rewrite (7.4) as

$$F_{n+2} - 2F_{n+1} + F_n = 0 \tag{7.5}$$

Multiply (7.5) throughout by x^{n+2} and make the summation from $n = 0$ to ∞. Then (7.5) becomes

$$\sum_{n=0}^{\infty} F_{n+2}x^{n+2} - 2\sum_{n=0}^{\infty} F_{n+1}x^{n+2} + \sum_{n=0}^{\infty} F_n x^{n+2} = 0 \tag{7.6}$$

Let $f(x) = \sum_{n=0}^{\infty} F_n x^n$. (7.6) becomes,

$$(f(x) - xF_1 - F_0) - 2xf(x) + 2xF_0 + x^2 f(x) = 0$$

Hence,

$$f(x) = \frac{x}{1 - 2x + x^2} = \frac{x}{(1-x)^2}$$

Investigating $f(x)$, we get

$$\begin{aligned} f(x) &= x(1-x)^{-2} \\ &= x\sum_{n=0}^{\infty}(n+1)x^n \\ &= x(1 + 2x + 3x^2 + 4x^3 + \cdots) \\ &= (x + 2x^2 + 3x^3 + \cdots) \\ &= \sum_{n=1}^{\infty} nx^n = \sum_{n=0}^{\infty} nx^n \end{aligned}$$

$$\therefore F_n = n, \quad n \geq 0.$$

Example 7.6 Solve

$$F_n - 5F_{n-1} + 6F_{n-2} = 0, \quad n \geq 2$$

where $F_0 = 0, \; F_1 = 1$.

Solution: Consider the recurrence relation equivalent to the given as,

$$F_{n+2} - 5F_{n+1} + 6F_n = 0, \quad n \geq 0$$

Multiplying throughout by x^{n+2} and taking sum from 0 to ∞

$$\sum_{n=0}^{\infty} F_{n+2}x^{n+2} - 5\sum_{n=0}^{\infty} F_{n+1}x^{n+2} + 6\sum_{n=0}^{\infty} F_n x^{n+2} = 0 \tag{7.7}$$

Let $f(x) = \sum_{n=0}^{\infty} F_n x^n$

(7.7) becomes

$$(f(x) - F_1x - F_0) - 5x(f(x) - F_0) + 6x^2 f(x) = 0$$

By using initial conditions $F_0 = 0$, $F_1 = 1$,

$$\begin{aligned} f(x) &= \left(\frac{x}{1 - 5x + 6x^2}\right) \\ &= \frac{x}{(1 - 2x)(1 - 3x)} \\ &= \frac{1}{1 - 3x} - \frac{1}{1 - 2x} \end{aligned}$$

Using inverse of $f(x)$, we get

$$F_n = (3^n - 2^n), \quad n \geq 0$$

Example 7.7 Solve the recurrence

$$F_n = 2F_{n-1} + 1, \quad F_0 = 0$$

Solution: Rewrite the above recurrence relation as,

$$F_{n+1} - 2F_n - 1 = 0$$

The linear recurrence which is not homogeneous. Multiplying the equation through by x^{n+1}, and summing from $n = 0$ to ∞, we get,

$$\sum_{n=0}^{\infty} F_{n+1}x^{n+1} - 2\sum_{n=0}^{\infty} x^{n+1}F_n - \sum_{n=0}^{\infty} x^{n+1} = 0 \tag{7.8}$$

Let

$$\sum_{n=0}^{\infty} F_n x^n = f(x).$$

Then (7.8) becomes

$$(f(x) - F_0) - 2xf(x) = \sum_{n=0}^{\infty} x^{n+1}$$

$$\begin{aligned}(1-2x)f(x) &= \frac{x}{1-x} \\ f(x) &= \frac{1}{(1-2x)(1-x)} = \frac{1}{1-2x} - \frac{1}{1-x}\end{aligned}$$

Using inverse of $f(x)$, we get

$$\begin{aligned}F_n &= 1.2^n - 1 \\ &= 2^n - 1\end{aligned}$$

Remark 7.8 From the above example we get the definition of a non homogeneous recurrence relation and the same generating function method of solving it.

Example 7.9 Show that a Fibonacci number F_n is even if and only if n is a multiple of 3.

Solution: The proof is by induction on k where $n = 3k$. For $k = 1$, $F_1 = 1$, $F_2 = 1$ which are odd numbers. So F_3 is even.

Assume for $n = 3k$ and it is true that F_{3k} is even. As F_{3k-1} and F_{3k-2} and F_{3k} are related by,

$$F_{3k} = F_{3k-1} + F_{3k-2},$$

both F_{3k-1} and F_{3k-2} must be odd.

Now induction step is that

$$F_{3(k+1)} = F_{3k+3} = F_{3k+2} + F_{3k+1}$$

and to prove F_{3k+3} is even. Now $F_{3k+2} = F_{3k+1} + F_{3k}$ being odd and $F_{3k+1} = F_{3k} + F_{3k-1}$ being odd F_{3k+3} is even. Hence, the result.

Example 7.10 Let $p_k(n)$ denote the number of partitions of the integer n into atmost k-parts, then show that

$$\sum_{n=0}^{\infty} p_k(n)x^n = \prod_{i=1}^{k} \frac{1}{(1-x^i)}$$

Solution: Let us take an example that $n = 10$ and $k = 3$. Then, it can written as
$1 + 9, 2 + 8, 3 + 7, 4 + 6, 5 + 5$ (2-parts partitions)
$1 + 1 + 8, 1 + 2 + 7$ and so on (3-parts partitions)
10 (1-part partition)
$\sum_{n=0}^{\infty} p_k(n)$ counts the total. The right hand side product is

$$\prod_{i=1}^{k} \frac{1}{(1 - x^i)} = (1 + x + x^2 + \cdots)(1 + x^2 + x^4 + x^6 + \cdots) \cdots (1 + x^k + x^{2k} + \cdots)$$

The problem is to determine the coefficient of x^n in the product. That is we need to choose t terms such that each x^{ik_i} comes from different terms and $1k_1 + 2k_2 + \cdots + tk_t = n$. A term from the ith parenthesis is of the form x^{ik_i}, that is $i + i + \cdots + i$, (k_i copies of i). Hence, we obtain the coefficient to be partition of n having atmost k partitions.

By the same assignment as above, we can see that each time, the coefficient of x^n from the product on the right hand side, gives the partition of n into atmost k parts and conversely.

Remark 7.11 In the above example, if $p(n)$ denotes the number of partitions of n, then

$$\sum_{n=0}^{\infty} p(n)x^n = \prod_{k=1}^{\infty} \frac{1}{1 - x^k}.$$

We now see few examples where we use exponential generating functions for solving recurrence relations.

Example 7.12 Solve the recurrence relation

$$F_{n+1} = (n + 1)(F_n - n + 1), \text{ if } n \geq 0$$

where $F_0 = 1$.

Solution: Let $f(x) = \sum_{n=0}^{\infty} F_n \frac{x^n}{n!}$ be the exponential generating function for the given recurrence relation.

Multiply the given recurrence relation throughout by $\frac{x^{n+1}}{(n+1)!}$ and sum from $n = 0$ to ∞. Then

$$\sum_{n=0}^{\infty} F_{n+1}\frac{x^{n+1}}{(n+1)!} = \sum_{n=0}^{\infty} F_n(n+1)\frac{x^{n+1}}{(n+1)!} - \sum_{n=0}^{\infty}(n+1)(n-1)\frac{x^{n+1}}{(n+1)!}$$

$$\sum_{n=0}^{\infty} F_{n+1}\frac{x^{n+1}}{(n+1)!} = \sum_{n=0}^{\infty} F_n\frac{x^{n+1}}{(n)!} - \sum_{n=0}^{\infty}(n-1)\frac{x^{n+1}}{(n)!}$$

$$(f(x)-1) = xf(x) - x^2e^x + xe^x$$

$$f(x) = \frac{1}{1-x} + xe^x = \sum_{n=0}^{\infty} x^n + \sum_{n=0}^{\infty}\frac{x^{(n+1)}}{n!}$$

Hence, the coefficient of $\frac{x^n}{n!}$ in $f(x)$ will be $n! + n$

Example 7.13 Determine the number of ways to color the squares of a $1 \times n$ diving board using colours white, black and red if an even number of squares are coloured black.

Solution: Let F_n be the number of ways of coloring satisfying the given conditions in the problem. Let $F_0 = 1$. Each such coloring is nothing but a permutation of three objects W (white), R (red) and B (black) where we allow repetition. Here, 'B' must appear even number of times. Then using the exponential generating function for the occurrence of each color as $\sum_{n=0}^{\infty}\frac{x^n}{n!}$, we get the product exponential function as

$$f(x) = \left(\sum_{n=0}^{\infty}\frac{x^{2n}}{(2n)!}\right)\left(\sum_{n=0}^{\infty}\frac{x^n}{n!}\right)^2$$

where the first term is for the even occurrences of 'black' color.

Hence,

$$\begin{aligned} f(x) &= \left(\frac{e^x + e^{-x}}{2}\right)(e^{2x}) \\ &= \frac{1}{2}(e^{3x} + e^x) \\ &= \frac{1}{2}\sum_{n=0}^{\infty}(3^n + 1)\frac{x^n}{n!} \end{aligned}$$

Hence, $F_n = \frac{3^n+1}{2}, n \geq 0$.

7.2 Other Methods of Solving Recurrence Relations

We have seen how generating functions can be used in solving recurrence relations. We will now discuss two more methods of solving a recurrence equation. We will stick on only to linear recurrence relations throughout. The choice of the method of solving a recurrence may vary from problem to problem. The first method that we discuss now is called the direct substitution method.

1. **Direct method**

   In this case we start with the initial term of the sequence given by the initial conditions specified by the recurrence relation problem. Then we proceed to generate a few more terms using the relation. If possible guess the possible pattern of the nth term of the sequence. Then, we may check whether the guessed solution is correct, using some proof technique learned earlier. This method is also called as 'forward substitution' method.

Example 7.14 Let c be a circle. The construction is to draw diameters on the circle. If n diameters are drawn, find the number of sectors formed on the circle.

Solution: Let F_n be the number of sectors formed in the circle when n diameters are drawn. The problem is to find F_n.
We have, for
$n = 1, F_1 = 2$
$n = 2, F_2 = 4$
$n = 3, F_3 = 6$
$n = 4, F_4 = 8$

When one diameter is drawn, the circle is divided into 2 sectors. Each additional diameter will split the existing two sectors into two. Hence the recurrence relation that F_n will satisfy will be $F_{n+1} = F_n + 2$.

Now from the initial terms written above, the guessed value for F_n will be $2n$. It can be shown that $F_n = 2n$ for $F_{n+1} = F_n + 2$ by mathematical induction.

For $n = 1$, $F_2 = F_1 + 2 = 4 = 2n$. Assume for $n = k$ that $F_k = 2k$. Consider $F_{k+1} = F_k + 2 = 2k + 2 = 2(k + 1)$.

Hence, $F_n = 2n$.

Example 7.15 Solve the recurrence relation $F_{n+1} = 3F_n - 4$ where $F_1 = 5$.

Solution: We approach for the solution to this recurrence relation by what is known as direct method of substitution. The method is,

$$\begin{aligned}
F_{n+1} &= 3F_n - 4 \\
&= 3(3F_{n-1} - 4) - 4 \\
&= 3^2 F_{n-1} - 4(1+3) \\
&= 3^2(3F_{n-2} - 4) - 4(1+3) \\
&= 3^3 F_{n-2} - 4(1+3+3^2) \\
&= 3^4 F_{n-3} - 4(1+3+3^2+3^3) \\
&\vdots \\
&= 3^n F_1 - 4(1+3+3^2+\cdots+3^{n-1}) \\
&= 3^n F_1 - 4\left(\frac{3^n - 1}{3-1}\right) \\
&= 3^n F_1 - 2(3^n - 1)
\end{aligned}$$

If $F_1 = 5$, then $F_{n+1} = 3^{n+1} + 2$. i.e., $F_n = 3^n + 2$.

2. **Characteristic Equation Method**

In the direct substitution method, we did not use any big method of reaching the solution. We simply either guessed the solution or, iteratively substituted forward to reach the solution. However such methods may not be always useful. For example, for the recurrence relation $a_{n+2} - 4a_{n+1} + 3a_n = 0$ with $a_1 = 1, a_2 = 1$, the above method is not suitable for the known reasons of complication. We need a general method to solve such recurrences. The motivation comes from methods of solving differential equations. Consider the differential equation

$$\frac{d^2y}{dx^2} - 4\frac{dy}{dx} + 4y = 0.$$

This can also be written differently as

$$D^2y - 4Dy + 4y = 0.$$

Here, $\frac{d}{dx}$ or D is known as the differential operator.
The solution for the differential equation

$$D^2y - 4Dy + 4y = 0$$

is solved using the characteristic equation

$$m^2 - 4m + 4 = 0$$

The roots of this equations are 2, 2 and the solution is $y = C_1e^{2x} + C_2xe^{2x}$.

We follow the same method for solving recurrence relations. We understand the underlying auxiliary equation by defining a recurrence operator T from a vector space V to a vector space V. The elements of V are functions from integers to set of complex numbers. Let

$$T(g(n)) = g(n+1), T^2(g(n)) = g(n+2), \ldots, T^k(g(n)) = g(n+k)$$

where k is any positive integer.

Example 7.16 Let $g(n) = 2n + 3$ and $g(n) \in V$ where $T : V \to V$. Consider the application $2T^2 - 3T + 3$ to $g(n)$. That is,

$$\begin{aligned}(2T^2 - 3T + 3)(g(n)) &= 2T^2(g(n)) - 3T(g(n)) + 3g(n)\\ &= 2g(n+2) - 3g(n+1) + 3g(n)\\ &= 2(2(n+2)+3) - 3(2(n+1)+3) + 3(2n+3)\\ &= 4n + 8.\end{aligned}$$

Already we have defined both homogeneous and non homogeneous recurrence relations. We only need to know how to write the recurrence relation using the operator T.

Example 7.17 Let $\{F_n\}_{n=0}^{\infty}$ be a sequence that satisfies the recurrence relation $F_{n+1} = 3F_n$ where $F_0 = 1$. Write the sequence in terms of the operator T and hence solve for F_n.

Solution: Let $g(n) = F_n$ for $n \geq 0$. Then $g(0) = 1$ and

$$\begin{aligned}T(g(n)) &= 3g(n)\\ i.e.\ (T-3)(g(n)) &= 0\end{aligned}$$

Since T is an operator similar to D, the solution to $g(n) = c.3^n$ where c is a constant. Now the initial condition $F_0 = 1$, gives $c = 1$. Hence, $g(n) = 3^n\ (= F_n)$.

Remark 7.18 From the above discussion and examples, now on we will proceed with examples, where we solve a linear recurrence relation with constant coefficient by what is known as characteristic equation method.

Example 7.19 Solve the recurrence relation $F_{n+2} - 4F_{n+1} + 3F_n = 0$ by characteristic equation method given the fact that $F_1 = 1$ and $F_2 = 2$.

Solution: Let $F_n = g(n)$ for $n \geq 0$. Then $g(1) = 1$, $g(2) = 2$. Then the recurrence relation in terms of T will be

$$\begin{aligned} T^2(g(n)) - 4T(g(n)) + 3g(n) &= 0 \\ (T^2 - 4T + 3)(g(n)) &= 0 \end{aligned} \tag{7.9}$$

The auxillary equation is

$$(T^2 - 4T + 3) = (T - 3)(T - 1)$$

Its roots are 3, 1 and are distinct and real. Hence, the solutions of

$$(T - 3)(g(n)) = 0 \text{ and } (T - 1)(g(n)) = 0$$

$$g(n) = C_1 3^n \text{ and } g(n) = C_2(1^n)$$

Hence, the general solution is

$$g(n) = C_1 3^n + C_2.$$

Substituting $F_1 = 1$ and $F_2 = 2$, we get

$$g(n) = \frac{3^{n-1} + 1}{2}.$$

Example 7.20 Solve the recurrence relation $F_{n+2} - 4F_{n+1} + 3F_n = 3^n$ with the initial conditions $F_1 = 1$ and $F_2 = 3$

Solution: The given recurrence relation is a non homogeneous recurrence relation. We proceed by the same way of solving a non homogeneous differential equation.

First consider the homogeneous recurrence relation,

$$F_{n+2} - 4F_{n+1} + 3F_n = 0.$$

By the previous example, its solution will be $F_n = C_1 3^n + C_2$ where C_1 and C_2 are constants.

Now consider the given recurrence relation whose right hand side is 3^n. As the underlying homogeneous recurrence has 3^n as its solution, we must try with $F_n = C_3 n 3^n$ as a possible solution for the given recurrence relation.

Substituting it in the recurrence relation

$$\begin{aligned} C_3(n+2)3^{n+2} - 4C_3(n+1)3^{n+1} + 3C_3n3^n &= 3^n \\ C_3(9n+18-12n-12+3n) &= 1 \\ 6C_3 &= 1 \\ C_3 &= \frac{1}{6} \end{aligned}$$

Hence the solution will be

$$F_n = C_1 3^n + C_2 + \frac{1}{6}n3^n$$

where

$$C_1 = -\frac{1}{12} \text{ and } C_2 = \frac{3}{4}$$

Hence the complete solution will be

$$\begin{aligned} F_n &= -\frac{1}{12}3^n + \frac{3}{4} + \frac{1}{6}n3^n \\ &= \frac{3^n(2n-1)}{12} + \frac{3}{4} \\ &= \frac{3^{n-1}(2n-1)}{4} + \frac{3}{4} \end{aligned}$$

Example 7.21 Solve the recurrence relation $F_{n+2} - 3F_{n+1} + F_n = 0$ $n \geq 1$ given $F_0 = 1$ and $F_1 = 2$

Solution: Let $g(n) = F_n$. Then writing the recurrence relation in terms of the operator T, we get,

$$\begin{aligned} T^2(g(n)) - 3T(g(n)) + g(n) &= 0 \\ (T^2 - 3T + 1)g(n) &= 0 \end{aligned}$$

The roots of $T^2 - 3T + 1 = 0$ are α, β where $\alpha = \frac{3+\sqrt{5}}{2}$ and $\beta = \frac{3-\sqrt{5}}{2}$. Hence,

$$g(n) = C_1\left(\frac{3+\sqrt{5}}{2}\right)^n + C_2\left(\frac{3-\sqrt{5}}{2}\right)^n$$

Given $g(0) = 1, g(1) = 2$

$$C_1 + C_2 = 1$$

$$\frac{3+\sqrt{5}}{2}C_1 + \frac{3-\sqrt{5}}{2}C_2 = 2$$

$$C_1 = \frac{1+\sqrt{5}}{2\sqrt{5}}, \quad C_2 = \frac{\sqrt{5}-1}{2\sqrt{5}}$$

$$\therefore \quad F_n = \left(\frac{1+\sqrt{5}}{2\sqrt{5}}\right)\left(\frac{\sqrt{5}+3}{2}\right)^n + \left(\frac{\sqrt{5}-1}{2\sqrt{5}}\right)\left(\frac{3-\sqrt{5}}{2}\right)^n$$

Example 7.22 Solve the recurrence relation $F_n - 2F_{n-1} - F_{n-2} + 2F_{n-3} = 0, n \geq 3$ with the initial conditions $F_0 = 1$, $F_1 = 2$ and $F_2 = 0$.

Solution: Let $F_n = g(n)$. Then the auxiliary equation in terms of the operator T will be

$$T^3 - 2T^2 - T + 2 = 0$$

$$i.e. (T-2)(T+1)(T-1)g(n) = 0$$

Then, the solution is,

$$g(n) = C_1(-1)^n + C_2 + C_3 2^n$$

Using initial conditions we get,

$$C_1 + C_2 + C_3 = 1$$

$$C_1 - C_2 + 2C_3 = 2$$

$$C_1 + C_2 + 4C_3 = 0.$$

Solving the above by a little computation we get, $C_1 = 2$, $C_2 = \frac{-2}{3}$, $C_3 = \frac{-1}{3}$.

Hence,

$$F_n = 2(-1)^n - \frac{2}{3} - \frac{1}{3}2^n.$$

Example 7.23 Solve the following difference equation,

$$F_n - F_{n-1} = 3n^2 - 5n^3, \ n \geq 1, \ F_0 = 2.$$

Solution: Consider the underlying homogeneous recurrence relation

$$F_n - F_{n-1} = 0.$$

Let $g(n) = F_n$. Then the recurrence can be written down as,

$$F_{n+1} - F_n = 0$$

and

$$\begin{aligned} T(g(n)) - g(n) &= 0 \\ (T - 1)g(n) &= 0. \end{aligned}$$

Hence, the solution for this will be $g_1(n) = C_1(1)^n = C_1$.

Now consider the equation

$$F_n - F_{n-1} = 3n^2 - 5n^3.$$

The solution will be of the form

$$g_2(n) = C_2 \sum_{k=1}^{n} (3k^2 - 5k^3)$$

Hence, the solution will be

$$\begin{aligned} g(n) &= C_1 + C_2 \sum_{k=1}^{\infty} (3k^2 - 5k^3) \\ g(0) &= 2 \Rightarrow C_1 = 2 \end{aligned}$$

Let $C_2 = 1$, then

$$\begin{aligned} g(n) = 2 + \frac{3n(n+1)(2n+1)}{6} \\ = -5 \left(\frac{n(n+1)}{2} \right)^2 \end{aligned}$$

(Note $C_2 \neq 0$).

7.3 Some Sorting Methods

Sorting numbers, strings, symbols etc. occurs very often in algorithms. By sorting we mean an ordering or positioning some data of information for computational purposes. The method of sorting differs for different data types. Also there are different techniques for sorting. Commonly used sorting algorithms are for numbers so that the numbers are put in ascending or descending order. There is also an ordering in a collection of strings or words called dictionary ordering. This dictionary ordering is totally dependent on the alphabet ordering over which the strings are made.

For example one needs to order only two numbers, then simple comparison and exchange positions will work. But if the input is a set of numbers whose cardinality is more than 2, then there are different techniques to sort these numbers. The first method that is presented below is known as "Bubble Sort" method.

Bubble Sort

Let $\{n_1, n_2, \ldots, n_t\}$ be the set of numbers to be sorted. The method goes in steps by comparing adjacent numbers and swapping if required. That is the algorithm compares n_1 and n_2. If $n_1 > n_2$, then swap n_1 and n_2. Update the input incorporating the change. Now a comparison of second and third number is done and update the input as before. t steps of comparisons, updates the input of t numbers so that the largest number is in the tth position.

Now in the updated list the above said procedure is once again carried out for $(t-1)$ numbers. The updated list of numbers will have t^{th} and $(t-1)^{th}$ positions filled as required. This is done $(t-1)$ times to completely obtain a sorted list.

When we discuss the computational steps, the comparisons made are to be counted. If a_t is the number of comparisons made, then

$$a_t = (t-1) + a_{t-1}.$$

This will be the recurrence relation for the count of comparisons. One can easily see that

$$a_t = \frac{t(t-1)}{2}.$$

The best case running time occurs when the input numbers are already sorted. The worst case running time occurs when the input numbers are completely reversed. So, we say safely on an average case, its complexity will be $O(t^2)$.

Insertion Sort

The name of the method suggests a technique of inserting numbers at the appropriate positions. In the beginning, the first number n_1 of the input list $n_1, n_2, \ldots, n_t$ is taken and it is sorted. Now the numbers are read in sequence. In the first step consider n_1 and n_2, compare them and sort their positions. Update the list where the first two numbers are in order. Now as a next step, consider the next unvisited number which is the third one here. Now compare this number with the previous number. Now positions are swapped or remain unchanged based on the comparison. If they are swapped, then the highest value number from the first three numbers is now in the third position. But the number in the second position may be smaller than the first one. Hence another comparison is needed with the first number to sort the first three numbers. In the next pass the number in the fourth position is compared with the first three numbers as before. The correct index or position of the fourth number is found and it is inserted. The procedure continues till the input list is completely sorted.

Example 7.24 Let the numbers be,

$$14, 13, 15, 4, 5$$

First step: $\boxed{14}, \boxed{13}, 15, 4, 5$
14 and 13 are compared. As $14 > 13$ the numbers are swapped. Now the new updated list is

$$13, \boxed{14}, \boxed{15}, 4, 5$$

Second step: Now comparing 15 and 14, as $15 > 14$, there will be no swapping and hence the list now will be,

$$13, 14, \boxed{15}, \boxed{4}, 5$$

Third step: Comparing 4 and 15, $15 < 4$ and hence they swap their positions. Since there is a swapping step, 4 will be compared with 14 and successively with 13. Each time comparison shifts the position to the first which is the correct position for 4. After this step, the list looks as below.

$$4, 13, 14, 15, 5$$

Fourth step: The number 5 is compared with 15, 14, 13 and 4 and it is inserted between 4 and 13. Now the updated list is sorted.

The following is the algorithm for insertion sort. We can understand the recursion and hence its complexity from a brief analysis.

Insert-Sort Algorithm
Ins-Sort (M)
1. for $i = 1$ to t
2. $\quad key = M[i]$
3. $*$ $M[i]$ to be inserted in the correct position in the sorted list
$M[1]M[2]\ldots M[i-1]$ $*$
4. $j = i - 1$
5. while $(j > 0$ and $M[j] > key)$
6. $\quad M[j+1] = M[j]$
7. $\quad j \leftarrow j - 1$
8. $M[j+1] \leftarrow key$

Here '$*$' indicates that it is a comment statement and hence takes only no time.
Step '1' runs for t times
Step '2' runs for $(t-1)$ times
Step 3 is a comment statement and is run only once
Step '4' runs for $(t-1)$ times

Step 5, 6, 7 are within a loop. So Step 5 runs $\sum_{j=2}^{t} n_j$ times and hence Step 6 runs $\sum_{j=2}^{t}(n_j-1)$ times. Step 7 also runs the same number of times as Step 6. Step 8 runs for $(t-1)$ times.

Clearly the recursion occurs in the while loop. Suppose step i takes c_i units for each iteration. So the total time units will be,

$$A(t) = c_1 t + c_2(t-1) + c_0 0 + c_4(t-1) + c_5 \sum_{j=2}^{t} n_j$$
$$+ c_6 \sum_{j=2}^{t}(n_j - 1) + c_7 \sum_{j=2}^{t}(t_j - 1) + c_8(t-1).$$

In the best case, the input array is already sorted. So the computation of $A(t)$ will be of the form $\ell t + m$ where ℓ, m are constants. So we say $A(t)$ is a linear function of t.

In the worst case, the list is reversely sorted. So every recursive step in 5, 6 and 7 are run for all possible indices in the loop. That is

$$\sum_{j=2}^{t} j = \frac{(t+1)t}{2} - 1$$
$$\sum_{j=2}^{t}(j-1) = \frac{t(t-1)}{2}$$

Hence it is not difficult to see that $A(t)$ will be of the form $\ell t^2 + mt + k$.

So we say that the time complexity of insertion sort is in $O(t^2)$.

In simple terms, one can understand that in the jth iteration, the list or array from index 1 to $j-1$ is sorted. So we understand the recurrence relation involved in calculating the number of time steps will be,

$$A(t) = A(t-1) + t,$$

where 't' denotes the time needed to insert the last element in the list of sorted $(t-1)$ elements.

Merge Sort

Let $A = \{n_1, n_2, \ldots, n_t\}$ be the list of input numbers. Now in merge sort, we apply a technique called 'divide-and-conquer' technique. Briefly the technique is to divide a list of t-numbers into two lists of each having $\frac{t}{2}$ elements each. Then sort the two divided sublists. Finally merge the two sorted sublists to get one final sorted list. We have a pseudocode of the merge sort as below.

Merge-Sort (A, m, n)

if $m < n$

$i = \frac{m+n}{2}$

Merge-Sort(A, m, i)

Merge-Sort$(A, i + 1, n)$

Merge(A, m, n)

Here Merge(A, m, n) is the procedure that sorts the numbers from index m to index n. The sub-program 'merge-sort' sorts the subarrays before merging.

If $A(t)$ is the number time units totally taken to sort the 't' numbers, then the recursion will be

$$A(t) = 2\,A\left(\frac{t}{2}\right) + t.$$

We saw that a recurrence relation will be an expression describing a function in terms of smaller input values. Many sorting algorithms have their running time dependent on recursive call of subproblems and hence a recurrence, relation on its running time is encountered.

Exercise

1. Solve the recurrence relation $F_n = 5F_{n-1} - 6F_{n-2}, n \geq 2$ where $F_0 = 1$, $F_1 = -2$.
2. Solve the recurrence relation $F_n - 6F_{n-1} + 7F_{n-2} = 1$ where $F_0 = 0$, $F_1 = 1$.
3. Find the recurrence relation for the number of ways of decorating a n-feet long rod such that 4 color stickers of the following sizes are to be used. Red stickers are of size 2 feet long, yellow, green and orange stickers are of size 1 feet long each.
4. A student has to climb up a staircase of n steps to reach his classroom. At each step, the student climbs either one step or two steps up. Find the recurrence relation to the number of ways of climbing the n steps and solve the recurrence relation.
5. Let F_n be the number of ways to pay a 'n rupees' bill using 10 rupee coins, five rupee coins and one rupee coin only. Find the ordinary generating function for F_n. Solve for F_n explicitly.
6. Solve the recurrence relation with the initial condition $F_0 = 2$, $F_n = (F_{n-1})^2, n \geq 1$.
7. Determine the number a_n of a n digit (in base 10) numbers with each digit being odd. Also a condition that digit 1 and 3 must occur even number of times.
8. Find F_n in terms of n in the following recurrence relations.

(i) $F_{n+2} - 4F_{n+1} + 4F_n = 0$, $F_1 = 1$, $F_2 = 2$

(ii) $F_{n+2} - 4F_{n+1} + 4F_n = 2^n$, $F_1 = 1$, $F_2 = 2$

9. Solve the following recurrence relations by direct method of substitution.

   (i) $F_n - F_{n-1} = n^2$ where $F_0 = 2$
   (ii) $F_n - F_{n-1} = n2^n$ where $F_0 = 1$
   (iii) $F_n - F_{n-1} = n + n^2$ where $F_0 = 1$

10. Solve the following recurrence relations by characterstic equation method.

   (i) $F_{n+2} - 5F_{n+1} + 6F_n = 0$ where $F_0 = 0$, $F_1 = 1$
   (ii) $F_{n+2} + F_{n+1} - 12F_n = 0$ where $F_0 = 0$, $F_1 = 2$
   (iii) $F_{n+4} + 2F_{n+3} - 12F_{n+2} + 14F_{n+1} + 5F_n = 0$ where $F_0 = 0$, $F_1 = 2$, $F_2 = 1$, $F_3 = 1$

11. Write down the following recurrence relations in terms of the operator T.

   (i) $F_{n+2} - F_{n+1} - 3F_n = 0$
   (ii) $F_n - F_{n-1} - 3F_{n-2} = 2^{n+1} - n^2$
   (iii) $F_{n+3} - 5F_{n+1} + F_n = 2^n$

12. For each of the recurrence relations written in terms of the operator T, find its general solution.

   (i) $(T - 2)(T + 3)g(n) = 0$
   (ii) $(T^2 - 5T + 4)g(n) = 0$
   (iii) $(T^3 - 3T^2 + 3T + 1)g(n) = 0$

13. Solve the following non homogeneous equations.

   (i) $(T - 5)(T + 4)g(n) = 2^n$
   (ii) $(T^2 - 2T + 1)g(n) = 4n$
   (iii) $(T - 2)(T - 3)^2 g(n) = 2n + 3$

14. Find the number of 10 digit positive integers that satisfy the following two conditions.

   (i) Each digit is either 1 or 2
   (ii) There exist exactly two consecutive 1's

15. Find the number of ways that the numbers 1, 2, ..., 2004 can be arranged in a row such that, except for the first number, each number differs from some number on its left by 1.

16. A circle C is given. Chords are drawn inside the circle which split the interior region of the circle into smaller regions. Find F_n, the largest number of regions that can be formed by n chords.
17. Let F_n be the number of expressions for the positive integer n as sum of positive integers. For example,

$$3 = 0 + 3, 3 = 1 + 2,\ \ 3 = 2 + 1,\ \ 3 = 1 + 1 + 1.$$

So, $F_3 = 4$. We can check F_4 to be 8.

(i) Show that $F_1 = 1$ and $F_n = 1 + \sum_{k=1}^{n-1} F_k$ for $n \geq 2$.
(ii) Also, prove from (i) that $F_1 = 1$, $F_n = 2(F_{n-1})$, $n \geq 2$.
(iii) Solve for F_n.

18. Consider the sequence $\{a_n\}_{n=0}^{\infty}$ where $a_0 = 1$, $a_{n+1} = (n+1)(a_n - n + 1)$. Find the direct formula for a_n using exponential generating funtion.
19. Solve the following recurrence relation using generating function method.

$$a_n - a_{n-1} - 2a_{n-2} = 2^n, n \geq 2, a_0 = 2, a_1 = 1$$

20. Solve for $\{a_n\}_{n=0}^{\infty}$ if

$$\sum_{t=0}^{n} a_t a_{n-t} = 1$$

where $a_0 = 1$.
21. Find a simple expression for the generating function

$$\mathcal{B}(x) = \sum_{n=0}^{\infty} B_n \frac{x^n}{n!}$$

where $B_0 = 1$, $B_1 = 1$,

$$B_{n+1} = \sum_{n=0}^{\infty} C(n,t) B_t$$

for $n > 1$. B_n's are the Bell numbers.
22. Solve for a_n explicity if

$$a_{n+1} = 2a_n + 3^n, n \geq 0, a_0 = 15$$

23. Let a_n denote the number of strings over $\{0, 1\}$ that do not contain the substring '11'. Find a recurrence relation for a_n.

Inclusion Exclusion Principle

8

8.1 The Principle of Inclusion-Exclusion

This section is concerned with events which are defined in terms of other events $A_1, A_2, \ldots, A_k$. Of the k events, one, two or more can occur simultaneously. This means there can be overlapping events. For example if A_1 and A_2 are two events, then $A = A_1 \cup A_2$ denotes that either A_1 or A_2 or both can occur. Here $A = A_1 \cup A_2 \cup \cdots \cup A_k$ and the problem is to find $|A|$. We can in the beginning find the cardinality of A by some direct method. But what is obtained is only an approximate counting. We need to subtract off an overcounting, as the events $A_1, A_2, \ldots, A_k$ are overlapping. This method will be named as 'method of Inclusion-Exclusion' or 'Principle of Inclusion-Exclusion' (PIE). PIE is one of the important and basic tools in enumerative combinatorics.

Example 8.1 There are 14 students in a class who are good in mathematics and 17 students who are good in physics. Four students are good in both mathematics and physics. How many students are good atleast in one of the subjects?

Solution: The situation can be described using a Venn-diagram.

Clearly $14 + 17 - 4 = 27$ are good in at least one of the subjects (Fig. 8.1).

The situation that the students are tested for their knowledge in three subjects becomes more complicated but manageable.

Example 8.2 Suppose in a class, there are 14 students who are good in mathematics, 17 students are good in physics and 18 students are good in chemistry. Four students are good in maths and physics, three students are good in maths and chemistry, and five students are good in physics and chemistry. There is one student who is good in all the three subjects. How many students are good in at least one of the subjects?

R. Rama, *Topics in Combinatorics and Graph Theory*,
https://doi.org/10.1007/978-3-031-74252-1_8

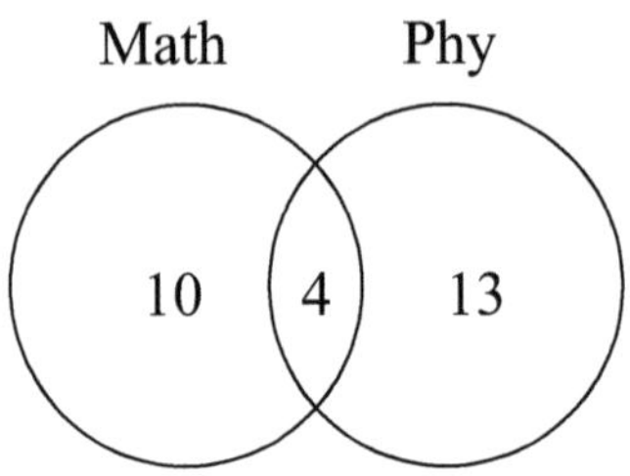

Fig. 8.1 .

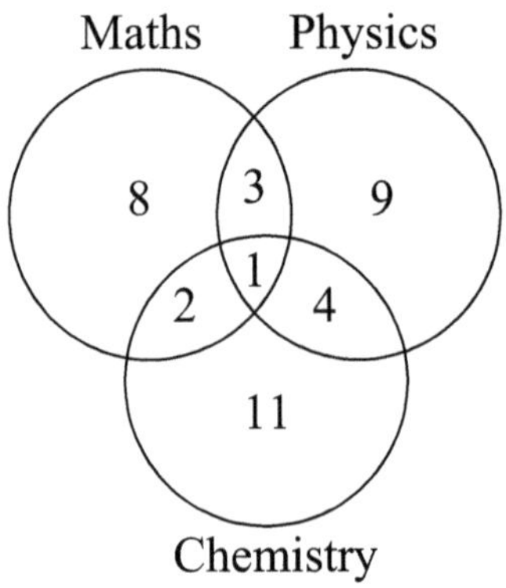

Fig. 8.2 .

Solution: Adding the number of students who are good in maths, physics and chemistry $14 + 17 + 18 = 49$ results in an over counting. We have to subtract the number of students who are good in two subjects i.e. the modified count will be $49 - 4 - 3 - 5 = 37$. This number is also an approximate value as this does not include one student who is good in all the three subjects, but the student has been subtracted three times (once for each pair of subjects). It means, this student together with 37 is the required answer. By Venn-diagram it means (Fig. 8.2).

Remark 8.3 From the previous example, it is clear to analyse the situation, by repeatedly including and excluding the events for the correct counting.

Theorem 8.4 *Let $A_1, A_2, \ldots, A_n$ be subsets of some finite set A. Then*

$$|A_1 \cup A_2 \cdots \cup A_n| = \sum_{j=1}^{n} (-1)^{j-1} \sum_{i_1, i_2, \ldots, i_j} |A_{i_1} \cap A_{i_2} \cap \cdots \cap A_{i_j}| \tag{8.1}$$

where $(i_1, i_2, \ldots, i_j)$ ranges all j-element subsets of an n-element set X.

Proof By looking at the left hand of (8.1) we can see that the union counts each element of A_i exactly once. We need to show that each element on the LHS is counted exactly once on the RHS of (8.1) also. Let $x \in A_1 \cup A_2 \cup \cdots \cup A_n$. Let A be a subset of indices. Let A satisfy the condition that $x \in A_i$ if and only if $i \in A$. Let the cardinality of A be t. Clearly $t \geq 1$. Then $x \in A_{i_1} \cap A_{i_2} \cap \cdots \cap A_{i_t}$ if and only if $(i_1, i_2, \ldots, i_t) \subseteq A$. It means that $x \in A_{i_1} \cap A_{i_2}$, $x \in A_{i_2} \cap A_{i_3} \cap A_{i_4}$ and so on. That is the RHS of (8.1) counts x once for each of these subsets with alternating signs. Hence x is counted

$$t - C(t, 2) + C(t, 3) - C(t, 4) + \cdots + (-1)^{t-1} C(t, t) = 1 \tag{8.2}$$

times. That is one time. To see this subtract 1 from both sides of (8.2),

$$1 - t + C(t, 2) - C(t, 3) + \cdots + (-1)^t C(t, t) = 0$$

which is true due to binomial theorem. □

Example 8.5 Suppose there are 200 graduate students in total. In this 100 students opted for logic and 70 students opted for computability courses. If there are 50 students who opted for both the courses, how many students take neither of the course?

Solution: Let
p: be the property that the student has opted for logic
q: be the property that the student has opted for computability.
Now $N = 200$, $N(p) = 100$, $N(q) = 70$, $N(pq) = 50$. Then by PIE

$$\begin{aligned}\overline{N} &= N - [N(p) + N(q)] + N(pq)\\ &= 80\end{aligned}$$

Therefore there are 80 students who do not opt for both the courses.

Example 8.6 Find the number of numbers between 1 and 10^4 that are not divisible by 6, 7 or 8.

Solution: Let
p: the property that n is divisible by 6
q: the property that n is divisible by 7
r: the property that n is divisible by 8.
Then $N(p) = \lfloor \frac{10^4}{6} \rfloor = 1666$,

$$N(q) = 1428, N(r) = 1250$$

$$N(pq) = \lfloor \frac{10^4}{6.7} \rfloor = 238, N(qr) = 178,$$

$$N(pr) = \lfloor \frac{10^4}{lcm(6,8)} \rfloor = \lfloor \frac{10^4}{24} \rfloor = 416$$

$$N(pqr) = \lfloor \frac{10^4}{lcm(6,7,8)} \rfloor = \lfloor \frac{10^4}{168} \rfloor = 59$$

Then by PIE, the number of numbers satisfying p, q, r will be

$$[N(p) + N(q) + N(r)] - N(pq) - N(pr) - N(qr) + N(pqr) \tag{8.3}$$

But we need the numbers which do not satisfy p, q and r. Hence it will be

$$\overline{N} = 10^4 - \text{expression (8.3)} = 6429$$

Example 8.7 Let a bag contain 7 balls of 7 different colours and let there be three different boxes B_1, B_2, B_3 (say). Find the number of ways of placing the balls into the three different boxes so that each box contains at least one ball.

Solution: Let N be the total number of ways to distribute the balls into the boxes and $N = 3^7$.

Let

p: the property that B_1 has no balls.
q: the property that B_2 has no balls.
r: the property that B_3 has no balls.

The problem is to find

$$N - (N(p) \cup N(q) \cup N(r)$$

$$N(p) = N(q) = N(r) = 2^7$$

$$N(pq) = 1^7 = N(qr) = N(pr)$$

$$N(pqr) = 0$$

$\therefore$ The required count will be, by PIE

$$3^7 - [3.2^7 - 3.1^7 + 0] = 1806.$$

Example 8.8 How many integers in the set $\{1, 2, 3, \ldots, 180\}$ have atleast one prime divisor in common with 180?

Solution: The prime divisors of 180 are 2, 3 and 5. i.e. $180 = 2^2 \cdot 3^2 \cdot 5$. Let E_i be numbers between 1 to 180 that have divisor i, $i = 2, 3, 5$. The problem is to find $|E_2 \cup E_3 \cup E_5|$. We use PIE for the calculations.

$$|E_2| = \lfloor \frac{180}{2} \rfloor = 90,\ |E_3| = \lfloor \frac{180}{3} \rfloor = 60$$

$$|E_5| = \lfloor \frac{180}{5} \rfloor = 36,\ |E_2E_3| = \lfloor \frac{180}{6} \rfloor = 30$$

$$|E_2E_5| = \lfloor \frac{180}{10} \rfloor = 18,\ |E_3E_5| = \lfloor \frac{180}{15} \rfloor = 12$$

$$|E_2E_3E_5| = \lfloor \frac{180}{30} \rfloor = 6$$

therefore

$$|E_2 \cup E_3 \cup E_5| = 132.$$

Example 8.9 How many integers in the set $\{1, 2, \ldots, 180\}$ are relatively prime to 180?

Solution: From the previous example it will be $180 - 132 = 48$. Hence we have the following.

If n is an integer and if $p_1, p_2, \ldots, p_t$ are the prime factors of n, then the number of integers which are less than 'n' and relatively prime to 'n' can be found by PIE. That is known as Euler's ϕ-function and is denoted by $\phi(n)$.

$$\phi(n) = n\left[1 - \frac{1}{p_1}\right]\left[1 - \frac{1}{p_2}\right]\cdots\left[1 - \frac{1}{p_t}\right].$$

8.2 Euler's Phi-Function

Prime Numbers

The numbers $1, 2, 3, 4, \ldots$ are called natural numbers. Number theory will deal with such numbers, properties and relationships between certain subsets of these. Natural numbers can be grouped as below:

Even Numbers: 2, 4, 6, 8, 10, ...

Odd Numbers: 1, 3, 5, 7, ...

Squares: 1, 4, 9, 16, ...

Composite Numbers: 4, 6, 8, 9, 10, 12, 14, ...

Prime Numers: 2, 3, 5, 7, 11, 13, 17, ...

Fibonacci Numbers: 1, 1, 2, 3, 5, 8, ...

Many of these are already well known subsets of natural numbers. Here after number means natural number.

Example 8.10 Is it possible to find numbers x, y and z such that $x^2 + y^2 = z^2$? Such a triplet will be called as 'Pythagorean Triplet'. This is a property of numbers. Not every triplet satisfies this property. For example (3, 4, 5) is a Pythagorean Triplet.

Definition 8.11 A prime number p is a number whose only factors are p and 1.

Example 8.12 17, 19 are prime numbers. This is a property of numbers. Naturally, we want to see in the whole set of natural numbers, is the subset containing all the primes infinite? If so how to list them? How to find whether a given number is prime or not?

Before we proceed further, let us understand integer divisibility.

Definition 8.13 Let m, n be two integers such that $m \neq 0$. We say 'm divides n' if there exists an integer k, such that $n = km$. We say m is a factor of n or n is a multiple of m.

Notation 8.14 The standard notation is, if we write '$m|n$', then we mean m divides n. If we write $m \not| n$, then we mean 'm does not divide n'.

Example 8.15 (i) For the subset 'E' of even numbers, 2 divides m, where $m \in E$. i.e., $2|m$. (ii) For the subset 'O' of odd numbers 2 does not divide m', where $m' \in O$. $2 \not| m'$.

Remark 8.16 We include 0 wherever required. For any $n \in \mathbb{N}$, we see $n|0$.

We state the following simple theorem without proof.

Theorem 8.17 *Let n_1, n_2, n_3, p and q be integers. If $n_3|n_1$ and $n_3|n_2$, then $n_3|(pn_1 + qn_2)$.*

We now see what a division algorithm is. When two numbers are multiplied, we get another number. Multiplication is nothing but repeated addition. Likewise division is repeated subtraction. So we have the following divison theorem which proposes and defines division of an integer by another number.

Theorem 8.18 *Let m and n be two numbers with $m > 0$. Then there exists unique numbers q and r such that $n = qm + r$, $0 \leq r < m$. If $m \not| m$, then $r \neq 0$.*

Proof Since division is a repeated subtraction, we define the following sequence for m and n as below.

$$n - m, n - 2m, n - 3m, n - 4m, \ldots$$

Clearly each element is distinct. Choose the smallest nonnegative number in this sequence, which is r here. Let it be, $n - qm = r$.

The elements r, and q can be shown to be unique, by using method of contradiction. □

Remark 8.19 The above theorem is true for any two integers and it is also called as division algorithm.

Example 8.20 If $m = 5, n = 73$, then $73 = 14 \times 5 + 3$. Here $q = 14, r = 3$.

The following properties can be verified easily.

1. If m, n, s and t are integers with $m, s \neq 0$ such that $m|n$, $s|t$, then $ms|nt$.
2. If m and n are positive integers and $m|n$, then $m \leq n$.
3. If m, n and s are integers such that $ms|ns$, then $m|n$.
4. If m and n are integers such that $m|n$ and $n|m$, then $m = \pm n$.

Divisors

By division algorithm, for any two integers m and n, we have $n = qm + r$ where $0 \leq r < m$. When $r = 0$, we say that m divides n or m is a divisor of n.

Definition 8.21 If a number t is a divisor of two integers m and n. We call t as a common divisor. For $m, n \neq 0$, the number of common divisors of m and n will be a finite set. The number in the set of divisors of m, n which has the maximum value is called as the greatest common divisor.

Notation 8.22 For $m, n \in \mathbb{N}$, $div(m, n) = d$ means that d is a divisor of m, n; $gcd(m, n) = d$ means that d is the greatest common divisor of m and n.

Definition 8.23 For any integers m, n if $gcd(m, n) = 1$, then we call m and n as relatively prime to each other.

Example 8.24 The $gcd(4, 10) = 2$. Then $2 = s_1 4 - s_2 10$ where $s_1 = 3, s_2 = 1$.

In general for any two integers m and n, we have

$$m = 1 \cdot m + 0 \cdot n$$
$$n = 0 \cdot m + 1 \cdot n$$

A linear combination set will be $\{s_1 m + s_2 n | s_1, s_2 \text{ are integers}\}$. This set is not empty. By well-ordering property it has a least d such that $d = s_1 m + s_2 n$. The following theorem presents the idea that the gcd of two integers m, n can be represented as a linear combination of m and n, provided both are not zero.

Theorem 8.25 *Let m, n be two integers, both not zero. Then the $gcd(m, n) = d = s_1 m + s_2 n$ where s_1 and s_2 are some integers.*

Remark 8.26 The idea of gcd can be extended to k numbers. The above theorem will be true for k numbers.

Definition 8.27 Let $m_1, m_2, \ldots, m_t$ be integers such that not all of them are zero. Then

(i) The gcd of $m_1, m_2, \ldots, m_t$ is the largest number that is a divisor of all the t numbers
(ii) If $gcd(m_1, m_2, \ldots, m_t) = 1$, then the t numbers are said to be mutually prime to each other.
(iii) The t numbers are said to be pairwise relatively prime, if any two numbers a_i, a_j are relatively prime for $i \neq j$.

Remark 8.28 Relatively prime pairs are also called as co-prime pairs. The following properties can be proved easily.

1. For any integer s, the $gcd(sm, sn) = sgcd(m, n)$.
2. If $gcd(s, m) = gcd(s, n) = 1$, then $gcd(s, mn) = 1$.
3. For any integer s, $gcd(m, n) = gcd(m, -n) = gcd(m, n + sm)$
4. If for integers m, n and t, $t|mn$, and t and n are co-prime, then $t|m$.

The definition and properties of divisors gave a basic understanding on the integers and divisibility. Now by Euclidean Algorithm we come across a method that determines the gcd of two input numbers m and n. We have seen that the $gcd(m, n)$ can be written down as a linear combination of m and n. That is, $gcd(m, n) = s_1 m + s_2 n$, s_1, s_2 are integers. We also noted that $gcd(m, n) = gcd(m, -n)$. We state the following lemma which follows directly from properties of gcd.

Lemma 8.29 *Let m and n be two integers. If $m = pn + r$ where p and r are integers, then $(m, n) = (n, r)$.*

The following is called the Euclidean algorithm which states that the gcd of two integers m and n is the last non-zero remainder obtained by repeated division.

Theorem 8.30 *Let m and n be two positive integers where $m \geq n$. Define*

$$r_i = r_{i+1}q_{i+1} + r_{i+2} \text{ where } 0 \leq r_{i+2} < r_{i+1}$$

Starting with $r_0 = m$, $r_1 = n$, the successive applications of division algorithm for all $i = 0, 1, 2, \ldots, t-2$ will terminate in $r_t = gcd(m, n)$. Here $r_{t+1} = 0$.

Proof The proof is straightforward. However, let us write down the steps obtained by applying of division algorithm

$$\begin{aligned} r_0 &= r_1 q_1 + r_2 \;\; 0 \leq r_2 < r_1 \\ r_1 &= r_2 q_2 + r_3 \;\; 0 \leq r_3 < r_2 \\ &\ldots\ldots \\ r_{t-2} &= r_{t-1} q_{t-1} + r_t \;\; 0 \leq r_t < r_{t-1} \\ r_{t-1} &= r_t q_t \end{aligned}$$

The final step will be the one where the remainder term is zero as all the remainders are integers and every remainder in the next step being less than the remainders in the previous step. So,

$$gcd(m, n) = gcd(n, r_2) = gcd(r_2, r_3) = \cdots = gcd(r_t, 0) = r_t.$$

□

Example 8.31 Find the gcd of 11468 and 3137 using Euclidean Algorithm.

Solution:

$$\begin{aligned} 11468 &= 3137 \times 3 + 2057 \\ 3137 &= 2057 \times 1 + 1080 \\ 2057 &= 1080 \times 1 + 977 \\ 1080 &= 977 \times 1 + 103 \\ 977 &= 103 \times 9 + 50 \\ 103 &= 50 \times 2 + 3 \\ 50 &= 3 \times 16 + 2 \\ 3 &= 2 \times 1 + 1 \\ 2 &= 2 \times 1 \\ gcd(11468, 3137) &= 1. \end{aligned}$$

Remark 8.32 Note that 1 can be expressed as a linear combination of 11468 and 3137.

Product of Primes

We will now see that every number can be represented as product of primes. Such a representation is called as 'prime decomposition' of the number. Given a number, finding such a decomposition is a difficult problem. For example consider the number 1425. It can be written as $3 \cdot 5^2 \cdot 19$. We consider $3 \cdot 19 \cdot 5^2$ is the same as $3 \cdot 5^2 \cdot 19$. The order of the prime powers does not matter. So we will see that such a decomposition is unique for each number.

Proposition 8.33 *Let n be a natural number. Then it has a prime decomposition.*

Proof If $n = 1$, then it is the null product of prime. If n is prime, then n is the prime decomposition of itself. Suppose n is composite. Then $n = n_1 \cdot n_2$ where $n_1, n_2 < n$. Then it can be established by induction that n_1 and n_2 are products of primes. So n has a prime decomposition. □

Now the problem is given a number n,

(1) Is the prime decomposition unique?
(2) Does there exist a polynomial time algorithm to perform the decomposition?

We will see first Question (1). The following theorem is known as the Fundamental theorem of arithmetic.

Theorem 8.34 *Every positive integer $n(\neq 1)$ can be represented uniquely as a product of primes.*

Proof We know for any positive integer n there is a prime decomposition. If possible let n have two decompositions

$$n = p_1 p_2 \dots p_k,$$
$$n = q_1 q_2 \dots q_t.$$

As $p_1 | n = q_1(q_2 \dots q_t)$, we have either $p_1 = q_1$ or $p_1 | (q_2 \dots q_t)$. Note that $p_1 = q_i$ for some i. Now cancel p_1 and its equivalent term q_i. Repeating this procedure, we see that the two representations are the same. □

Question (2) above still is an open problem except a polynomial time algorithm using quantum computers being proposed by Peter Shor.

Integers Modulo n

Numbers can be grouped based on certain property. For example we can make two sets of numbers as prime number set and compositive number set. We also saw several divisibility properties of numbers. Congruence is a property of divisibility. Congruency, groups numbers based on their divisibility with respect to a particular number n. We directly give the following definition.

Definition 8.35 Let a, b, n be integers. If $n|(a-b)$, then we say that a and b are congruent with respect to n. We also call it as a and b are congruent modulo n. We write down here the standard notation which is $a \equiv b \bmod n$.

Remark 8.36 As $n|(a-b)$, $a = tn + b$ for $t \in \mathbb{Z}$.

Example 8.37

$$21 \equiv 5 \bmod 8$$
$$21 \equiv 4 \bmod 17$$
$$21 \equiv 3 \bmod 9$$

We now state the following simple properties of congruence relation and prove some of them.

Theorem 8.38 *Let a, b, c and d be any integers. Let n be a positive integer. Then*

(1) If $a \equiv b \bmod n$ then $b \equiv a \bmod n$.
Proof*: By $a \equiv b \bmod n$. We mean $a = tn + b$. That is $(a - b) = tn$ and $(b - a) = (-t)n$. So $b \equiv a \bmod n$.*
(2) If $a \equiv b \bmod n$ then $b \equiv c \bmod n$ then $a \equiv c \bmod n$. That is the relation congruence is transitive.
Proof is left as an exercise.
(3) If $a \equiv b \bmod n$ then $(a + c) \equiv (b + c) \bmod n$.
Proof*: By definition, $a \equiv b \bmod n$ means $a = tn + b$. Now,*

$$(a + c) = tn + (b + c)$$

which means $(a + c) \equiv (b + c) \bmod n$.
(4) If $a \equiv b \bmod n$ then $(a - c) \equiv (b - c) \bmod n$.
Proof is left as an excercise.

(5) *If* $a \equiv b \bmod n$ *then* $ac \equiv bc \bmod n$.
Proof: *By definition* $a = tn + b$. *So* $ac = (tc)n + bc$. *So* $ac \equiv bc \bmod n$.
(6) *If* $a \equiv b \bmod n$ *then* $c \equiv d \bmod n$, *then* $ac \equiv bd \bmod n$.
Proof is left as an excercise.

Definition 8.39 Let m and n be two positive integers. The least common multiple m, n is the smallest positive integer that is a multiple of both.

Notation 8.40 We denote lcm of m, n as $lcm(m, n)$ or $< m, n >$.

Example 8.41 $< 3, 9 >= 9$, $< 7, 6 >= 42$.

Theorem 8.42 *If* $a \equiv b \bmod n$ *and* $a \equiv b \bmod m$ *then* $a \equiv b \bmod (lcm(n, m))$.

Proof Since $a \equiv b \bmod n$, $a = tn + b$ and since $a \equiv b \bmod m$, $a = \bar{t}m + b$. That is $(a - b) = tn$, $(a - b) = \bar{t}m$. That is $lcm(n, m)|(a - b)$. Hence

$$a \equiv b \bmod (lcm(n, m)).$$

□

Now we can see that '≡' relation is defined on set of numbers. It can be seen that it is an equivalence relation. We now define what is called as residue classes.

Suppose that n is a number and a, b are two given integers. By writing $a \equiv b \bmod n$, we mean $a = tn + b$, $0 \leq b < n$. We can call 'b' as residue of a when divided by n. Also it may be that $a \equiv 0 \bmod n$, $a \equiv 1 \bmod n$, ..., $a \equiv (n - 1) \bmod n$.

Definition 8.43 A residue class modulo n is a set of integers such that every integer is congrument to some b mod n for one b between 0 and $(n - 1)$. We denote this class by $[b]$.

Thus every integer belongs to one residue class and there will be 'n' residue classes totally with respect to modulo n.

Example 8.44 The 5 residue classes of modulo 5 will be [0], [1], [2], [3] and [4]. Also every integer in $\mathbb{N}$ belongs to exactly one residue class. So $[0] \cup [1] \cup [2] \cup [3] \cup [4] = \mathbb{N}$ and $[i] \wedge [j] = \phi$, if $i \neq j$.

In general, the complete residue classes of modulo n are $[0], [1], \ldots, [n - 1]$.

Let $a, b \in \mathbb{Z}$ and $n \in \mathbb{Z}$. We write $a \equiv b \bmod n$ which means $n|(a - b)$. That is $(a - b) = tn$. By writing $n\mathbb{Z}$. We mean the subsets of $\mathbb{Z}$ consisting of all multiples of n.

Definition 8.45 The set $\mathbb{Z}|n\mathbb{Z}$ denotes the integer modulo n residue classes of the integer n.

Example 8.46

$$\begin{aligned}\mathbb{Z}/2\mathbb{Z} &= \{\{... - 2, 0, 2, 4, ...\},\\ &\quad \{... - 3, -1, 1, 3, 5, 7, ...\}\}\\ \mathbb{Z}/3\mathbb{Z} &= \{\{... - 3, 0, 3, 6, ...\},\\ &\quad \{... - 2, 1, 4, 7, ...\}\\ &\quad \{... - 1, 2, 5, 8, ...\}\}.\end{aligned}$$

Definition 8.47 (1) By writing $(a + n\mathbb{Z}) +_n (b + n\mathbb{Z})$ we mean it is equal to $(a + b) + n\mathbb{Z}$.
(2) By writing $(a + n\mathbb{Z}) \times_n (b + n\mathbb{Z})$ we mean $(a \times b) + n\mathbb{Z}$.

Example 8.48 $(1 + 3\mathbb{Z}) +_3 (2 + 3\mathbb{Z}) = (1 + 3 \cdot t) + (2 + 3 \cdot t')$ for t, t' being multiple of 3. So it is $3 + 3(t + t')$. So $(1 + 3\mathbb{Z}) + (2 + 3\mathbb{Z}) = 3 + 3(t + t') = 3(t + t' + 1)$. So it belongs to $3\mathbb{Z}$. Similarly $(1 + 3\mathbb{Z}) \times_n (2 + 3\mathbb{Z}) = 2 + 3\mathbb{Z}$.

Euler's ϕ Function

The function that counts the number of positive integers less than a given number n and relatively prime to n is called the Euler's ϕ function. It is denoted by $\phi(n)$. It can be shown by the method of inclusion-exclusion that

$$\phi(n) = n\left[1 - \frac{1}{p_1}\right]\left[1 - \frac{1}{p_2}\right]\cdots\left[1 - \frac{1}{p_t}\right]$$

where $p_1, p_2, \ldots, p_t$ are the distinct prime divisors of n.

Example 8.49 $\phi(9) = 6$ as the numbers 1, 2, 4, 5, 7, 8 are relatively prime to 9. So we conclude that the number elements in a least residue system will be $\phi(n)$. We state the following theorem without proof.

Theorem 8.50 *If $x_1, x_2, \ldots, x_{\phi(n)}$ is a least residue system modulo n and if t and n are relatively prime, then $tx_1, tx_2, \ldots, tx_{\phi(n)}$ is a least residue system modulo n.*

8.3 Derangements

Derangement of a set is a permutation where no element occupies its original place. For example the some derangements of the set $\{a, b, c, d\}$ are $\{b, a, d, c\}$, $\{d, c, b, a\}$ and so on.

The number of derangements of a set with n-elements is called the nth derangement number or renontres number or the subfactorial of n. It is denoted by D_n or $!n$. One should understand the expression $x!y$ properly. If it is derangement, then it is $x(!y)$. The $!n$ satisfies the recurrence

$$!n = n(!(n-1)) + (-1)^n$$
$$!n = (n-1)(!(n-1)+!(n-2))$$

In general,

$$!n = n!\sum_{k=0}^{n} \frac{(-1)^k}{k!}$$

For example,

$$!3 = 3!\sum_{k=0}^{3} \frac{(-1)^k}{k!} = 6\left[\frac{1}{1} - \frac{1}{1} + \frac{1}{2} - \frac{1}{6}\right] = 2$$

Theorem 8.51 *For any integer $n \geq 0$, $(!n =)D_n = n!\sum_{k=0}^{n} \frac{(-1)^k}{k!}$.*

Proof This can be proved by the PIE. First $(n!)$ gives the number of arrangements of the n elements in all possible ways. From this one has to subtract the number of arrangements in which each element appears in its original place. Hence the new count will be $n! - C(n,1)(n-1)!$. Then in the next step, we must add back the number of permutations in which each set of two elements stay in their original places, as we subtracted them twice. Now the modified count will be

$$n! - C(n,1)(n-1)! + C(n,2)(n-2)!$$

The modification continues and alternate terms subtract and add to give the following.

$$n! - C(n,1)(n-1)! + C(n,2)(n-2)! + \cdots + (-1)^n \tag{8.4}$$

Now

$$C(n,k) = \frac{n!}{k!(n-k)!}$$

Equation (8.4) becomes

$$n! - \frac{n!}{1!} + \frac{n!}{2!} - \cdots + (-1)^n \frac{n!}{n!}$$
$$= n!\left[1 - \frac{1}{1!} + \frac{1}{2!} - \cdots + (-1)^n \frac{1}{n!}\right]$$
$$= n!\sum_{k=0}^{n} \frac{(-1)^k}{k!}.$$

□

Example 8.52 A meeting was attended by 10 people who left their shoes outside the meeting hall. After the meeting ended the people came out and due to power failure could not identify their respective shoes properly. When they went back home, none of them wore their own shoes. Find the number of ways that this could happen.

Solution: The problem is clearly a derangement problem of a set of size $n = 10$. Hence the direct answer will be

$$!10\left[1 - \frac{1}{1} + \frac{1}{2!} - \frac{1}{3!} + \cdots - \frac{1}{10!}\right].$$

8.4 Counting Combinations by PIE

Counting arrangements of n objects in a specified manner is a general problem in arrangement. There may or may not be restrictions in such arrangements. But one can always consider a specific arrangement where some elements occur only in certain positions.

Suppose the numbers $\{1, 2, 3, \ldots, n\}$ are to be permutated, what we get an assignment of the numbers $1, 2, \ldots, n$ to positions $1, 2, \ldots, n$ such that each number is assigned one position between 1 and n. For example for the numbers 1, 2, 3, 4 one permutation will be 4 1 2 3. It means the number 4 is given the first position, the number 1 is given the second position, and so on. There is no restriction here on the arrangement. Clearly the number of permutations will be $n!$.

Let us consider a generalisation of the above situation. Let $A_1, A_2, \ldots, A_n$ be the subsets of $S = \{1, 2, \ldots, n\}$ where each $|A_i| \geq 0$, $1 \leq i \leq n$ and $A_i \cap A_j \neq \emptyset$. Consider permutations $a_1a_2 \ldots a_n$ of the set S such that a_i is not in A_i. That is a_i can be any number in $\overline{A}_i$. Let A be the set of all such permutations. The problem is to find $|A|$.

Example 8.53 Let $S = \{1, 2, 3, 4, 5\}$ and $A_1 = \{1\}$, $A_2 = \{2, 3\}$, $A_3 = \{3, 4\}$, $A_4 = \{4\}$, $A_5 = \{1, 5\}$. Here

$$A = \left\{ a_1a_2a_3a_4a_5 \,\middle|\, \begin{array}{l} a_1 \neq 1, \quad a_3 \neq 3, 4,\ a_5 \neq 5 \\ a_2 \neq 2, 3,\ a_4 \neq 4, \end{array} \right\}$$

Some permutations are

$$2\,1\,5\,3\,4,\quad 3\,4\,1\,5\,2,\quad 4\,5\,2\,3\,1.$$

Example 8.54 Let $S = \{1, 2, 3, 4, 5, 6\}$ and

$$\begin{aligned} A_1 &= \{1, 2, 3, 4\} \\ A_2 &= \{2, 3, 4, 5\} \\ A_3 &= \{3, 4, 5, 6\} \\ A_4 &= \{4, 5, 6, 1\} \\ A_5 &= \{5, 6, 1, 2\} \\ A_6 &= \{6, 1, 2, 3\} \end{aligned}$$

We now draw a grid of size 6×6 as below. In the grid the symbol '$*$' indicates 'not to be used' and 'blank' indicates 'can be selected'.

Table 8.1 6×6 grid

	A_1	A_2	A_3	A_4	A_5	A_6
1	*			*	*	*
2	*	*			*	*
3	*	*	*			*
4	*	*	*	*		
5		*	*	*	*	
6			*	*	*	*

Clearly in forming $a_1a_2a_3a_4a_5a_6$, $a_1 \in \{5, 6\}$, $a_2 \in \{1, 6\}$, $a_3 \in \{1, 2\}$, $a_4 \in \{2, 3\}$, $a_5 \in \{3, 4\}$, $a_6 \in \{4, 5\}$, with a condition each number is chosen only once against a position. So the permutations are

$$6\,1\,2\,3\,4\,5, \quad 5\,6\,1\,2\,3\,4.$$

We use the principle of inclusion-exclusion for counting the above event which is explained in the next section.

8.5 Rook Polynomials

Consider a $m \times m$ grid of squares. Let the rows be numbered from 1 to m in a top down way. Columns are numbered from left to right as $1, 2, \ldots, m$. We defined a board $\mathcal{B}$ to be a subset of this $m \times m$ grid. The problem is to understand how to place k non-attacking rooks in $\mathcal{B}$. By non-attacking rooks we mean no two rooks should share a row or column. If $k = 0$, we understand that there is only way of placing $k(= 0)$ rooks in $\mathcal{B}$. Let us understand the way of placing k non-attacking rooks in a board $\mathcal{B}$. That is, if s_k denotes the number of ways of placing exactly 'k' non-attacking rooks in a board $\mathcal{B}$, then we can write

$$g_{\mathcal{B}}(x) = \sum_{k \in [n]} s_k x^k \tag{8.5}$$

and this polynomial $g_{\mathcal{B}}(x)$ is defined for any board $\mathcal{B}$ of size $n \times n$, $1 \leq n \leq m$.

The subproblem of placing k non-attacking rooks in a board $\mathcal{B}$ is to be understood carefully.

Let us now go back to our problem of forming permutations with forbidden positions. Refer Table 8.1. The $*$ positions indicate the 'not-to-be-used' indices in forming permutations on the input set S. Also the problem was to count the possible permutations of the elements of S with the forbidden set condition. We use the method of inclusion-exclusion for this counting.

We already know what is derangement. We also saw permutations with forbidden sets. For any permutation $a_1 a_2 \ldots a_n$ of set $S = \{1, 2, \ldots, n\}$, one can associate placement of non-attacking rooks in a $n \times n$ grid. That is we say the rook positions are $(1, a_1)$, $(2, a_2)$, $\ldots$, (n, a_n) i.e., (i, a_i) corresponds to the position i^{th} row a_i^{th} column in the grid. So the forbidden set A_1 for each i corresponds to the squares on the grid where the non-attacking rooks cannot be put. For example, consider the example of the previous section. The 5×5 grid with the forbidden subsets A_1, A_2, A_3, A_4 and A_5 is drawn as below

	A_1	A_2	A_3	A_4	A_5
1	$*$				$*$
2		$*$			
3		$*$	$*$		
4			$*$	$*$	
5					$*$

$*$ indicates forbidden positions. For each set A_j, we say that a set B_j corresponds to the placement of a rook in the jth column which contains a '$*$', that is the column position is indicated by A_j. So, $\overline{A}_j$ indicates the set of squares in the jth column that are not marked with '$*$'. This description is for all j, $1 \leq j \leq n$.

Hence if $F(A_1, A_2, \ldots, A_n)$ denote the set of all permutations of S with $A_1, A_2, \ldots, A_n$ as forbidden sets, then

$$|F| = |\overline{B}_1 \cap \overline{B}_2 \cdots \cap \overline{B}_n|$$

Using the principle of inclusion exclusion we can write

$$\begin{aligned} &|\overline{B}_1 \cap \overline{B}_2 \cap \cdots \cap \overline{B}_n| \\ &\quad = n! - \sum |B_i| + \sum |B_i \cap B_j| - \cdots + (-1)^n |\overline{B}_1 \cap \overline{B}_2 \cap \cdots \cap \overline{B}_n| \end{aligned}$$

Here $|B_i|$ stands for the number of ways to place n non-attacking rooks in row i in the positions (i, t_i) where $t_i \in A_i$. It can be done in $|A_i|$ ways. Rest of the rooks are sent to remaining $(n-1)$ positions in a non-attacking manner. Hence

$$|B_i| = |A_i|(n-1)!, 1 \leq i \leq n.$$

Let

$$|s_1| = \left|\bigcup_{i=1}^{n} A_i\right|$$

and

$$\sum |B_i| = s_1(n-1)!$$

$|B_i \cap B_j|$ corresponds to the number of ways to place n non attacking rows where rook in row i and rook in row j occupy the '$*$ed' location indicated by the sets B_i and B_j respectively. Remaining $(n-2)$ rooks can be placed in $(n-2)!$ ways in a non-attacking fashion. Hence

$$\sum |B_i \cap B_j| = s_2(n-2)!$$

where 's_2' indicates the number of ways of placing two non-attacking rooks on the $*$ed positions indicated by B_i and B_j, $1 \leq i, j \leq n$. In general

$$\sum |B_{i_1} \cap B_{i_2} \cdots \cap B_{i_t}| = s_t(n-t)!$$

for $i_k \in \{1, 2, \ldots, n\}$ and $1 \leq t \leq n$. We state the following result which counts the number of ways of placing n non attacking rooks in a $n \times n$ grid.

Theorem 8.55 *The number of ways of placing n non-attacking rooks in a $n \times n$ grid with forbidden grid positions is given by*

$$n! - s_1(n-1)! + s_2(n-2)! - \cdots + (-1)^n s_n.$$

This is the count of number of permutations of S with the forbidden sets $A_1, A_2, \ldots, A_n$.

Let $\mathcal{B}$ is a grid of size $n \times n$. We now marked on $\mathcal{B}$ with $*$'s indicating forbidden sets $A_1, A_2, \ldots, A_n$. Suppose $s_t(\mathcal{B})$ denote the number of ways of placing 't' non attacking rooks on $\mathcal{B}$. Here t will vary from 1 to n. We understand the calculation of $s_t(\mathcal{B})$ through different subsets of $\mathcal{B}$. Each of this subset will be a bound by itself.

Two subsets B_1 and B_2 of $\mathcal{B}$ are said to be disjoint if they do not share any squares on $\mathcal{B}$. Consider

$$g_{\mathcal{B}}(x) = s_0 + s_1 x + s_2 x^2 + \ldots \tag{8.6}$$

This $g_{\mathcal{B}}(x)$ is called the rooks polynomial. Here 's_k' stands for placing k-rooks on $\mathcal{B}$. Suppose $\mathcal{B}$ has some '$*$'s marked on it. One can consider $\mathcal{B}$ to contain subsets with '$*$'s and those without '$*$'s. Let $B_1, B_2, \ldots, B_\ell$ be the subsets of $\mathcal{B}$ with '$*$'s on them. Then $s_k(B_i)$ denotes the number of ways of placing 'k' rooks on B_i in a non-attacking way. So we can write a polynomial $g_{B_i}(x)$ as in equation (8.6) for every i, $1 \leq i \leq m$. Then their product gives the required rooks polynomial for the board $\mathcal{B}$.

Exercise

1. Let there be 15 chairs for 15 students in a hall where a lecture is delivered. Find the number of the students sit in the chair in which they did not sit in the previous class.
2. Find the number of derangements of the integers 10–20 both inclusive such that

   (1) 10, 11, 12, 13, 14 are in a derangement
   (2) 15–20 are in derangement

3. 20 students in a class are given a class test. The teacher collects the answer papers, corrects and distributes back the papers one to each so that none of them get their own paper. In how many ways it can happen?
4. Find the rooks polynomial for the following boards.

(a) 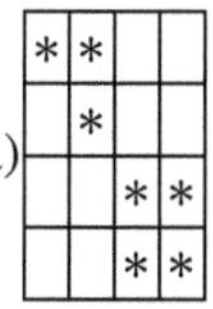 (b) 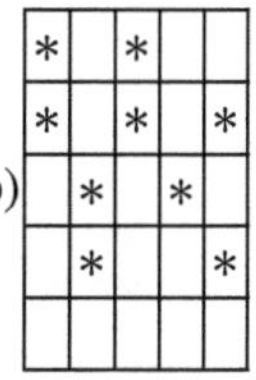

5. Find the number of ways of rearranging the symbols in the word FCBBCF so that no symbol appear in its original position.
6. Find the number of permutations of $\{1, 2, \ldots, 7\}$ in which atleast one even integer in its natural position.
7. Find the number of words of English alphabet that do not contain the sequence words 'DULA', 'PART' and 'TABLE' in them. The words are constructed using all 26 letters exactly once.
8. Find the positive integers which are less than 150 which are relatively prime to 150.
9. A box contains 20 red balls, 20 blue balls, 20 green balls and 20 orange balls. You are allowed to draw one ball in every minute. How long will it take to ensure that you have atleast 12 balls of same color drawn by you?
10. Find the number of positive integers less than or equal to 500 that are not divisible by any of the following numbers $\{3, 6, 25\}$.
11. Find the number of 'onto' functions from a set $\{a, b, c\}$ to itself.

Partial Order and Lattices 9

9.1 Partial Order Relation

We have seen what a set is and how to handle the elements of a set. For example in the chapter on 'Algebraic structures' we saw how to look for some structure among the elements of the set. Throughout we had the definition of binary operations defined between any two elements of the set. Also looked at relating the elements as 'xRy' which means x and y are related by a relation R. The relation R was looked at, from the following three aspects.

(i) R being reflexive
(ii) R being symmetric
(iii) R being transitive

Then we went on to define an equivalence relation R. Logically we can write down the expression as follows:

xRy	: x and y are related by R.
$\neg(xRy)$	: x and y are not related.
xRx	: x is related to x. It means R is reflexive.
$\sim(xRx)$	: R is irreflexive.
$xRyRz$	: R is transitive.
$(xRy) \wedge (yRx)$	: symmetric.
$(xRy) \wedge (yRx) \Rightarrow x = y$	: Antisymmetric

R. Rama, *Topics in Combinatorics and Graph Theory*,
https://doi.org/10.1007/978-3-031-74252-1_9

Definition 9.1 Let $(S, \leq)$ be a set with the binary relation $\leq$. If the following conditions are satisfied, then $(S, \leq)$ is said to be a partially ordered set or poset (in short).

(i) $\forall x \in S, x \leq x$ (reflexive).
(ii) $(x \leq y) \wedge (y \leq x) \Rightarrow x = y$ (Antisymmetry).
(iii) $(x \leq y) \wedge (y \leq z) \Rightarrow x \leq z$ (Transitivity).

The relation '$\leq$' is either read as 'less than or equal to' or 'x is included in y'. The relation '$\leq$' is properly called as order relation.

Remark 9.2 (i) The relation 'inclusion' or 'less than' need not be defined for each pair of elements $x, y \in S$.
(ii) The statement $\neg(x \leq y)$ does not imply $x > y$.

Example 9.3 Let S be a set and $\mathbb{P}(S)$ be the power set of S. Consider $(\mathbb{P}(S), \subseteq)$. Here $A \subseteq B$, $A, B \in \mathbb{P}(S)$ means that the subset A is included as a subset of B which may or may not be proper. $(\mathbb{P}(S), \subseteq)$ is a poset.

Example 9.4 Let $(\mathbb{P}(S), \subset)$ be the given set $\mathbb{P}(S)$, with a '*proper*' inclusion between two sets A and B. Is $(\mathbb{P}(S), \subset)$ a poset? Note that there is no set S' such that $S \subset S'$.

Remark 9.5 The reader is advised to refer the topic 'well-ordering' of a set and axiom of choice.

Example 9.6 Consider $(\mathbb{R}, \leq)$. This is both partially ordered and totally ordered set.

Definition 9.7 A totally ordered set is called as a chain. So, in a chain $(S, \leq)$, for any two elements $x, y \in S$, either $x \leq y$ or $y \leq x$.

Example 9.8 $(\mathbb{R}, \leq)$ is a chain, $\leq$ is usual 'less than or equal to'.

Remark 9.9 In a poset $(S, \leq)$, any two elements $x, y \in S$, are either comparable like, either $x \leq y$ or $y \leq x$, or they are incomparable.

Notation:

(i) For two elements x and y that are comparable, we denote it by $x \sim y$
(ii) For two elements x, y that are not comparable, we denote by $\neg(x \sim y)$.

9.2 Hasse Diagrams

A poset can be diagramatically represented using the methods of graph theory. Each element of the poset will be a vertex of the graph and the edges and paths are defined in such a way that there is a 1–1 correspondence between the representation of the poset as a graph and the poset. That is from the graph, writing down the poset description can be done uniquely.

Definition 9.10 Let $(S, \leq)$ be a poset. If for any two elements $x, y \in S$, $x \leq y$ and there is no $z \in S$ such that $x \leq z \leq y$, then we say that x is covered by y in S and is denoted by $x \sqsubset y$.

In $x \sqsubset y$, y is also called the upper bound for x. In a Hasse diagram, there will be an edge between two vertices x and y if $x \sqsubset y$. The vertex 'x' is plotted below the vertex 'y'. So, by $x \leq y$, we mean one can traverse from x to y via edges in the upward direction. However, the Hasse diagram is not a directed graph (Figs. 9.1, 9.2 and 9.3).

Fig. 9.1 Hasse diagram

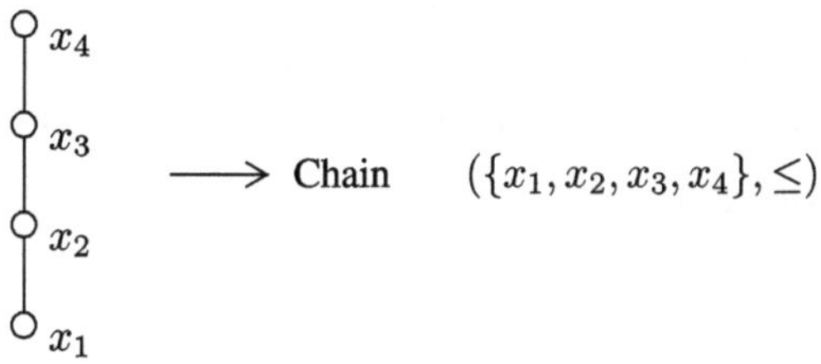

Fig. 9.2 Another Hasse diagram

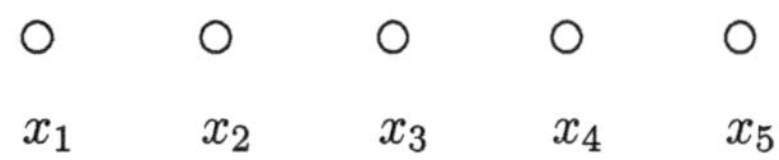

Fig. 9.3 Not a chain

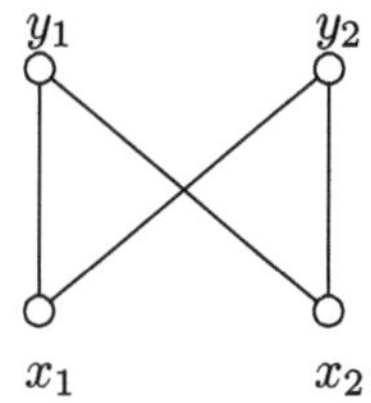

Fig. 9.4 Power set

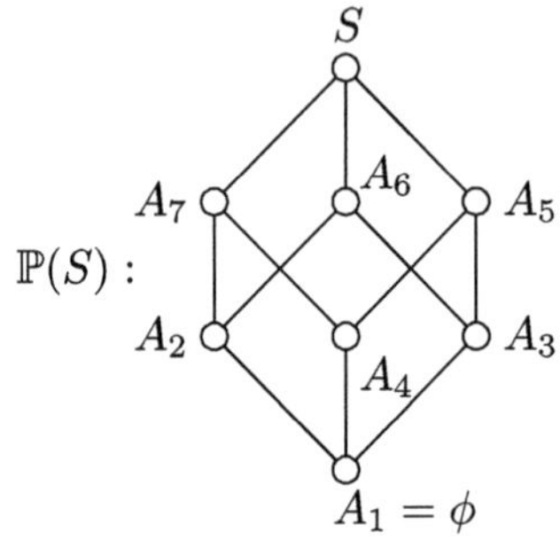

Example 9.11 (*Hasse diagrams*)

(a)
(b) $(S, \leq)$ is a poset which is totally unordered or an anti-chain, where $S = \{x_1, x_2, \ldots, x_5\}$.
(c) $((x_1, x_2, y_1, y_2), \leq)$ where $x_2 \leq y_1, x_1 \leq y_1, x_1 \leq y_2, x_2 \leq y_2$. It is not a chain (Why?)
(d) This is the Hasse diagram corresponding to a $(\mathbb{P}(S), \subseteq)$ where $S = \{a, b, c\}$. It is not a chain. Here $A_1 = \phi$, A_2, A_3, A_4 are 1-element subsets A_5, A_6, A_7 are 2-element subsets.

Definition 9.12 Two posets $(S, \leq)$ and $(T, \leq')$ having the same Hasse diagram are said to be isomorphic.

Definition 9.13 Let S be a set. If in S, for any two elements x and y if $x \leq y$, we say that x is contained in y or x is less than or equal to y. We can also say that either 'y contains x' or y is greater than or equal to x. This is known as 'duality principle' in lattice theory. We call '$\leq$' as dual of '$\geq$'. If $(S, \leq)$ is a poset, then $(S, \geq)$ is also a poset.

It is clear that a poset has some ordering of elements of a set, using a binary relation. Also the following facts are realizable ones.

Fact 1: Any poset $(S, \leq)$ has an element e such that $e \leq x, \forall\, x \in S$, then e is called the least or null element of S.

Remark 9.14 A poset may or may not contain e. Figure ??c above, this lattice has no e. However, if it is there then it must be unique.

Fact 2: Any poset, we can define an element 'F' such that for all $x \in S, x \leq F$. F is called the full or universal element and if S has one, then it is unique.

Definition 9.15 Let $\tilde{S}$ be a subset of S where $(S, \leq)$ is a poset. We now define least upper bound (l.u.b) and greatest lower bound (g.l.b) for $\tilde{S}$ in S

(i) An element $x \in S$ such that $y \leq x$ for all $y \in \tilde{S}$, then x is said to be an upper bound for $\tilde{S}$. x is said to be l.u.b if there does not exist an element x' $(\neq x)$ in S such that $x' \leq x$ and x' is an upper bound of $\tilde{S}$.
(ii) An element being a greatest lower bound of $\tilde{S}$ can be identically defined.

Remark 9.16 The g.l.b and l.u.b are computed from the binary relation defined on the set. The l.u.b may or may not belong to $\tilde{S}$.

Next we define what are known as minimal and maximal elements of a poset $(S, \leq)$.

An element σ is said to be a minimal element of a poset $(S, \leq)$ if $\forall\, x \in S,\ \sigma \leq x \Rightarrow \sigma = x$.

An element τ is said to be a maximal element of $(S, \leq)$ if $\forall\, x \in S\ \tau \leq x \Rightarrow \tau = x$.

Example 9.17 (i) We know $(\mathbb{Z}, \leq)$ is totally ordered but does not have a maximal element.
(ii) If $S = \{a, b, c\}$ then we know $(\mathbb{P}(S), \subseteq)$ is a poset. The maximal element is 'S' itself and the minimal element is 'ϕ'.
(iii) This poset has two minimal and two maximal elements.

We know that a chain is totally ordered. Any totally ordered S with a property that every nonempty subset of S has a minimal element is called well ordered.

Example 9.18 $(\mathbb{Z}, \leq)$ is not well ordered as it has no minimal element.

Fact: All finite totally ordered sets are well ordered.

Example 9.19 Let $S = \{a\}$. S^* is the set of strings $\{\lambda, a, aa, aaa, \cdots\}$ where λ is the empty string and minimal (S^*, lx) is well ordered where 'lx' is the lexicographic ordering. The strings on any subset of S^* can be arranged in the increasing order of their length. Hence every subset of S^* is totally ordered and has a minimal string.

Well ordering principle is one by which every set can be well ordered. And also assume that every chain in S has an upper bound. That is let S be the set. If $S \neq \phi$, then there is a $c_0 \in S$. c_0 is maximal then, $\{c_0\}$ is well-ordered, otherwise we can build a chain from c_0 as $c_0 < c_1 < c_2 \ldots$. This chain will eliminate the maximal element. Then it has a upper bound say t_0. We already saw this t_0 need not be in the set $\{c_0, c_1, c_2, \cdots\}$. From t_0, build a new chain recursively as $t_0, t_1, t_2, \ldots$ and this chain will reach a maximal element say s_0. Continue building chains of the set S. We hope to reach through every element of S by this process. If there is a $\bar{C}$ which goes on and on without an upper bound, then for all $x \in S$, there is an element on the chain say 'p' so that 'x and p' are incomparable. But this cannot happen as $\bar{C}$ must have upper bound (say) $\bar{P}$ so that $\forall\, x \in \bar{C}, x \leq \bar{P}$. Hence a chain cannot continue without an upper bound. Hence the elements of 'S' can be well-ordered. We now state Zorn's lemma without proof.

Lemma 9.20 (Zorn's lemma) *Let S be non empty poset such that every chain C in S has an upper bound in S. Then S has a maximal element.*

Remark 9.21 (i) A proof of Zorn's lemma will be to show that well-ordering principle and Zorn's lemma are equivalent. We already discussed how well-ordering takes care of the maximal element.
(ii) The proof of Zorn's lemma can be given using compactness theorem of K. Gödel which says that 'A set of well formed formula is satisfiable' if and only if it is finitely satisfiable.

9.3 Lattices

In the earlier section, we saw several interesting aspects of partially ordered sets. It will be interesting to observe that posets are endowed with some more structures which arise naturally in algebra. This is due to the fact that in a poset, we already have a binary relation, we denote this in posets generally as '$\leq$' which we denoted as $*$ in groups. So we now briefly see how to look at a poset with special properties as a semi group.

Let $(S, *)$ be an idempotent semigroup where for all $a, b, c \in S$

(i) $a * a = a$
(ii) $a * b = b * a$
(iii) $a * (b * c) = (a * b) * c$

That is '$*$' with $\leq$ has the following transformation.

(i) $a * a = a$ means $a \leq a$ (symmetry).
(ii) $a * b = b * a$ means $a = a \leq b = b \leq a = b$ (irreflexive).
(iii) $a * (b * c) = (a * b) * c$ means $a \leq (b \leq c) = (a \leq b) \leq c$. So $a \leq b$ and $b \leq c$ means $a \leq c$ (transitive).

Remark 9.22 So a semilattice is one in which every pair of elements $\{x, y\}$ are incomparable.

Definition 9.23 (i) For any two elements $\{x, y\}$, the $l.u.b\{x, y\}$ is called the join of x and y and it is denoted by $x \vee y$.
(ii) For any two elements $\{x, y\}$, the $g.l.b\{x, y\}$ is called the meet of x and y and it is denoted by $x \wedge y$.

Example 9.24 (i) In any chain any pair of elements have both meet and join.
(ii) The following posets are on which either have a meet or a join, but not both.

Definition 9.25 A lattice $(S, \vee, \wedge)$ is a nonempty set L where every pair of elements x and y has both meet and join in S and satisfying following conditions on $\vee$ and $\wedge$.

(i) $\forall x, y, z \in S, x \vee (y \vee z) = (x \vee y) \vee z, x \wedge (y \wedge z) = (x \wedge y) \wedge z$. These are known as associativity w.r.to $\vee$ and $\wedge$.
(ii) $\forall x, y \in S, x \vee y = y \vee x$ and $x \wedge y = y \wedge x$. These are known as commutative w.r.to $\vee$ and $\wedge$.
(iii) $\forall x, y, z \in S, x \vee (x \wedge y) = x$ and $x \wedge (x \vee y) = x$. These are known as absorption laws.

Remark 9.26 Basically the set S defined above is a poset in which every pair of elements has a g.l.b and l.u.b in S (Figs. 9.5 and 9.6).

Example 9.27 1. Consider the Hasse diagram in Fig. 9.7.
It is a lattice.
2. Figure 9.8 is not a lattice as the pair (x_2, x_6) has three upper bounds x_3, x_4, x_5 but no l.u.b.

Example 9.28 Let S be a set. Let $\mathbb{P}(S)$ be the power set of S. Define for any two elements $A, B \in \mathbb{P}(S)$, $A \wedge B = A \cap B$ and $A \vee B = A \cup B$. Then $(\mathbb{P}(S), \vee, \wedge)$ is a lattice.

For example if $S = \{a, b, c\}$ refer Fig. 9.4 which is a lattice with ϕ as the least and S as the greatest element of the lattice $(\mathbb{P}(S), \vee, \wedge)$ (Fig. 9.9).

The following is a simple direct property of a lattice.

Proposition 9.29 *(i) Let $(S, \leq)$ be a poset in which every pair of elements x, y has g.l.b and l.u.b. Then $(x \leq y$ iff $x \vee y = y)]$ and $(x \leq y$ iff $x \wedge y = x)$. These are called laws of consistency.*

Fig. 9.5 A poset

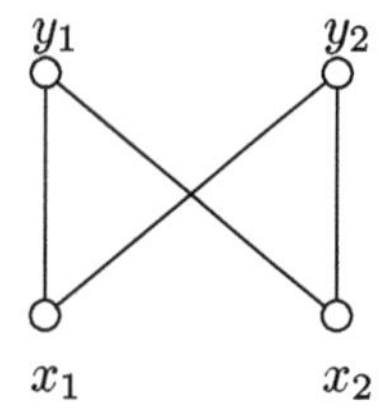

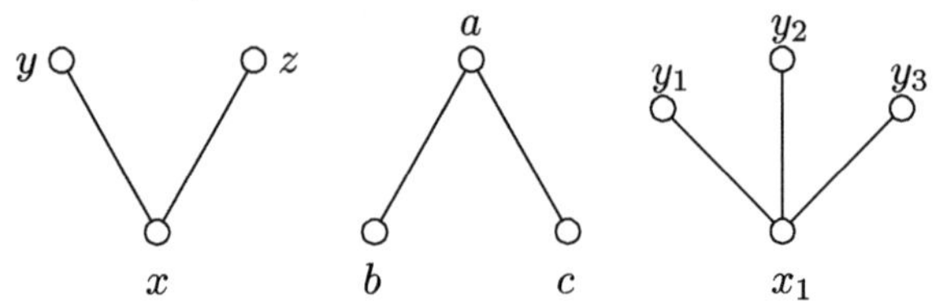

Fig. 9.6 Poset example

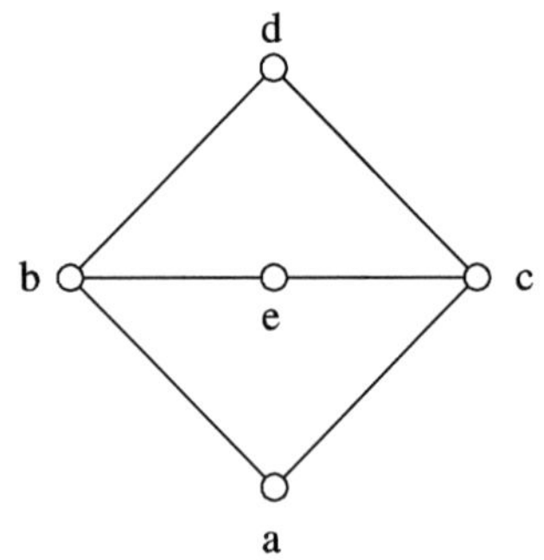

Fig. 9.7 A lattice

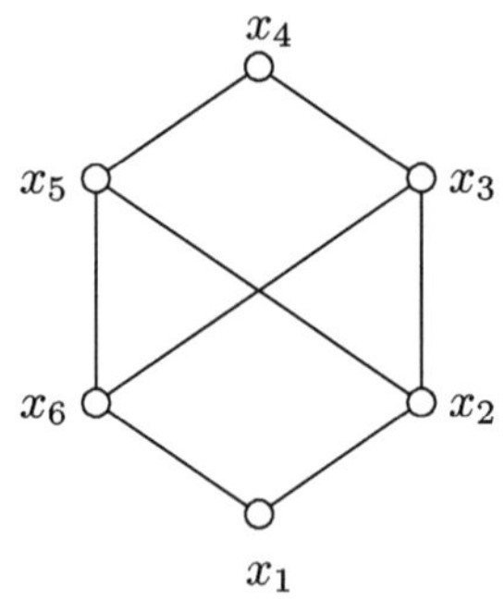

Fig. 9.8 Not a lattice

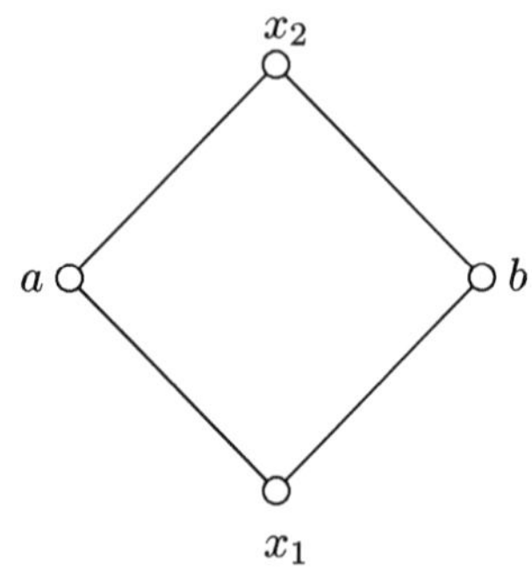

Fig. 9.9 Bounded lattice

(ii) *Conversely let* $(S, \vee, \wedge)$ *be a lattice. Define for every pair* $x, y \in S$, $x \leq y$ *iff* $x \vee y = y$ *(or)* $x \leq y$ *iff* $x \wedge y = x$. *Then* $(S, \leq)$ *will be a poset.*

9.3.1 Bounded Lattices

A lattice $(S, \vee, \wedge)$ is said to be bounded below if there is an element $e \in S$ such that $e \wedge x = e$ and $e \vee x = x \; \forall \, x \in S$. Here '$e$' is called the lower bound. A lattice $(S, \vee, \wedge)$ is said to be bounded above if there is an element $u \in S$ such that $u \wedge x = x$ and $u \vee x = u \; \forall \, x \in S$. Then '$u$' is called the upper bound. A bounded lattice is one which is both bounded above and below.

Example 9.30 (i) $(\mathbb{P}(S), \vee, \wedge)$ in Example 9.28 is a bounded lattice.
(ii) $([0, 1], \vee, \wedge)$ such that $x, y \in [0, 1], x \vee y = y, x \wedge y = x$ is a bounded lattice.
(iii) $((0, 1), \vee, \wedge)$ is an unbounded lattice.

Example 9.31 The lattice given below by means of Hasse diagram is bounded

9.3.2 Modular and Distributive Lattices

We have a simple lemma below which motivates to define a distributive lattice.

Lemma 9.32 *In any lattice* $(S, \vee, \wedge)$, *we have* $x \wedge (y \vee z) \geq (x \wedge y) \vee (x \wedge z)$ *for* $x, y, z \in S$.

Proof For $x, y, z \in \mathbb{Z}$,

$$\begin{aligned} x \wedge (y \vee z) &\geq x \wedge y \\ x \wedge (y \vee z) &\geq x \wedge z \end{aligned} \tag{9.1}$$

Combining the above two equation, we get

$$x \wedge (y \vee z) \geq (x \wedge y) \vee (x \wedge z)$$

□

Example 9.33 Consider the Hasse diagram in Fig. 9.10. Here $2 \wedge (3 \vee 4) \geq (2 \wedge 3) \vee (2 \wedge 4)$.

Definition 9.34 A lattice $(S, \vee, \wedge)$ is modular for all $x, y, z \in S$, $x \geq z$ imply $x \wedge (y \vee z) = (x \wedge y) \vee z$.

Example 9.35 Consider the following lattice. This lattice is not modular because,

$$\begin{aligned} 2 \wedge (3 \vee 5) &= 2 \wedge 1 = 2 \\ (2 \wedge 3) \vee 5 &= 3 \vee 5 = 1 \end{aligned} \tag{9.2}$$

Fig. 9.10 A lattice

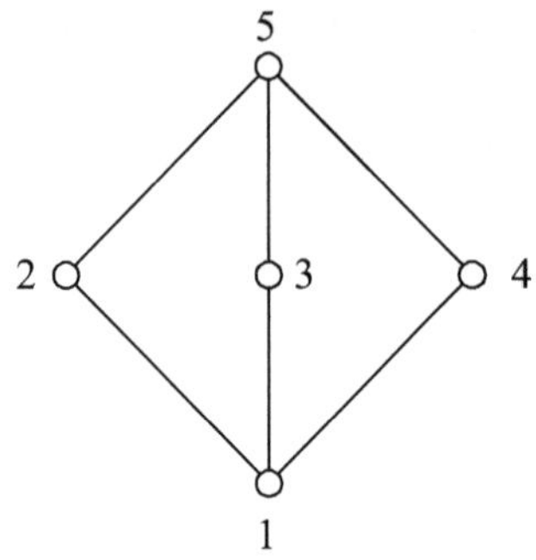

Lemma 9.36 *Let* $(S, \vee, \wedge)$ *be a lattice. For* $x, y, z \in S$, *for all* $x \geq z$, $x \wedge (y \vee z) \geq (x \wedge y) \vee z$.

Proof Since $x \geq z$, $x \wedge z = x$. Hence from the previous lemma the result follows. □

Definition 9.37 Let $(S, \vee, \wedge)$ be a lattice. It is said to be distributive if for all $x, y, z \in S$, $x \wedge (y \vee z) = (x \wedge y) \vee (x \wedge z)$.

Theorem 9.38 *Every distributive lattice is modular.*

Proof Since the given lattice is distributive we have $\forall x, y, z$ $x \wedge (y \vee z) = (x \wedge y) \vee (x \wedge z)$. If $x \geq z \Rightarrow x \wedge (y \vee z) = (x \wedge y) \vee z$. Hence the lattice is modular (Figs. 9.11, 9.13, 9.14, 9.15, 9.16, 9.17, 9.18, 9.19, 9.20 and 9.21). □

Remark 9.39 The converse of the above theorem is in general not true.

Fig. 9.11 Not a modular lattice

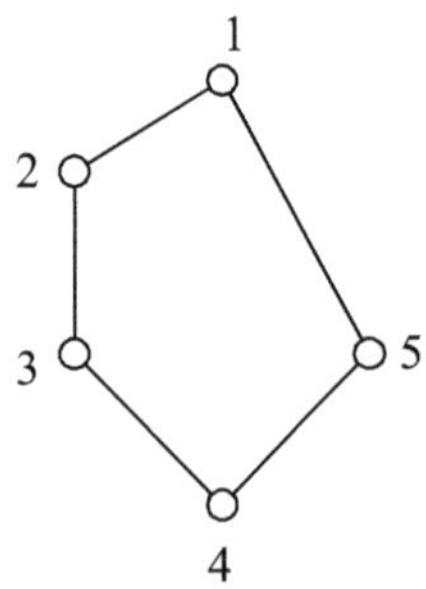

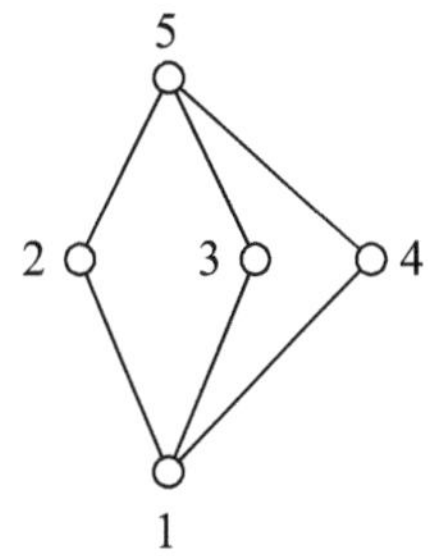

Fig. 9.12 Modular lattice

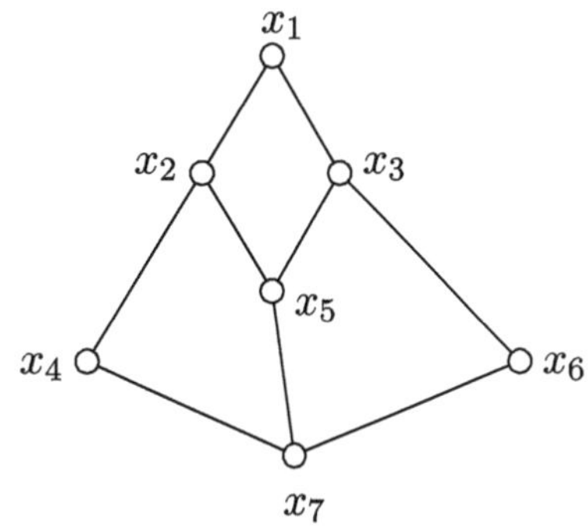

Fig. 9.13 Complete lattice

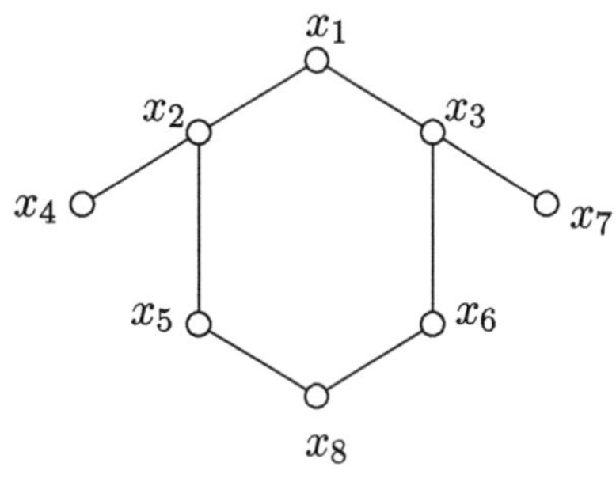

Fig. 9.14 A lattice

Example 9.40 Consider the lattice given in Fig. 9.12.
This lattice is not distributive, but it is modular.

9.4 Basic Results

Definition 9.41 A lattice $(S, \vee, \wedge)$ is said to be a complete lattice if any subset of S has glb and lub.

Remark 9.42 By Zorn's lemma, a complete lattice must have a unique least and greatest element. Let us denote them by 'e' and 'u' respectively.

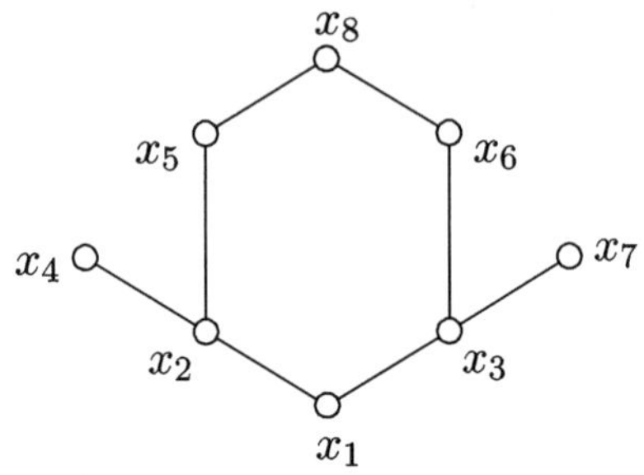

Fig. 9.15 Dual lattice

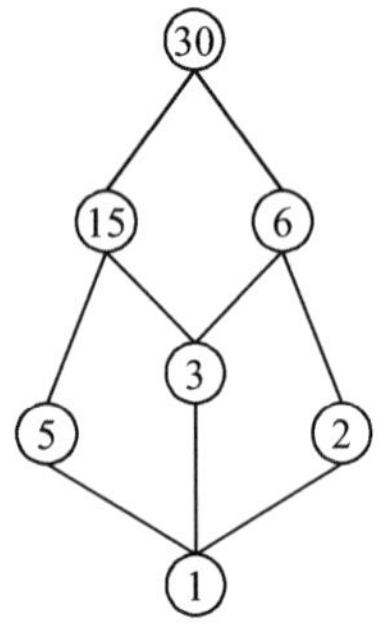

Fig. 9.16 A poset

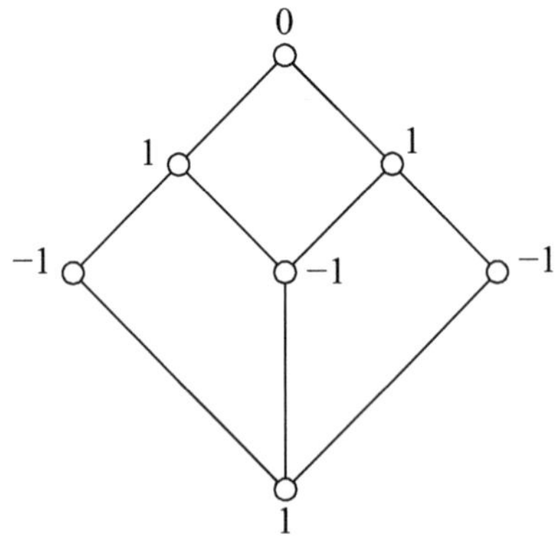

Fig. 9.17 A poset

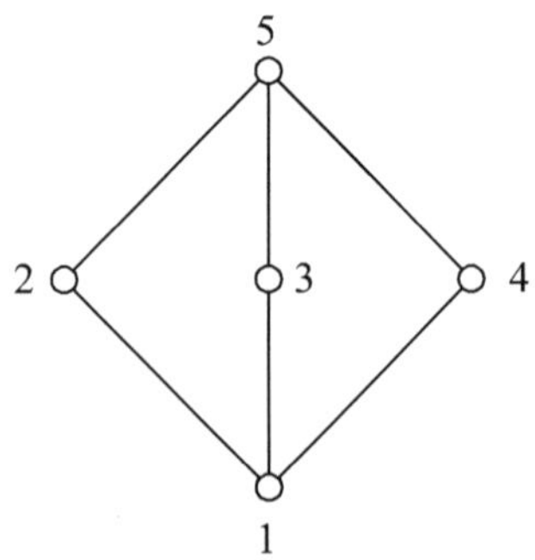

Fig. 9.18 A Hasse diagram

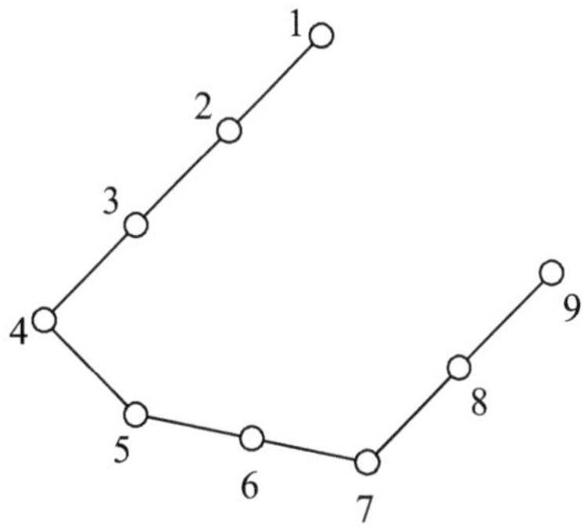

Fig. 9.19 A Hasse diagram

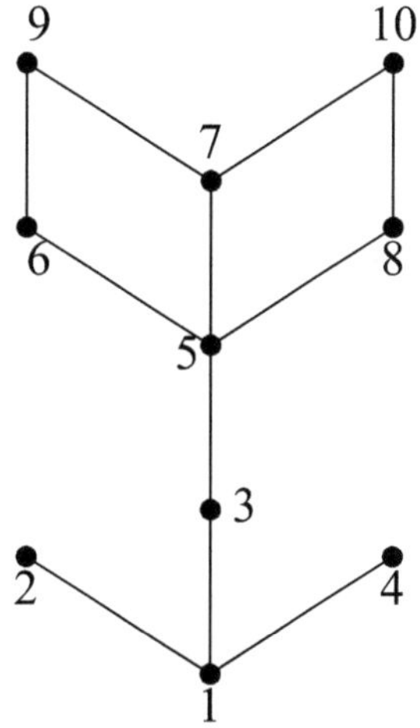

Fig. 9.20 A Hasse diagram

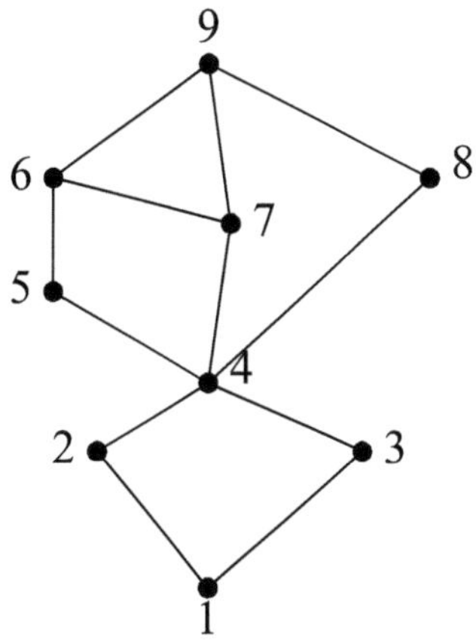

Fig. 9.21 A Hasse diagram

Example 9.43 The lattice $(\mathbb{P}(S), \vee, \wedge)$ where $\mathbb{P}(S)$ is the power set of a finite set S and $\vee$ and $\wedge$ are as defined in Example 9.28.

Proposition 9.44 *Let $(S, \leq)$ be a poset. If S has greatest element 'u' and if every nonempty subset $\tilde{S}$ of S also has glb in S, then $(S, \vee, \wedge)$ with usual '$\vee$' and '$\wedge$' definition is a complete lattice.*

Remark 9.45 The proposition holds true for least element also.

We now look for some simple properties of complete lattices.

Example 9.46 Let $(S, \wedge, \vee)$ be a lattice with $\bar{\mathbb{N}} = \mathbb{N} \cup \{0\}$. Then $(S, \leq)$ is a poset and $(S, \vee, \wedge)$ is complete with '0' as the lower bound. Note that $(S, \leq)$ is also a bounded lattice.

Example 9.47 Consider the following lattice $L = \{x_1, x_2, x_3, x_4, x_5, x_6, x_7\}$ where the operations $\wedge$ and $\vee$ are given by the following Cayley table.

$\vee$	x_1	x_2	x_3	x_4	x_5	x_6	x_7
x_1	x_1	x_1	x_1	x_1	x_1	x_1	x_1
x_2	x_1	x_2	x_1	x_2	x_2	x_1	x_2
x_3	x_1	x_1	x_3	x_1	x_3	x_3	x_3
x_4	x_1	x_2	x_1	x_4	x_2	x_1	x_4
x_5	x_1	x_2	x_3	x_2	x_5	x_3	x_5
x_6	x_1	x_1	x_3	x_1	x_3	x_6	x_6
x_7	x_1	x_2	x_3	x_4	x_5	x_6	x_7

$\wedge$	x_1	x_2	x_3	x_4	x_5	x_6	x_7
x_1	x_1	x_2	x_3	x_4	x_5	x_6	x_7
x_2	x_2	x_2	x_5	x_4	x_5	x_7	x_7
x_3	x_3	x_5	x_3	x_7	x_5	x_6	x_7
x_4	x_4	x_4	x_7	x_4	x_7	x_7	x_7
x_5	x_5	x_5	x_5	x_7	x_5	x_7	x_7
x_6	x_6	x_7	x_6	x_7	x_7	x_6	x_7
x_7	x_7	x_7	x_7	x_7	x_7	x_7	x_7

The corresponding Hasse diagram will be

This is a complete lattice which is finite. The binary relations are $x_2 \leq x_1$, $x_3 \leq x_1$, $x_5 \leq x_2, x_5 \leq x_3, x_4 \leq x_2, x_7 \leq \{x_4, x_5, x_6\}$.

We state the following theorem which will be useful for us to define what are known as 'dual' lattices.

Theorem 9.48 *Let $(S, \vee, \wedge)$ be a lattice where $(S, \leq)$ is the corresponding poset. For any two elements $x, y \in S$. We have the following statements being equivalent*

(i) $x \leq y$
(ii) $x \wedge y = x$
(iii) $x \vee y = y$.

Proof Follows directly from the definitions of $\leq, \wedge, \vee$. □

Definition 9.49 Let $(S, \vee, \wedge)$ be a lattice. Consider the corresponding poset $(S, \leq)$. '$\leq$' satisfies the following property: for any $x, y \in S, x \leq y$ iff $x \wedge y = x$.

Remark 9.50 (i) We have seen in the previous theorem that $x \leq y$ is also equivalent to $x \vee y = y$.
(ii) $\leq$ is antisymmetric iff $x \leq y$ and $y \leq x$ imply $x = y$ iff $x \wedge y = x$ and $x \wedge y = y$ imply $x = y$.
(iii) $\leq$ is transitive iff $x \leq y$ and $y \leq z$ imply $x \leq z$ iff $x \wedge y = x$ and $y \wedge z = y$ imply $(x \wedge y) \wedge z = x \wedge z$ and $x \wedge (y \wedge z) = (x \wedge y)$. It means $(x \wedge z) = x$.

The points given in the above remark also imply that $\wedge$ and $\vee$ are both commutative and associative. If in any formula involving elements of S and $\vee, \wedge$ if we replace $\vee$ by $\wedge$ and $\wedge$ by $\vee$ we get a new expression involving elements of S, which evaluate new bounds for the variables in the formula. Hence we can define duality in lattices.

In $(S, \geq)$ we have $\vee'$ and $\wedge'$ defined as follows. For $x, y \in S$, if $x \vee y = c, x \wedge y = d$, then in dual lattice $(S, \vee', \wedge')$ $x \vee' y = d, x \wedge' y = c$. So if F is a valid expression involving elements of S, $\wedge$ and $\vee$ then its dual F is obtained by replacing in F every $\wedge$ by $\vee$ and every $\vee$ by $\wedge$. Recall that we have defined earlier in Logic, what are 'well-formed formulas' and their negations.

Definition 9.51 Let $(S, \vee, \wedge)$ be a lattice and let $(S, \leq)$ be the corresponding poset. The lattice $(S, \vee', \wedge')$ corresponding to the poset $(S, \geq)$ is called the dual lattice.

Example 9.52 Consider the lattice described by the following Hasse diagram.

Its dual will be

Example 9.53 Find the dual of the lattice given in Example 9.47.

We state the following theorem without proof.

Theorem 9.54 *The dual of every complete lattice is complete in which the glb and lub are reversed.*

9.5 Mobius Inversion

Let n be a positive integer and consider the set $S = \{1, 2, \ldots, n\}$. 2^S denotes the power set of S and we say for two subsets A, B of S, $A \subseteq B$ if A is a subset of B. It is not difficult to see $(2^S, \subseteq)$ is a poset. Define a real-valued function

$$f : 2^S \to \mathbb{R}.$$

For any poset $(A, \leq)$ and functions f, g defined from A to $\mathbb{R}$,

$$\text{if} \quad f(x) = \sum_{y \leq x} g(x), \quad \text{then}$$

the problem now is to find g using f.

For example, let $f : 2^S \to \mathbb{R}$, such that

$$f(A) = \sum_{T \subseteq A} g(T) \tag{9.3}$$

where A is a subset of S. Now the problem is to find g in terms of f. It is a type of inversion and Mobius inversion will help to find a solution.

We have some definitions that are useful in understanding Mobius inversion.

Definition 9.55 Let $(A, \leq)$ be a poset. Then an incidence algebra of this poset is the set

$$I(A) = \{f : A \times A \to \mathbb{R} / f(x_1, x_2) = 0 \text{ unless } x_1 \leq x_2\}$$

Note that f maps pairs (x_1, x_2) to $\mathbb{R}$. Here $x_1, x_2 \in A$.

$I(A)$ is closed under scalar multiplication. That is if $\alpha \in I(A)$, then $c\alpha \in I(A)$, where 'c' is a scalar. Also $I(A)$ is closed under addition and multiplication. Here for any $f, g \in I(A)$, multiplication as defined to be $fg(x, y) = \sum_{x_1 \leq z \leq x_2} f(x_1, z)g(z, y_2)$.

For $x_1, x_2 \in A$,

$$
\begin{aligned}
e(x_1, x_2) &= 1 \text{ for } x_1 = x_2 \\
&= 0 \text{ for } x_1 \neq x_2
\end{aligned}
$$

we can call 'e' to be the multiplicative identity.

Definition 9.56 The function given by

$$
\zeta(x_1, x_2) = \begin{cases} 1 & \text{if } x_1 \leq x_2 \\ 0 & \text{if } x_1 \nleq x_2 \end{cases}
$$

is called the zeta function.

Clearly ζ belongs to the incidence algebra of a poset $(A, \leq)$.

Mobius Function: A function μ in $I(A)$ defined as below is called the Mobius function.

$$
\begin{aligned}
\mu(x_1, x_2) &= 1 \text{ if } x_1 = x_2 \\
&= 0 \text{ if } x_1 < x_2 \\
&= -\sum_{x_1 \leq s < x_2} \mu(x_1, s)
\end{aligned}
$$

Remark 9.57 Consider

$$
\begin{aligned}
\mu\zeta(x_1, x_2) &= \mu(x_1, x_2)\zeta(x_1, x_2) \\
&= \begin{cases} 1 & \text{for } x_1 = x_2 \\ 0 & \text{otherwise} \end{cases} \\
&= e(x_1, x_2)
\end{aligned}
$$

Proposition 9.58 *A function $f \in I(A)$ is invertible if and only if $f(x, x) \neq 0$ for all $x \in A$.*

Remark 9.59 The inverse of ζ is μ.

Example 9.60 Consider the number $n = 30$. Its divisors are 1, 2, 3, 5, 6, 10, 15 and 30. One can draw a Hasse diagram for the divisors of 30 as below.

For the diagram, we mark the calculation of $\mu(x, y)$ as below. Here '1' is chosen to be the least element of the poset.

Example 9.61 Let A and B be two subsets of $S = \{1, 2, \ldots, n\}$. We note that $(2^S, \subseteq)$ is a poset. Then

$$\mu(A, B) = (-1)^{|B|-|A|} \text{ if } A \subseteq B$$
$$= 0 \quad A \nsubseteq B.$$

Solution: Consider $A = \emptyset$. Then $\mu(\emptyset, B) = (-1)^{|B|}$. This can be proved by induction on $|B|$. If $|B| = 0$, then it is trivially true as $\emptyset$ is the least element of $(2^S, \subseteq)$. Assume the result to be true for $|B| = k$ elements. Then $|\emptyset - B| = |B|$ and for every subset $C \subseteq B$, $\mu(\emptyset, C) = (-1)^{|C|}$. Consider

$$\mu(\emptyset, B) = -\sum_{C \subset B} \mu(\emptyset, C)$$
$$= -\sum_{C \subset B} (-1)^{|c|}$$
$$= (-1)^{|B|}$$

By replacing $\emptyset$ by any set A, it can be shown that

$$\mu(A, B) = (-1)^{|B|-|A|} \text{ if } A \subseteq B$$
$$= 0 \text{ otherwise}$$

Inclusion-Exclusion formula can be obtained using Mobius inversion.

Let $S_1, S_2, \ldots, S_n$ be subsets of a finite set S. Define $f(A)$ to be the number of elements of S that belong to sets S_i such that i is forbidden index. The forbidden index set is $A \subset \{1, 2, \ldots, n\}$. That is for any element $x \in S$, x is counted by f with a conclusion that $x \notin S_i$ for every $i \in A$. Clearly $x \in S_j$ for every $j \notin A$. Then the function

$$g(A) = \sum_{C \subseteq A} f(C)$$

enumerates the elements of S satisfying the above condition. That is

$$f(A) = \sum_{C \subseteq A} (-1)^{|A|-|C|} g(C)$$

So the count of $\bar{S}_1 \cap \bar{S}_2 \cdots \cap \bar{S}_n$ will be

$$|\bar{S}_1 \cap \bar{S}_2 \cdots \cap \bar{S}_n| = \sum_{C \in 2^S} (-1)^{n-|C|} g(C)$$
$$= \sum_{\bar{C} \in 2^S} (-1)^{|\bar{C}|} g(\bar{C})$$
$$= \sum_{\bar{C} \in 2^S} (-1)^{|\bar{C}|} \left| \bigcap_{i \in \bar{C}} S_i \right|$$

This is equivalent to the inclusion-exclusion formula given in ??.

Exercise

1. Show that the following sets are posets.

   (a) $(\mathbb{N}^+, \times_f)$ where $x \times_f y$ means x and y have a common factor.
   (b) $(\mathbb{R}, \leq)$, $\leq$ has the usual arithmetic meaning.
   (c) $(\mathbb{A}, \leq)$ where $\mathbb{A}$ is the set of people in a city where $x \leq y$ means x and y are the same person or x is related to y.

2. Let $S = \{1, 2, 3, 4, 5, 6, 8, 9, 12, 15, 18\}$. Let $\div$ be the binary relation such that $x \div y$ means x divides y. Show that $(S, \div)$ is a poset. Draw its Hasse diagram. Write down all its subsets which are '*chains*'.
3. Let $D_n = \{m \mid m \text{ is a divisor of } n\}$. Show that $(D_{16}, \leq)$ where for $d, \bar{d} \in D_{16}$, $d \leq \bar{d}$ means d is a divisor of $\bar{d}$, is a poset. Draw its Hasse diagram.
4. Draw the Hasse diagram for the poset $S = \{a, b, c, d, e, f, g, h\}$ with

   (a) $a < b, a < d, b < e, c < e, f < g, d < g, d < h, e < a, e < g$.
   (b) $a < c, c < e, e < g, b < d, d < f, f < h$.

5. Draw all the posets of a set with 4 elements which are non-isomorphic.
6. Let $S = \{1, 2, 3, 4\}$. Show that $\mathbb{P}(S)$ is a poset. Identify the minimal and maximal elements of the subset $\{1, 3\}$.
7. Let there be three distinct balls to be distributed in three boxes. Enumerate all possible distributions by indicating only the nonempty boxes. Take this set and define an ordering, so that the following Hasse diagram is reached.
8. Extract the maximal and minimal chains from the following Hasse diagram. Define length of a finite chain C to be $l(C) = |C| - 1$ and rank of a finite poset S to be, $r(S) = \max\{|C| - 1 \mid C \text{ is a chain }\}$. Find $l(c)$ and $r(S)$ for the above poset.
9. Find the glb and lub of $\{2, 3, 4\}$, $\{3, 5, 6\}$, $\{1, 3, 5, 7\}$, $\{9, 10\}$ in the following Hasse diagram.
10. Check whether the following Hasse diagram form a lattice or not. Justify your answer.
11. Given $T = T_1 \times T_2$ of two posets $(T_1, \leq)$ and $(T_2, \leq)$ with $\leq$ being dictionary ordering, show that $(T, \leq)$ is a poset.
12. Prove that every finite lattice is a complete lattice.
13. Let $(S, \leq)$ be a poset and let $A \subseteq S$. Can $x \in S$ be the least upper bound of A? Justify your answer.
14. Consider $\mathbb{Z}^+$, the set of positive integers. Consider the set $(\mathbb{Z}^+ \times \mathbb{Z}^+, \sim)$ such that $(x, y) \sim (u, v)$ if x divides u and y divides v. Is $(\mathbb{Z}^+ \times \mathbb{Z}^+, \sim)$ a poset? Justify your answer.

15. Let $\leq$ be a relation on the set $S = \{1, 2, 3, 6, 7, 8\}$ with $x \leq y$ if $\frac{x}{y}$ is odd. Is $(S, \leq)$ a poset? Justify your answer.
16. Does there exist a lattice without minimum element? Justify your answer.
17. Let $S = \{a_1, a_2, a_3, a_4\}$. How many lattices can be formed using 'S' with some ordering '$\leq$'. List all of them.

Polya's Theory

10

10.1 Permutation Group

In this section we define groups. Basically every monoid is also a semigroup. Every semigroup is a groupoid. Now, we consider a monoid with one more property. So, a group is a set with a binary operation $*$ defined on the elements of the set. In monoids and semigroups also, an operation is applied on any two elements of the set S and the nature of the output is studied in S itself. For example for any $a, b \in S, a * b \in S$ is known as closure property. Groups of finite order are useful in coding theory, and cryptography.

Definition 10.1 A group is a set S with a binary operation $*$ such that

1. for any two elements $a, b \in S, a * b \in S$. That S is said to satisfy $*$-closure.
2. for any three elements $a, b, c \in S$,

$$(a * b) * c = a * (b * c).$$

   That is, $*$ is associative.
3. there exists an element $e \in S$ called the identity element w.r.to $*$. That is, for any $a \in S$, $a * e = e * a = a$.
4. for any $a \in S$, there exists an element $\overline{a} \in S$ such that $a * \overline{a} = \overline{a} * a = e$. $\overline{a}$ is called the inverse of a and is usually denoted by a^{-1}.

Notation: We call $(S, *)$ as a group which imply that S is a group w.r.to the binary operation $*$.

R. Rama, *Topics in Combinatorics and Graph Theory*,
https://doi.org/10.1007/978-3-031-74252-1_10

Example 10.2 1. The set $\mathbb{Z}$ with the usual + operation is a group. We know that $(\mathbb{Z}, +)$ is a monoid. Since for every $n \in \mathbb{Z}$, $-n \in \mathbb{Z}$ such that $n + (-n) = 0$, 0 being the identity, $(\mathbb{Z}, +)$ is a group.
2. The set $\mathbb{R}$ w.r.to + is a group.
3. The set $\mathbb{R}\backslash\{0\}$ is a group w.r.to multiplication.
4. Let $(M_{2\times 2}(\mathbb{N}), +)$ is a group where $M_{2\times 2}(\mathbb{N})$ is the set of 2×2 matrices with every entry being in $\mathbb{N}$. + is the usual matrix addition. Clearly the matrix $\begin{bmatrix} 0 & 0 \\ 0 & 0 \end{bmatrix}$ is the identity.
For any matrix $\begin{bmatrix} n_1 & n_2 \\ n_3 & n_4 \end{bmatrix}$, $\begin{bmatrix} -n_1 & -n_2 \\ -n_3 & -n_4 \end{bmatrix}$ is the inverse.
The next example is a special type of group which is more popular. Before we go with what are known as symmetric groups, we define abelian group.

Definition 10.3 A group $(S, *)$ is said to be an abelian group or commutative group if for any $x, y \in S$, $x * y = y * x$.

All the four groups defined in the previous examples are abelian groups.
Table

Example 10.4 (*Symmetric groups*) Let $S = \{1, 2, 3\}$ and let $f : S \to S$ be a mapping which is bijective. For example, if

$$f(1) = 2$$
$$f(2) = 3$$
$$f(3) = 1$$

is a mapping. It is something like distributing three distinct balls in 3 boxes such that no box is empty. The number of ways of doing this will be 3! = 6. Here there are exactly 6 distinct mappings that map the elements of S to itself in a bijective manner. We may also write the above mapping as

$$\begin{pmatrix} 1 & 2 & 3 \\ 2 & 3 & 1 \end{pmatrix}$$

which is called a permutation. Let us denote the 6-distinct permutations in a set denoted by S_3. The following table gives all the members of S_3.

Remark 10.5 $(S_n, \circ)$ can be shown to be a group called the permutation group.

mapping	permutation
e	$\begin{pmatrix}1&2&3\\1&2&3\end{pmatrix}$
f_1	$\begin{pmatrix}1&2&3\\1&3&2\end{pmatrix}$
f_2	$\begin{pmatrix}1&2&3\\3&2&1\end{pmatrix}$
f_3	$\begin{pmatrix}1&2&3\\2&1&3\end{pmatrix}$
f_4	$\begin{pmatrix}1&2&3\\2&3&1\end{pmatrix}$
f_5	$\begin{pmatrix}1&2&3\\3&1&2\end{pmatrix}$

Now $(S_3, \circ)$ is a group where $\circ$ is the composition of functions. The composition table w.r.t. to the set S_3 is as below:

$\circ$	e	f_1	f_2	f_3	f_4	f_5
e	e	f_1	f_2	f_3	f_4	f_5
f_1	f_1	e	f_4	f_5	f_2	f_3
f_2	f_2	f_5	e	f_4	f_3	f_1
f_3	f_3	f_4	f_5	e	f_1	f_2
f_4	f_4	f_3	f_1	f_2	f_5	e
f_5	f_5	f_2	f_3	f_1	e	f_4

From the above table, we see that e is the identity element of S_3. Also entries in the table where e is present indicate the inverses. For example, 5th row last column element is e and so f_4 is the inverse of f_5 and vice versa. Hence $(S_3, \circ)$ is a group but not commutative. It is called the symmetric group.

Remark 10.6 Similar to $(S_3, \circ)$, $(S_n, \circ)$ can be defined. The operation table given above is also called as Cayley table.

Example 10.7 Show that S_n has $n!$ elements.

Solution: By combinatorial argument it can be shown that $|S_n| = n!$.

Next we see a definition which talks about the cardinality of a set S in $(S, *)$.

Definition 10.8 Let S be any set. For any group $(S, *)$, the cardinality of $(S, *)$ is the number of elements in S. It is also called as order of the group $(S, *)$.

The order can be finite or infinite. For example $(\mathbb{Z}, +)$ is infinite order group whereas $(S_3, \circ)$ is finite order group.

Canonical Form of Permutations

When we write each k cycle by writing its largest value in the first and then writing the product in the increasing of their first element, then what we get is called the canonical form product of permutations. We observe that such a representation is unique of its nature.

We state the following theorem without proof.

Theorem 10.9 *Let $n_1, n_2, \ldots, n_k$ be integers with each $n_i \geq 0,\ 1 \leq i \leq k$. Let $\sum_{i=1}^{k} i \cdot n_i = m$. That is in n_i's, there may be a repetition. Then the number of permutations of length m with n_i cycles of length i where $1 \leq i \leq m$ will be $\frac{m!}{n_1!n_2!\cdots n_k!\cdot 2^{n_1}2^{n_2}\cdots 2^{n_k}}$*

Notation: If a m-permutation p has n_1 cycles of length 1, n_2 cycles of length 2, n_3 cycles of length 3 and so on, n_k cycles of length k, then we denote $(n_1, n_2, n_3, \ldots, n_k)$ as p-type permutation.

Example 10.10 The number of permutations having one cycles is of $(\underbrace{0, 0, 0, \ldots, 0}_{(n-1)\text{ times}}, 1)$-type and is equal to $(n-1)!$.

Definition 10.11 The number of m-permutations with t-cycles are also called as $s(m, t)$, where $s(m, t)$ is the Stirling number of first kind without sign. If $c(n, t)$ denote the number of such permutations, then

$$s(m, t) = (-1)^{m-t} c(m, t)$$

By the recurrence relation given earlier, we have the following theorem.

Theorem 10.12 *Let m and t be two positive integers such that $m \geq t$. Then*

$$c(m, t) = c(m-1, t-1) + (m-1)c(m-1, t)$$

Stirling Numbers

There are two types of Stirling numbers. They are Stirling number of first kind and Stirling number of second kind

Stirling Numbers of First Kind

Consider the product

$$x(x+1)(x+2)\cdots(x+m-1).$$

The coefficients of x^t, $0 \le t \le m$, in the product is known as 'Stirling Numbers'. The product expands as a polynomial in x and it can be written as

$$p(x) = \sum_{t=0}^{m}(-1)^{m-t}s(m,t)x^t.$$

The numbers $s(m,t)$ are known as unsigned Stirling numbers of first kind.

The following recursion is satisfied by Stirling numbers of first kind.

$$s(m,t) = s(m-1,t-1) - (m-1)s(m-1,t)$$

This recursion is given with the initial condition

$$s(m,0) = 0,\ m \ge 0$$

Given that

$$\begin{aligned}\sum_{t\ge 0} s(m,t)x^t &= P_m(x)\\ &= (x+m-1)P_{m-1}(x)\\ &= xP_{m-1}(x) + (m-1)P_{m-1}(x)\\ &= \sum_{t\ge 0} s(m-1,t)x^{t+1} + \sum_{t\ge 0}(m-1)s(m-1,t)x^t\end{aligned}$$

Comparing the coefficient of x^t on both sides

$$s(m,t) = s(m-1,t-1) + (m-1)s(m-1,t).$$

By using this recurrence relation for various sample values of m and t we have $s(m,t)$ values as

$m \backslash t$	0	1	2	3	4	5	6
0	1	0	0	0	0	0	0
1		1	1	2	6	24	120
2			1	3	11	50	274
3				1	6	35	225
4					1	10	85

Where do these Stirling numbers of first kind seen?

Definition 10.13 Let S be an n-element set. Any permutation can be viewed as a function $f : S \to S$. That is the permutation 312 for a set $S = \{1, 2, 3\}$ can be viewed as $f : S \to S$ such that $f(1) = 3$, $f(2) = 1$, $f(3) = 2$.

Usual compositions of functions can be defined on sets having the same cardinality which are usually called as product of permutations. The group with a multiplication operation will be called as permutation group which will be seen later. In permutations one can define what are known as cycles. Before we give the following simple result, by f^2 we mean $f^2 = f \circ f = ff$ is the product permutation.

Proposition 10.14 *Let $f : S \to S$ where S is a set having n elements. Let $a \in S$ be any element. Then there exists a positive integer $1 \le i \le n$ such that $f^i(a) = a$.*

The proof can be given using PHP. We have the definition of k-cycles where $k \le n$.

Definition 10.15 Let f be a function from the set S to S such that $\#S = n$. Let $a \in S$ and let k be the least positive integer such that $f^k(x) = x$. Then the elements a, $f(a)$, $f^2(a)$, $\cdots$, $f^k(a)$ form a k-cycle or cycle of length k of f.

The following result is very useful and popular.

Proposition 10.16 *All permutations can be decomposed into the disjoint union of their cycles.*

Example 10.17 Consider a permutation 321465, i.e., $f(1) = 3$, $f(2) = 2$, $f(3) = 1$, $f(4) = 4$, $f(5) = 6$, $f(6) = 5$. Then the (31), (65) are 2-cycles. (2) and (4) are 1-cycles or fixed points of f. By writing the product

$$(31)(2)(4)(65),$$

we mean, we have written a cycle decomposition of (321465).

10.2 Burnside's Lemma

Multiplication of permutations is nothing but a composition of permutations. That is for any two permutations π_1 and π_2, $\pi_1 \circ \pi_2$ is nothing but composition of two bijective functions from $X = \{1, 2, \ldots, n\}$ to itself. So, $\pi_1 \circ \pi_2(a) = \pi_1(\pi_2(a))$ for $a \in \{1, 2, \ldots, n\}$. We know S_n is the set of permutations of X and $|S_n| = n!$.

We now define action of a group G on a set X. It is defined as a function $f : X \to X$ with the following properties.

(i) $f(g(x)) = f \circ g(x), x \in X, f, g \in G$
(ii) $e(x) = x$ for all $x \in X$.

'e' is an element of G and by condition (ii) above 'e' maps any element $x \in X$ to itself. But it may not be true for every $f \in G$.

Definition 10.18 If for an element $g \in G$, G acting on a set X, $g(x) = x$, then 'g' is said to fix $x \in X$. Let

$$F(g) = \{x \in X \mid g(x) = x\}$$

denote the elements of X fixed by $g \in G$. The set $O(x) = \{g(x) | g \in G\}$ will be called as the orbit of the element $x \in X$.

The set $T(x) = \{g \in G | g(x) = x\}$ is called the stabilizer set of an element $x \in X$. This is the set of elements of G which fix an element $x \in X$. The following lemma is simple and direct from the definitions.

Lemma 10.19 *For all* $x \in X$,

$$|T(x)||O(x)| = |G|.$$

Remark 10.20 Above lemma is known as orbit counting lemma.

Let G be a group acting on X and '$\sim$' be a relation defined on X as $x \sim y$ if and only if there exists an element $g \in G$ such that $g(x) = y$, for all $x, y \in X$. It is not difficult to see that $\sim$ is an equivalence relation. Now we give Burnside lemma which counts the number of equivalence classes of X.

Burnside Lemma: The number of equivalence classes of $\sim$ is equal to $\dfrac{1}{|G|}\sum_{g \in G} |F(g)|$.

Proof Let $S = \{(f, x) \in G \times X | f(x) = x\}$.
Now one can get the cardinality of S either by looking at the sum

$$\sum_{g \in G} |F(g)| \text{ or}$$

by looking at those $x \in X$ such that

$$\sum_{x} |T(x)|.$$

Here $\sum_{g \in G} |F(g)|$ is a summation over 'g' where g fixes some elements of X.

Consider $\sum_{x \in X} |T(x)| = \dfrac{\sum_{\text{orbits}} |\text{number of orbits}||G|}{|\text{number of orbits}|}$.
Here we can see that $T(x)$ is a subgroup of G and hence

$$|T(x)| = \frac{|G|}{|\text{number of orbits of } X|}$$

Hence the number of equivalence classes is equal to

$$\frac{1}{|G|}\sum_{g \in G} |F(g)|.$$

It is nothing but number of orbits. □

Example 10.21 (*Beads Necklace Symmetry*) The problem is to arrange in a circular fashion, n beads where some are black in color and some are white in color. Here n is prime. Two such necklaces are considered to be the same if one can be seen as the other by means of circular rotation. Otherwise they are different. How many different necklaces can be made from 'n' beads for any n?

Solution:
Number of necklaces that can be got by using 'n' beads is 2^n and let K be the set of these arrangements. In these two necklaces $\alpha, \beta \in K$ are identical if α can be seen to be the same by rotating β. Otherwise they are different. For example $bwww$ will be same as $wbww$, $wwbw$, $wwwb$. Such necklaces are to be counted only once. By Burnside's lemma,

$$t|G| = \sum_{g \in G} |F(g)|$$

where 't' is the number of orbits. Let us understand that we want to find 't'. Before we proceed, let us understand about the arrangements of beads in detail

(i) A necklace that is completely white or completely black means that every position of the bead (white or black) is fixed by every element of G.
(ii) Consider the necklaces are not fixed by any $g \in G$ except the identity element of G say 'e' of G. Now we must understand G as a group

$$G = (\mathbb{Z}_n, +) \text{ where } \mathbb{Z}_n = \{0, 1, 2, \ldots, n-1\}$$

and + is addition modulo n operation. Let the necklaces be denoted by $(x_1, x_2, \ldots, x_n)$.

Consider a necklace such that

$$(x_1, x_2, \ldots, x_n) = g(x_1, x_2, \ldots, x_n), g \in G.$$

Then

$$g(x_1, x_2, \ldots, x_n) = (x_1, x_2, \ldots, x_n).$$

Iterative use of 'g' and the definition of G, imply that

$$tg(x_1, x_2, \ldots, x_n) = (x_1, x_2, \ldots, x_n).$$

If 'n' is prime, then we see

$$(x_1, x_2, \ldots, x_n) = fg^{-1}g(x_1, x_2, \ldots, x_n)$$

for $t = fg^{-1}$. That is

$$(x_1, x_2, \ldots, x_n) = (x_{1+f}, x_{2+f}, \ldots, x_{n+f})$$

Hence the chosen necklaces are same which is a contradiction. So when

$$\begin{cases} g = e, \ F(g) = 2^n \\ g \neq e, \ F(g) = 2. \end{cases}$$

Now

$$
\begin{aligned}
tn &= \sum_{g \in G} F(g) = F(e) + \sum_{\substack{g \neq e \\ g \in \mathbb{Z}_n^+}} F(g) \\
&= 2^n + \sum_{\substack{g \neq 0 \\ g \in (\mathbb{Z}_n^+)}} 2 = 2^n + 2(n-1) \\
\Rightarrow t &= \frac{2^n + 2(n-1)}{n} \text{ orbits}
\end{aligned}
$$

10.3 Equivalent Colorings

This section is a continuation of the previous section where we look for assignment of colors to the elements of a set X and identify non equivalent outputs under the action of a group G that identify symmetries in coloring. This section gives different examples to illustrate the concept of equivalent colorings.

Example 10.22 Let n students sit in a circular fashion and let there be $n!$ ways of seating. We can see that by cyclic rotation for every possible sitting, there are 'n' cyclic rotations of the sitting. This means each equivalence class will contain 'n' elements. Hence the number of equivalence classes will be $n!/n = (n-1)!$.

Example 10.23 Suppose we understand the arrangement of 8 chairs in a circular fashion. Let the chairs be only of two types.

Suppose the position of chairs are at equi-distant from each other. So we can see that there are 8 possible ways of placing the chairs. In the circular arrangement one can use

(i) Two chairs of type-1 and one chair of type-2.
(ii) Two chairs of type-2 and one chair of type-1.
(iii) All three chairs of same type.

The cyclic rotations are considered to be equivalent.

For example the arrangement

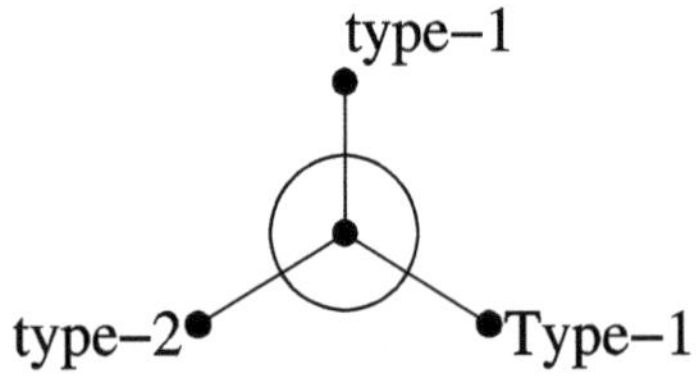

is equivalent to

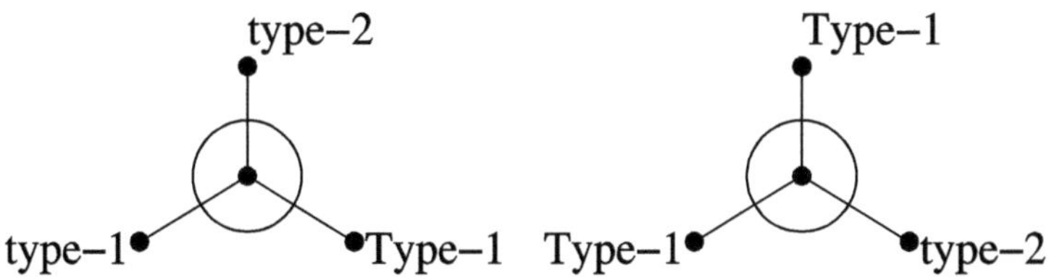

Hence the number of equivalent classes will be 4 which can be directly counted.

Suppose G is a group of permutations formed from the elements of X. Let ζ be the set of colors assigned to elements of X. Suppose $\zeta(1), \zeta(2), \ldots, \zeta(n)$ be the coloring of the elements of X. For example $\zeta = \{\text{blue, yellow}\}$ and $\zeta(i)$ may be blue if i is odd, $\zeta(i) =$ yellow when i is even. Then ζ assigned colors to every element of X. Consider a permutation $\sigma \in G$ such that

$$\sigma = \begin{pmatrix} 1 & 2 & 3 & \ldots & n \\ j_1 & j_2 & j_3 & \ldots & j_n \end{pmatrix}$$

Suppose G acts on X. Then action of σ on any color $s \in \zeta$ is denoted as $\sigma * s$ such that

$$\sigma * s = \begin{pmatrix} 1 & 2 & 3 & \ldots & n \\ s(j_1) & s(j_2) & s(j_3) & \ldots & s(j_n) \end{pmatrix}$$

Only requirement is that for every $\sigma \in G$ and for every $s \in \zeta, \sigma * s$ must also be in ζ. For any two permutations $\sigma_1, \sigma_2 \in G$

$$(\sigma_1 \circ \sigma_2) * s = \sigma_1 * (\sigma_2 * s).$$

Example 10.24 Consider $G = (\sigma, \sigma^2, \sigma^3, \sigma^4)$ where

$$\sigma = \begin{pmatrix} 1\,2\,3\,4 \\ 2\,3\,4\,1 \end{pmatrix}$$

$$\sigma^2 = \sigma \circ \sigma = \begin{pmatrix} 1\,2\,3\,4 \\ 3\,4\,1\,2 \end{pmatrix}$$

$$\sigma^3 = \sigma^2 \circ \sigma = \begin{pmatrix} 1\,2\,3\,4 \\ 4\,1\,2\,3 \end{pmatrix}$$

$$\sigma^4 = \begin{pmatrix} 1\,2\,3\,4 \\ 1\,2\,3\,4 \end{pmatrix}$$

G is a permutation group S_4. Suppose $\zeta = \{c_1, c_2\}$ are the colors assigned to 1, 2, 3 and 4. $\sigma^4 * \zeta = \gamma$. Then $\sigma, \sigma^2, \sigma^3$ are obtained by cyclic rotation of (1 2 3 4). Hence we can easily see the action of G under ζ.

If $c_1 = (r, b, b, r)$, then
$\sigma^4 * c_1 = (r, b, b, r)$
$\sigma * c_1 = (b, b, r, r)$
and so on.

If $c_2 = (r, b, r, b)$, then
$\sigma^4 * c_2 = (r, b, r, b)$
$\sigma^2 * c_2 = (r, b, r, b)$
and so on.

One can see that the colorings fall in one class by means of coloring function defined on the set X. Also the equivalence relation defined on X for counting the equivalent classes is any equivalence relation defined on X and G is any group acting on X. But in our present discussion we have a permutation group and the equivalence relation is induced by the permutations of G. Hence usage of Burnside lemma for counting equivalent coloring classes will not help.

Let us have one more example of coloring of permutation group.

Example 10.25 Let a square of unit length be given as below where the corners are well marked.

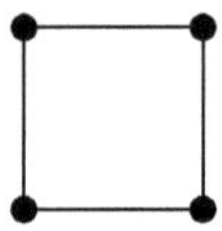

Find the number of ways we can construct squares by placing two flags, a black flag and a white flag so that the flagged squares are distinct. Suppose we allow counter clockwise rotation of a particular square about its center. We can say that the square is rotating and taking positions by $0°$, $90°$, $180°$ and $270°$.

Now a flagged square

α :

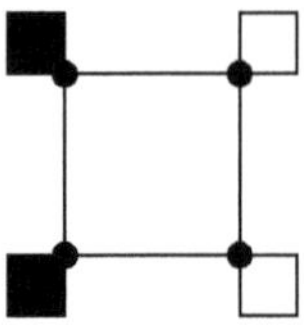

is same as the flagged square

β :

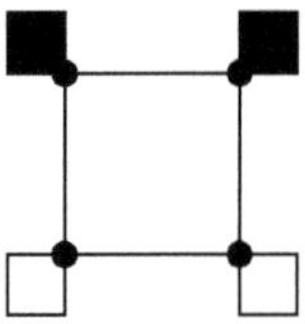

Here α when rotated by $90°$ we get β. We indicate the rotation by R and denote the rotation relation between α and β as $\alpha R \beta$. Here 'R' will be equivalence relation. There are 3 options of flagging the corners of the square. They are

1. All four corners get the same color flag.
2. Three corners are flagged with same color and one corner gets a different color.
3. Two corners get the same color flag and the other two corners get the other color flag.

Based on rotation, the flagged squares can be arranged as equivalent ones.

Any equivalence rotation partitions the set on which it is acting on. Since we look for equivalent coloring for a set, one has to observe operations on the set that induce equivalent colored elements of the set. That is if X is a set and ζ is the set of coloring assignments assigned to elements of X. Let R be an equivalence relation on X. For any two elements $\alpha, \beta \in X$, let $C(\alpha), C(\beta)$ be the coloring class where $C \in \zeta$. We say $C(\alpha) = C(\beta)$ if $\alpha R \beta$.

10.4 Cycle Index

We know for a set $X = \{1, 2, \ldots, n\}$ what a permutation group S_n is. We also know that any permutation in S_n can be written as the product of disjoint cycles. For example the permutation

$$\pi = \begin{pmatrix} 1\ 2\ 3 \\ 3\ 2\ 1 \end{pmatrix}$$

can be written as product of cycles (13) (2). Here (13) is a 2-cycle and (2) is cycle of length one.

A permutation $(x_1\ x_2\ \ldots\ x_n)$ is a permutation in S_n and it is a cycle of length n. For example $(2\ 3\ 4\ 1)$ is a cycle of length 4 and it is the same as $(3\ 4\ 1\ 2)$, $(4\ 1\ 2\ 3)$, $(1\ 2\ 3\ 4)$.

Let τ be a permutation written as a product of cycles. Let there be t_1 cycles of length 1, t_2 cycles of length 2 and so on t_k cycles of length k in this unique product. One can encode 'τ' by using the expression $x_1^{t_1} x_2^{t_2} \ldots x_k^{t_k}$. This encoding can be done for every element of a permutation group G. This leads to an encoding of the group G itself by summing up the individual encoding of each permutation in G. If 'k' is the length of the longest cycle in the cycle decomposition of any τ in G, then

$$P_G(x_1, x_2, \ldots, x_k) = \frac{1}{|G|} \sum_{\tau \in G} x_1^{t_1} x_2^{t_2} \ldots x_k^{t_k}$$

is called the cycle index of G.

Consider the permutation group S_n. Let us find all the permutations with their cycle decompositions or product of cycles. Let us do it in the following manner.

Suppose the integer partition of n is $1^{t_1} 2^{t_2} \ldots k^{t_k}$ and let us denote this as α. Clearly $t_1 + 2t_2 + \cdots + kt_k = n$.

We have the following theorem.

Theorem 10.26 *The number of permutations with cycle type α is*

$$\frac{n!}{\prod_{1\le j\le k} j^{t_j}(t_j!)}$$

Example 10.27 Consider $n = 7$.

$$\alpha = 1^0 2^2 3^1$$

Then the permutations may be $(1\ 2)\ (3\ 4)\ (5\ 6\ 7)$, $(1\ 3)\ (2\ 4)\ (5\ 6\ 7)$, $(1\ 4)\ (2\ 3)\ (5\ 6\ 7)$ and so on. Totally there are 7! such products. But all these 7! permutations are not distinct as rotational or cyclic symmetry will lead to the same permutation. For example in the cycle $(5\ 6\ 7)$, $(6\ 7\ 5)$, $(7\ 5\ 6)$ are falling under rotational symmetry.

So two permutations σ_1, σ_2 give rise to the same permutation if

(i) rotation of each cycle can be done and it can be done by in ℓ ways
(ii) also permute all the cycles of same size.

Let us use the following notation

$$Z_\alpha = \prod_{1\le j\le k} j^{t_j}(t_j!)$$

So in the above theorem number of permutations with cycle type α is $\frac{n!}{Z_\alpha}$.

Hence the cycle index polynomial for S_n will be

$$P_{S_n}(x_1, x_2, \ldots, x_k) = \frac{1}{n!}\sum_\alpha \frac{n!}{Z_\alpha} \prod_{1\le j\le k} x_j^{t_j}$$

where α is an integer partition of n.

$$= \sum_\alpha \prod_{1\le j\le k} \frac{x_j^{t_j}}{j^{t_j} t_j!}$$

Suppose G is a permutation group. i.e., $G \subseteq S_n$. Its cycle index polynomial is defined by

$$Z_G[y_1, y_2, \ldots] = \frac{1}{|G|}\sum_{g\in G}\prod_{i\in P} y_i^{m_i(\tau)}$$

where $m_i(\tau)$ = number of cycles of length i in τ.

Let σ_n be a permutation in S_n and let $\sigma_n = (1\ 2\ 3\ \ldots\ n)$, a n-cycle. Let G be the permutation group and $G =<\sigma_n>$, the group generated by σ_n. i.e., $G = \{\sigma, \sigma^2, \ldots, \sigma^{n-1}, \sigma^n = e\}$. For any $t \in \mathbb{N}$, the cycle of n under σ^k consists of

$$\{s \in \{1, 2, \ldots, n\} | s = at + bn, a, b \in \mathbb{Z}\} = \left\{\ell.gcd(t, n) | \ell = 1, 2, \ldots, \frac{n}{gcd(t, n)}\right\}.$$

Hence the cycle type of σ^k is $\underbrace{(e, e, \ldots, e)}_{gcd(t,n)}$.

Recall the cycles type of a permutation is a list consisting of length of the cycles in the product representation of the cycle and written as a s-tuple. Here s is the number of cycles in the product representation of the permutation. Now number of such t's such that $\frac{n}{gcd(t,n)} = e$ such that the cycle is $(e, e, e, \ldots, e)$ for σ^k is to be determined.

Consider the poset (divisors of $n, /$) the divisors of 'n' with partial ordering as divisibility '1'. Now consider

$$F_d = \left|\left\{t \in \{1, 2, \ldots, n\} \middle| \frac{n}{gcd(t, n)} = e\right\}\right|$$

$$H_d = \left|\left\{t \in \{1, 2, \ldots, n\} \middle| \text{'}e'\text{ divides } \frac{n}{gcd(t, n)}\right\}\right|$$

$$= e$$

Using Mobius inversion,

$$F_d = \sum_{\ell | e} \mu(\ell, e) = \sum_{\ell | e} \mu\left(\frac{e}{\ell}\right).\ell = \phi(e) \text{ (Euler } \phi)$$

$$Z_G[y_1, y_2, \ldots] = \frac{1}{n} \sum_{e|n} \phi(e) y_e^{\frac{n}{e}}$$

10.5 Polya's Theorem

We defined and discussed what a cycle index polynomial was.

Let G be a permutation group. Let $[k] = \{1, 2, \ldots, k\}$ denote set of k different colors. A coloring function assigns colors to elements of $X = \{1, 2, \ldots, n\}$. We already have seen what we mean by equivalent coloring. Two colorings C_1 and C_2 are said to be equivalent under the action of G if there exists a permutation $\tau \in G$ such that

$$C_1(\tau(x)) = C_2(x) \text{ where } \tau \in G$$

We denote this by $C_1 \underset{G}{\sim} C_2$. Let $\zeta_k(X)$ denote the set of all k-colorings of X. Let $G|\zeta_k(X)$ denote the set of equivalent colorings induced by G.

Example 10.28 Let $X = \{1, 2, \ldots, n\}$. Consider the permutation $\sigma = \big(2\ 3\ 4\ \ldots\ n\big)$. σ is a cyclic permutation and its length is n. Let $G =< \sigma >= \{\sigma, \sigma^2, \ldots, \sigma^{n-1}, e\}$ where $\sigma^n = e$. G is a permutation group. Now the k-colorings on X under the action of G gives the equivalent

k-colorings. This is nothing but the equivalence of necklaces with n beads formed using k-colors.

Now consider $C \in \zeta_k(x)$. Let $t_i = |\{x \in X | C(x) = i\}|$, $i = 1, \ldots, k$. That t_i gives the number of elements of X that are given color i. So that k-tuple $\alpha_C = (t_1, t_2, \ldots, t_k) \in \mathbb{N}^k$ is called the mode of a coloring $\tau \in \zeta$. Also $t_1 + t_2 + \cdots + t_k = |X| = n$. If $\tau \underset{G}{\sim} \bar{\tau}$, then $\alpha_C = \alpha_{\bar{C}}$. Let us do a more refined counting. For that consider $\alpha \in \mathbb{N}^k$ with $|\alpha| = n = |X|$. Let $\zeta_\alpha(x) = \{C \in \zeta_k(x) | \alpha_C = \alpha\}$ which is the colorings of X with mode α. The problem is to determine the number of G-equivalence classes of colorings of X with k-colors in the mode of α.

Definition 10.29 (*Power Sum Symmetric Polynomial*) A polynomial $g(x_1, \ldots, x_k)$ is said to be symmetric if

$$g(x_1, x_2, \ldots, x_k) = g(x_{w(1)}, x_{w(2)}, \ldots, x_{w(k)}\}$$

where $w \in S_k$.

Example 10.30 Consider the polynomial

$$g(x_1, x_2, x_3) = x_1^3 + x_2^3 + x_3^3 = p_3(x_1, x_2, x_3)$$

This is a 3rd power symmetric polynomial.

Consider an integer partition $\tau = (\tau_1, \tau_2, \ldots, \tau_m)$ and let

$$\begin{aligned} p_\tau(x_1, x_2, \ldots, x_k) &= (x_1^{\tau_1} + x_2^{\tau_1} + \cdots + x_k^{\tau_1} \\ &\quad (x_1^{\tau_2} + x_2^{\tau_2} + \cdots + x_k^{\tau_2} \\ &\quad (x_1^{\tau_k} + x_2^{\tau_k} + \cdots + x_k^{\tau_k} \\ &= \prod_{i=1}^{m} p_{\tau_i}(x_1, x_2, \ldots, x_k). \end{aligned}$$

These are called as power sum symmetric polynomial for the partition τ.

Consider now a permutation $\tau \in S_n$. $\mu(\tau)$ denote the cycle type of τ. That is $\mu(\tau) = (\tau_1, \tau_2, \ldots, \tau_m)$ where $\tau_1, \tau_2, \ldots, \tau_m$ are the size of the cycles of τ. Now for every permutation $\sigma \in S$ we try to associate a polynomial $p_{\tau(\mu)}$ in Polya enumeration theorem.

Lemma 10.31 *Let τ be any permutation. Then*

$$\sum_{\substack{\tau_1+\tau_2+\cdots+\tau_m=|X| \\ (\tau_1,\tau_2,\ldots,\tau_m)\in\mathbb{N}^m}} |\zeta_\alpha(x)^\tau| x_1^{\tau_1} x_2^{\tau_2} \ldots x_m^{\tau_m} = p_{\mu(\tau)}(x_1, x_2, \ldots, x_m).$$

Proof A τ-invariant coloring is one that is distinctly coloring the cycles of x. So,

$$\begin{aligned} p_{\mu(\tau)}(x_1, x_2, \ldots, x_m) &= \prod_{i=1}^{k} (x_1^{\mu_1} + \cdots + x_m^{\mu_k}) \\ &= \sum_{j_1=1}^{m} \sum_{j_2=1}^{m} \cdots \sum_{j_k=1}^{m} x_{j_1}^{\mu_1} x_{j_2}^{\mu_2} \ldots x_{j_k}^{\mu_k} \\ &\qquad \text{where } \mu(\tau) = (\mu_1, \mu_2, \ldots, \mu_k). \\ &= \sum_{(j_1, j_2, \ldots, j_k)\in[m]^k} x_{j_1}^{\mu_1} x_{j_2}^{\mu_2} \ldots x_{j_k}^{\mu_k} \end{aligned}$$

Each of the k-tuple $(j_1, j_2, \ldots, j_k)$ determine the coloring of x which is invariant under the action of G. This is because if we color all the elements $u \in x$ in the cycle of length μ_i with the color j_1 and if this coloring has shape α, then

$$x_{j_1}^{\mu_1} \ldots x_{j_k}^{\mu_k} = x^\alpha.$$

Hence

$$\begin{aligned} \sum_{\alpha} |\zeta_\alpha(x)\backslash G| x_1^{t_1} x_2^{t_2} \ldots x_k^{t_k} &= \frac{1}{|G|} \sum_{\alpha} \sum_{\tau\in G} (\zeta_\alpha(x))^\tau x_1^{t_1} x_2^{t_2} \ldots x_k^{t_k} \\ &= \frac{1}{|G|} \sum_{\tau\in G} p_{\mu(\tau)}(x_1, x_2, \ldots, x_m) \\ &\qquad \text{(by reversing the order of summation)} \end{aligned}$$

□

Polya Enumeration Theorem

Let x be a finite set and $G \subset S_n$ be a permutation group. Consider the summation

$$\sum_{\substack{t_1+t_2+\cdots+t_k=n \\ (t_1,t_2,\ldots,t_k)\in\mathbb{N}^k}} |\zeta_\alpha(x)\backslash G| x_1^{t_1} x_2^{t_2} \ldots x_k^{t_k}$$

where $|\zeta_\alpha/G|$ denotes the number of equivalence classes of $\zeta_\alpha(x)$ under the action of G. Here $\alpha = (t_1, t_2, \ldots, t_k)$. This summation is equal to

$$\frac{1}{|G|}\sum_{\tau\in G} p_{\mu(\tau)}(x_1, x_2, \ldots, x_k).$$

Proof By the orbit counting lemma

$$|\zeta_\alpha(x)\backslash G| = \frac{1}{|G|}\sum_{\tau\in G} |\zeta_\alpha(x)^\tau| \tag{10.1}$$

where $\zeta_\alpha(x)^\tau$ is the stabilizer of X under $\tau \in G$.

We use the above lemma and we get

$$\begin{aligned}
&\sum_{\tau\in G} |\zeta_\alpha(x)|G|x_1^{t_1}x_2^{t_2}\ldots x_k^{t_k} \\
&\quad = \frac{1}{|G|}\sum_{\alpha}\sum_{\tau\in G} |\zeta_\alpha(x)^\tau|x_1^{t_1}x_2^{t_2}\ldots x_k^{t_k} \\
&\quad = \frac{1}{|G|}\sum_{\tau\in G} p_{\mu(\tau)}(x_1, x_2, \ldots, x_k)
\end{aligned}$$

□

Example 10.32 Let $X = \{1, 2, 3, 4\}$.
Let $\tau = \left(1\ 2\ 3\ 4\right)$.
$G =< \tau >= \{e, \left(1\ 2\ 3\ 4\right), \left(1\ 3\right)\left(2\ 4\right), \left(4\ 3\ 2\ 1\right)\}$
So,

$$\frac{1}{|G|}\sum_{\tau\in G} p_{\mu(\tau)}(x_1, x_2, x_3, x_4)$$

(We assume there are 4-colors here).

$$\begin{aligned}
&= \frac{1}{4}\left[(x_1 + x_2 + x_3 + x_4)^4 + 2(x_1^4 + x_2^4 + x_3^4 + x_4^4)\right. \\
&\quad \left. +(x_1^2 + x_2^2 + x_3^2 + x_4^2)^2\right]. \\
&= x_1^4 + x_1^3x_2 + 2x_1^2x_2^2 + 3x_1^3x_2x_3 + 6x_1x_2x_3x_4 + x_2^4 + x_2^3x_3 + \cdots
\end{aligned}$$

So the number of necklaces with two beads of color 1, one bead of color 2, one bead of color 3 is 3 (verify). If one use all 4 colors different then 6 different necklaces can be made.

Exercise:

1. Show that every permutation on the set $X = \{1, 2, \ldots, n\}$ can be uniquely expressed as a product of transpositions.

2. Prove that the number of permutations of $X = \{1, 2, \ldots, n\}$ with odd number of cycles is equal to the number of permutations of X with even number of cycles.
3. Let σ and τ be permutations in S_3 such that

$$\sigma = \begin{pmatrix} 1 & 2 & 3 \\ 3 & 2 & 1 \end{pmatrix}$$

$$\tau = \begin{pmatrix} 1 & 2 & 3 \\ 2 & 1 & 3 \end{pmatrix}$$

Find $\sigma\tau(1), \sigma\tau(2), \sigma\tau(3)$.

4. Show that $(S_3, \circ)$ is a group. Write down explicitly the identity element of S_3. Also find the inverse of the permutation.

$$\tau = \begin{pmatrix} 1 & 2 & 3 \\ 3 & 1 & 2 \end{pmatrix}$$

5. Find the number of ways of arranging k distinct objects in a circular fashion.
6. Let D be a square with its corners labeled as a, b, c, d (i.e., corners distance) and its sides labeled as s_1, s_2, s_3, s_4. The following figure illustrates D.

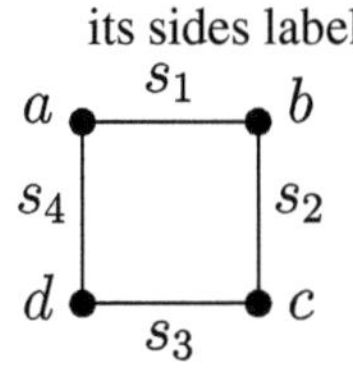

The rotational symmetry happens when D is rotated about its center through angles $0°, 90°, 180°$ and $270°$. Determine the rotation symmetry group of S_4.

7. In the square D considered in question 6, there are four reflection symmetrices about the diagonals 'ac' and 'bd'. This operation happens in space. Show that the reflection symmetries about the diagonal form a group.
8. Suppose the square D is such that the corners 1, 2, 3, 4 are colored using 3 colors red, blue and green.
(i) Determine the number of non equivalent colorings of the corners under the action of rotation symmetry group defined in question 6.
(ii) Determine the number of non equivalent colorings of the corners under the action of reflection symmetry group defined in question 7.
9. How many non equivalent ways are there to color the corners of a pentagon using 2 colors, red and blue?
10. A dihedral group on $\{1, 2, \ldots, n\}$ is a symmetry group. The symmetries are rotational symmetry and reflection symmetry. It is denoted as D_n and it is a group on n-sided regular polygon for $n \geq 3$.
(i) Show that D_n is an abelian permutation group.

(ii) Express each permutation in D_4 as a product of cycles. Write down their cycle types.
(iii) Find the cycle index of D_k, for $k \geq 3$.

11. Using the Burnside's Lemma find the number of non-isomorphic graphs on 5-vertices.
12. Find the cycle index polynomial $P_{S_5}(x_1, x_2, x_3, x_4, x_5)$ of K_5, the complete graph on 5-vertices.
13. Find x in the following equation

$$\begin{pmatrix} 1 & 2 & 3 & 4 \\ 4 & 1 & 3 & 2 \end{pmatrix}^2 x = \begin{pmatrix} 1 & 2 & 3 & 4 \\ 1 & 3 & 4 & 2 \end{pmatrix}$$

14. Check whether the following statements are true or false with justifications

(i) If τ is an odd cycle of a permutation, then τ^2 is also a cycle.
(ii) For $S = \{1, 2, 3, 4, \ldots, n\}$ any permutation $\sigma \in S_n$ can be written as a finite product of the following permutation.

15. Let $S = \{1, 2, 3, 4, 5\}$. Let $G = \{(1)(2)(3)(4)(5), (3\ 4\ 1)(2)(5), (2\ 3\ 4\ 5)(1)\}$. Find

(i) the orbit of 1, 3, 5
(ii) the fix of all the permutations in G
(iii) the stabilizer of every element of S.
(iv) verify Burnside Lemma.

More on Counting

11

11.1 Catalan Numbers

Catalan numbers are following sequence of numbers F_n that satisfy the recursive formula

$$\begin{aligned} F_0 &= 1 \quad F_1 = 1 \\ F_n &= F_0 F_{n-1} + F_1 F_{n-2} + \cdots + F_{n-1} F_0 \\ &= \sum_{i=1}^{n} F_{i-1} F_{n-i}, n \geq 1 \end{aligned}$$

The first few numbers are 1, 1, 2, 5, 14, 42, 132, 429, 1430, ...

Catalan numbers come out as answer to several combinatorial problems. We list some of them here.

Remark 11.1 We use for Catalan numbers notation F_n instead of C_n to avoid confusions in notations

Problem 1: Well formed sequences of parentheses

We define a string first over {(,)}. The set {(,)} has two elements. One is the left parenthesis '('and the other is the right parenthesis ')'. One can form strings by placing the parentheses one by the side of the other. For example (( )),)) ((are strings over {(,)}. Out of these two, only one is well formed where the other is not. We now have the following recursive definition of Well Formed Parentheses (WFP).

R. Rama, *Topics in Combinatorics and Graph Theory*,
https://doi.org/10.1007/978-3-031-74252-1_11

Definition 11.2 A sequence or a string of parentheses is WFP if it can be formed by applying the following sequence of rules.

1. The empty set or sequence is well formed.
2. If $'F'$ is a WFP, then (F) is a WFP.
3. If F and F' are WFP, then the concatenation FF' is WFP

Notation: The length of a WFP $'\alpha'$ is the number of parentheses occurring in it and it is denoted by $|\alpha|$.

Example 11.3 Find 2 WFPs of length 6 such that one is well formed and the other is not.

Solution:
$\alpha = (())()$ is WFP.
$\beta =)())()$ is not a WFP.
α satisfies the recursive rules and above, whereas β does not.

Remark 11.4 1. Any WFP will be of even length as there must be equal number of left and right parentheses.
2. If we replace every left parentheses by 0 and right parentheses by 1, what we obtain is a string over $\{0, 1\}$ that have equal number of 0's and 1's

We will first solve the recurrence relation satisfied by Catalan numbers using generating functions technique.

Example 11.5 Solve the recurrence relation given by

$$F_n = \sum_{i=1}^{n} F_{i-1}F_{n-i}, \quad F_0 = 1, n \geq 1 \tag{11.1}$$

Solution: We solve the recurrence relation by generating function method.

Let

$$f(x) = \sum_{n=0}^{\infty} F_n x^n \text{ with } F_0 = 1$$

be the generating function.

Multiplying the recurrence relation (11.1) throughout by x^n and taking the sum from 1 to ∞, we get

$$\begin{aligned}\sum_{n=1}^{\infty} F_n x^n &= \sum_{n=1}^{\infty} x^n \sum_{i=1}^{n} F_{i-1}F_{n-i} \\ &= \sum_{i\geq 0}\sum_{n\geq i-1} F_i F_{n-i+1} x^n \\ f(x) - F_0 &= x f^2(x) \\ &\quad \text{Assuming } F_{-1} = 0, \\ \Rightarrow x f^2(x) - f(x) + 1 &= 0, \quad \text{as } F_0 = 1\end{aligned}$$

We have to solve for $f(x)$.

$$f(x) = \frac{1+\sqrt{1-4x}}{2x}, \frac{1-\sqrt{1-4x}}{2x}$$

which of the two $f(x)$ we take to find F_n. The function $f(x)$ has a constant term which is 1. As $x \to 0$, $f(x) = \frac{1-\sqrt{1-4x}}{2x} \to \infty$ and we consider only

$$f(x) = \frac{1+\sqrt{1-4x}}{2x}$$

whose expansion will give a constant term.

Now expand $(1-4x)^{\frac{1}{2}}$ using binomial theorem, we get

$$(1-4x)^{\frac{1}{2}} = -2\sum_{n=0}^{\infty} \frac{C(2n-2, n-1)}{n} x^n$$

Hence $F_n = \frac{C(2n,n)}{n+1}$, $n \geq 0$.

Now we come back to our WFP problem.

Theorem 11.6 *The number of WFP of length* $2n$ *is the Catalan number* F_n

Proof We write down a direct method of reaching the recurrence relation satisfied by F_n as follows.

By recursive definition, for

	WFP	Number of ways
$n = 1$	()	1
$n = 2$	(()), ()()	2
$n = 3$	((())), (())(), ()()()	5
	()(()), (()())	
$n = 4$	(((()))), ...	14 (check)

Let us write down $\overline{F_0} = 1$, $\overline{F_1} = 1$, $\overline{F_2} = 2$, $\overline{F_3} = 5$, $\overline{F_4} = 14, \ldots, \overline{F_n}$ when $2n$ parentheses denotes the number of WFP with n-left parentheses and n-right parentheses.

Let us assume that we already have all WFP's of length $2(i-1)$ and their count is $\overline{F_{i-1}}$.

Any WFP will always begin with $'('$. As we have WFP, somewhere in the expression of WFP there must be a matching $')'$ for the left (. Hence the WFP is now of the form

$$(\overline{E_1})\overline{E_2}$$

where $\overline{E_1}$ and $\overline{E_2}$ are WFPs. Both $\overline{E_1}$ and $\overline{E_2}$ can now have $(n-1)$ matching parentheses. If $\overline{E_1}$ contains t-pairs, then $\overline{E_2}$ contains $(n-t-1)$ pairs. Here either $\overline{E_1}$ or $\overline{E_2}$ may be nil. So now the counting will begin as, $\overline{E_1}$ may contain 0-pair, $\overline{E_2}$ contains $(n-1)$ pairs $\overline{E_1}$ may contain 1-pair, $\overline{E_2}$ may contains $(n-2)$ pairs and so on. Summing up, we get the total number of n-balanced WFPs of length $2n$. That is,

$$\begin{aligned}
\overline{E_1} &= \overline{E_0 E_0} \\
\overline{E_2} &= \overline{E_1 E_0} + \overline{E_0 E_1} \\
\overline{E_3} &= \overline{E_2 E_0} + \overline{E_1 E_1} + \overline{E_0 E_1} \\
&\vdots \\
\overline{E_n} &= \overline{E_{n-1} E_0} + \overline{E_{n-2} E_1} + \cdots + \overline{E_0 E_{n-1}}
\end{aligned}$$

which is nothing but the same as the recurrence satisfied by the Catalan numbers. Hence,

$$\overline{E_n} = \frac{c(2n, n)}{n+1}$$

There are several counting problems which sum up to Catalan numbers. Some of the popular ones are polygon triangulation problem, non-crossing handshakes counting, tree counting problem, diagonal avoidance paths. Later in the graph theory chapters, we will discuss tree counting problem. □

11.2 Integer Lattice Points

Integer lattice is a set of points in the xy-coordinate plane where the co-ordinates of such points are always integers. For any two points (x_1, y_1), (x_2, y_2) with $x_1 \leq x_2$ and $y_1 \leq y_2$, a lattice path from (x_1, y_1) to (x_2, y_2) will consist of horizontal, vertical steps. One can use the notation 'H' to indicate horizontal step 'V' to denote the vertical step. For example, we can consider the plotting of lattice paths as below (Fig. 11.1).

Example 11.7 So the path from (2, 2) to (4, 4) is described by $VHVH$.

One always considers the shortest of such paths. No matter how one uses the 'H' steps and 'V' steps, the total number of such steps will be always '4' for all paths from (2, 2) to (4, 4).

In general a rectangular lattice path from (x_1, x_2) to (y_1, y_2) will always use $(y_1 - x_1)$ 'H' steps and $(y_2 - x_2)$ 'V' steps. For each such path, the sequence is unique. Once the 'H' steps are known, the option for 'V' steps is frozen. Hence out of $(y_1 - x_1) + (y_2 - x_2)$ steps, choose $(y_1 - x_1)$ 'H' steps in

$$C(y_1 - x_1 + y_2 - x_2, y_1 - x_1)$$

ways. Hence we have the following result.

Proposition 11.8 *The number of rectangular lattice paths from* (x_1, x_2) *to* (y_1, y_2) *is equal to the binomial coefficient*

$$C(y_1 - x_1 + y_2 - x_2, y_1 - x_1)$$

Remark 11.9 Each path is uniquely determined by a string over H and V. That is the string corresponding to the path in Example 11.7 will be '$VHVH$'.

Fig. 11.1 Lattice paths

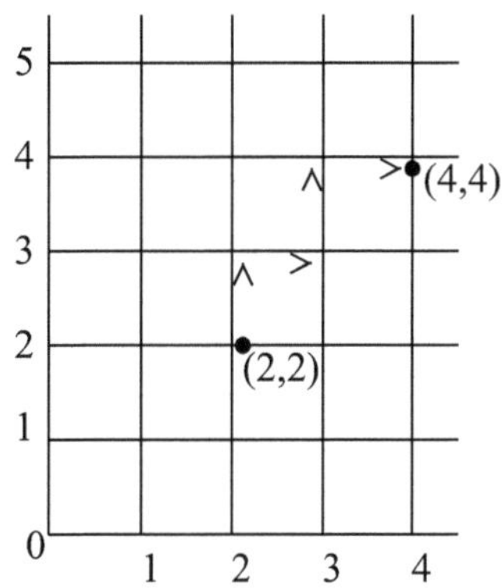

Suppose 'H' corresponds to '1' and 'V' corresponds to '0', then each path will uniquely correspond to a binary string of length equal to the length of the path.

Example 11.10 Suppose one starts at (0, 0) and end at (6, 6) using only H and V steps. Total number of such paths will be $C(12, 6)$. Suppose there is a restriction that one can move from (x, y) to $(x + 1, y)$ and (x, y) to $(x, y + 1)$ without crossing the line $y = x$ at every step. Then the total number of such restricted paths will not be $C(12, 6)$. One such path is marked in the following diagram (Fig. 11.2).

The path is $HVHVHVHVHVHV$.

The path '$HVHVVVHHVHVH$' is a bad path where this path crosses the line $y = x$.

Consider the bad path in Fig. 11.3

For any good path the first step cannot be a vertical one. Hence it should be a 'H' step. Also any prefix of a good path will contain more 'H' steps than 'V' steps and they can be equal. For example in the bad path above '$HVHVV$' is the substring that contains more

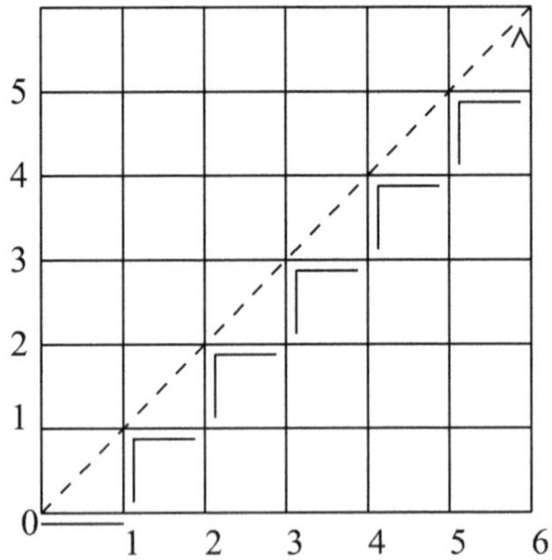

Fig. 11.2 One lattice path

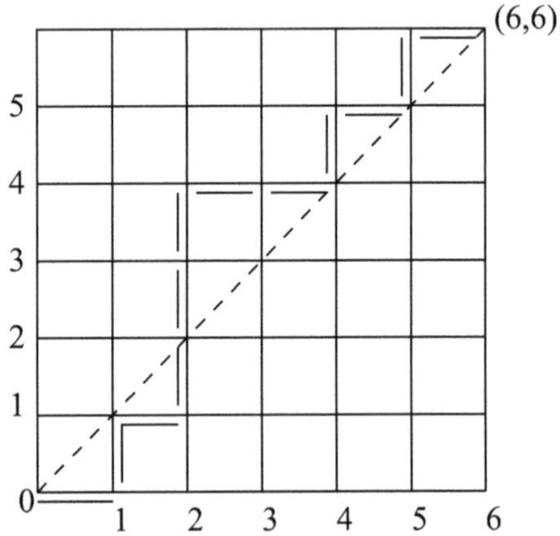

Fig. 11.3 Bad path

'V' steps than 'H' steps. So this tells the position where the path crossed the line $y = x$. Mark the whole bad path as $w_1 w_2$ where 'w_1' is a prefix indicating that crossing of $y = x$ line for the first time. Now construct a new path $w_1 \overline{w_2}$ where $\overline{w_2}$ is obtained from w_2 by swapping H with V and 'V' with H. Clearly this path will contain one H less than the previous count. Hence there is a one to one correspondence between bad paths $w_1 w_2$ with the newly constructed $w_1 \overline{w_2}$ paths. All such $w_1 \overline{w_2}$ paths will contain one 'V' more than that of $w_1 w_2$. Hence the total number of paths that do not cross the line $y = x$ will be

$$C(2n, n) - C(2n, n-1)$$

where the paths are constructed to move from (0, 0) to (n, n).

Definition 11.11 Catalan number C_n counts the number of lattice paths using only H and V steps that do not cross $y = x$.

Example 11.12 Find C_4.

Solution:

$$\begin{aligned} C_4 &= C(4, 2) - C(4, 1) \\ &= \frac{4!}{4} - \frac{4!}{3!} \\ &= 2 \end{aligned}$$

11.3 Eulerian Numbers

We know S_n is the set of permutations on the set $\{1, 2, \ldots, n\}$. We can denote the values in any permutation $\tau \in S_n$ by τ_i or $\tau(i)$, $1 \le i \le n$. As a single line notation we can express τ as product of cycles. For example if $\tau = \begin{pmatrix} 1 & 2 & 3 & 4 & 5 \\ 3 & 4 & 2 & 5 & 1 \end{pmatrix}$, we can write τ as $\begin{pmatrix} 1 & 3 & 2 & 4 & 5 \end{pmatrix}$ as a single cycle. Here $\tau(1) = 3$, $\tau(2) = 4$, $\tau(3) = 2$, $\tau(4) = 5$, $\tau(5) = 1$.

Definition 11.13 Let $\tau \in S_n$. We can define ascent and descent in τ as below.

(i) A descent of τ is a number $i \in \{1, 2, \ldots, n\}$ such that $\tau(i) > \tau(i+1)$. The set of all descents of τ is usually denoted as $des(\tau)$.
(ii) An ascent of τ is a number $i \in \{1, 2, \ldots, n\}$ such that $\tau(i) < \tau(i+1)$. The set of all ascents of τ is usually denoted as $asc(\tau)$.

Example 11.14 Consider S_3. The members of S_3 are

$$\begin{pmatrix}1\ 2\ 3\end{pmatrix},\begin{pmatrix}1\end{pmatrix}\begin{pmatrix}2\ 3\end{pmatrix},\begin{pmatrix}1\ 2\end{pmatrix}\begin{pmatrix}3\end{pmatrix},\begin{pmatrix}1\end{pmatrix}\begin{pmatrix}2\end{pmatrix}\begin{pmatrix}3\end{pmatrix},\\ \begin{pmatrix}1\ 3\ 2\end{pmatrix},\begin{pmatrix}1\ 3\end{pmatrix}\begin{pmatrix}2\end{pmatrix}$$

For $\tau = \begin{pmatrix}1\end{pmatrix}\begin{pmatrix}2\ 3\end{pmatrix}$, $|asc(\tau)| = 1$, $|dec(\tau)| = 1$
For $\tau = \begin{pmatrix}1\end{pmatrix}\begin{pmatrix}2\end{pmatrix}\begin{pmatrix}3\end{pmatrix}$, $|asc(\tau)| = 2$, $|dec(\tau)| = 0$
For $\ell = \begin{pmatrix}1\ 3\end{pmatrix}\begin{pmatrix}2\end{pmatrix}$, $|asc(\ell)| = 0$, $|dec(\ell)| = 2$

Definition 11.15 A number

$$E(n,k) = \|\{\tau \in S_n \,\big|\, |dec(\tau)| = k\}\|$$

is called as an Eulerian number. That is number of permutations in S_n with 'k' descents is the Eulerian number $E(n,k)$.

Example 11.16 Consider S_3. We can find the Eulerian numbers of the permutations of S_3 as in the following table.

	$\lvert dec(\tau)\rvert = 0$	$\lvert dec(\tau)\rvert = 1$	$\lvert dec(\tau)\rvert = 2$
$\tau:$	$\begin{pmatrix}1\end{pmatrix}\begin{pmatrix}2\end{pmatrix}\begin{pmatrix}3\end{pmatrix}$	$\begin{pmatrix}1\end{pmatrix}\begin{pmatrix}2\ 3\end{pmatrix}$	$\begin{pmatrix}1\ 3\end{pmatrix}\begin{pmatrix}2\end{pmatrix}$
		$\begin{pmatrix}1\ 2\end{pmatrix}\begin{pmatrix}3\end{pmatrix}$	
		$\begin{pmatrix}2\ 3\ 1\end{pmatrix}$	
		$\begin{pmatrix}3\ 1\ 2\end{pmatrix}$	

Hence $E(3,0) = E(3,2) = 1$
$E(3,1) = 4$.

Remark 11.17 One can see that 'dec' is a function from the set S_n to set of integers $\{1, 2, \ldots, n-1\}$. Also number of ascents of a permutation τ in S_n will be equal to the number $(n-1) - |dec(\tau)|$. Hence $E(n,k) = E(n, n-k-1)$.

We now prove a linear recurrence which resembles Pascal-identity of binomial coefficients.

Theorem 11.18

$$E(n,k) = (n-k)E(n-1,k-1) + (k+1)E(n-1,k)$$

for any k and $n \geq 1$.

Proof If $\tau \in S_n$ with 'k' descents, then removing 'n' from τ, results in a permutation $\overline{\tau} \in S_{n-1}$. This $\overline{\tau}$ will have either 'k' descents or $(k-1)$ descents.

Also if $\overline{\tau}$ is a permutation in S_{n-1} with 'k' or $(k-1)$ descents, one can insert 'n' into this $\overline{\tau}$ to form a permutation τ in S_n with k descents.

Assuming that there are $(n-1)$ students of different heights in a class. Also suppose that these students are linearly arranged so that there are $(k-1)$ descents in this arrangement. Now insert the tallest student into this arrangement so that the number of descents is either 'k' or $(k-1)$. This can be done in $(n-1-(k-1)) = (n-k)$ ways.

Identically if $\overline{\tau}$ is a permutation in S_{n-1} with 'k' descents, then forming a permutation τ in S_n with 'k' descents can be done in $(k+1)$ ways. Hence the recurrence. □

Example 11.19

$$E(4, 1) = 3E(3, 0) + 2E(3, 1)$$

In S_3 there are 4 permutations with '1' descent. They are

(i) $(1)(2\ 3), (1\ 2)(3), (2\ 3\ 1), (1\ 3\ 2)$
In $\overline{\tau} = (1)(2\ 3)$ with '1' descent, the permutations $(1)(2\ 3\ 4), (1)(3\ 2)(4)$ are in S_4 with '1' descent.

(ii) For $\overline{\tau} = (1\ 2)(3)$ with '1' descent, the permutations in S_4 are $(2\ 4\ 1\ 3)$ and $(1\ 2)(3)(4)$ with 1 descent.

(iii) For $\overline{\tau} = (2\ 3\ 1)$ with '1' descent we have the permutations $(2\ 3\ 4\ 1)$ and $(1\ 2\ 3)(4)$ are in S_4 with 1 descent.

(iv) For $\overline{\tau} = (3\ 1\ 2)$ with '1' descent in S_3, we have $(1\ 3)(2\ 4)$ and $(1\ 3\ 2)(4)$ in S_4 with 1 descent.

(v) For $\overline{\tau} = (1)(2)(3)$ with '0' descent in S_3, we have $(4\ 1\ 2\ 3)$, $(1)(2\ 4\ 3)$, $(1)(2)(3\ 4)$ are the permutations in S_4 with 1 descent.

This example illustrates the direct method of constructing permutations in S_4 with '1' descent from permutations in S_3 that have either '0' or '1' descent. This gives the recurrence procedure given in Theorem 11.18.

The following table of Eulerian numbers can be shown to be true using the above recurrence relation.

$n\backslash k$	0	1	2	3	4	5
1	1					
2	1	1				
3	1	4	1			
4	1	11	11	1		
5	1	26	66	26	1	
6	1	57	302	302	57	1

Fig. 11.4 Recurrence relation—weighted graph

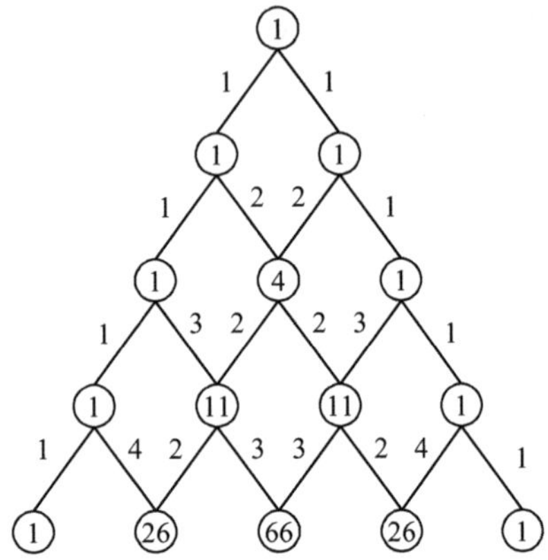

Clearly

$$E(4, 2) = 2E(3, 1) + 3E(3, 2).$$

Here one can think of 2 and 3 as weights of edges. Hence we can draw the following graph with weighted edges to illustrate the recurrence relation (Fig. 11.4).

One can associate polynomials with Eulerian numbers as seen below

$$\begin{aligned}
G_1(x) &= 1\\
G_2(x) &= 1 + x\\
G_3(x) &= 1 + 4x + x^2\\
G_4(x) &= 1 + 11x + 11x^2 + x^3\\
G_5(x) &= 1 + 26x + 66x^2 + 26x^3 + x^4
\end{aligned}$$

Assume $G_0(x) = 1$.

For some fixed value 'n'. We can have the generating function for the Eulerian numbers $E(n, k)$ as

$$\begin{aligned}
G_n(x) &= \sum_{\tau \in S_n} x^{|dec(\tau)|}\\
&= \sum_{t=0}^{n-1} E(n, t)x^t \qquad (11.2)
\end{aligned}$$

$$G'_n(x) = \sum_{t=1}^{n-1} tE(n, t)x^{t-1} \qquad (11.3)$$

From (11.2) we have,

$$
\begin{aligned}
(1+nx)G_n(x) &= (1+nx)\sum_{t=0}^{n-1} E(n,t)x^t \\
&= \sum_{t=0}^{n-1} E(n,t)x^t + \sum_{t=1}^{n} nE(n,t-1)x^t
\end{aligned} \tag{11.4}
$$

From (11.3) we have,

$$
\begin{aligned}
(x-x^2)G'_n(x) &= (x-x^2)\sum_{t=1}^{n-1} tE(n,t)x^{t-1} \\
&= \sum_{t=1}^{n-1} tE(n,t)x^t - \sum_{t=1}^{n-1} tE(n,t)x^{t+1} \\
&= \sum_{t=1}^{n-1} tE(n,t)x^t - \sum_{t=2}^{n} (t-1)E(n,t-1)x^t
\end{aligned} \tag{11.5}
$$

So,

$$
\begin{aligned}
&(1+nx)G_n(x) + x(1-x)G'_n(x) \\
&= \sum_{t=0}^{n-1} E(n,t)x^t + \sum_{t=1}^{n} nE(n,t-1)x^t \\
&\quad + \sum_{t=1}^{n-1} tE(n,t)x^t - \sum_{t=2}^{n} (t-1)E(n,t-1)x^t
\end{aligned} \tag{11.6}
$$

Coefficient of x^t will be

$$(n+1-t)E(n,t-1) + (t+1)E(n,t) \tag{11.7}$$

By the recurrence in Theorem 11.18 we have (11.7) to be $E(n+1,t)$. We can write the following theorem which gives a linear recurrence relation for the generating function $G_n(x)$.

Theorem 11.20 *If $G_n(x)$ is the generating function for the Eulerian numbers $E(n,t)$, we have*

$$G_{n+1}(x) = (1+nx)G_n(x) + (x-x^2)G'_n(x).$$

11.4 Narayana Numbers

A permutation $\tau \in S_n$ is a function f from $\{1, 2, \ldots, n\}$ to $\{1, 2, \ldots, n\}$. A permutation of the form

$$\begin{pmatrix} 1 & 2 & 3 & \ldots & n \\ \tau_1 & \tau_2 & \tau_3 & \ldots & \tau_n \end{pmatrix} \in S_n$$

can be written simply as '$\tau_1\tau_2 \ldots \tau_n$' with the understanding that $f(i) = \tau_i, 1 \leq i \leq n$. For example the permutation 4 3 1 2 $\in S_4$ stands for $\begin{pmatrix} 1 & 2 & 3 & 4 \\ 4 & 3 & 1 & 2 \end{pmatrix} \in S_4$. A sub-permutation of a permutation is of the form $\tau_i\tau_j\tau_k$ for $i < j < k$, $1 \leq i, j, k \leq n$. Here i, j, k need not be consecutive. If $\tau_i < \tau_k < \tau_j$, then we say that the sub-permutation $\tau_i, \tau_j\tau_k$ if of 132-pattern. If $\tau_k < \tau_i < \tau_j$, then we call the sub-pattern $\tau_i < \tau_j < \tau_k$ as 231-pattern. For example $\tau = \big(6\ 4\ 5\ 2\ 3\ 1\big)$ has 231-pattern as '452' as its subpermutation. We can understand the pattern avoidance by looking for permutations that avoid certain patterns. For example the permutation $\big(3\ 2\ 1\ 5\ 4\big)$ will not have a subpattern of the type '231'. We can draw a grid to show how the permutation function can be plotted. Also, from this plotting, one can easily look for subpermutations of certain patterns being present.

For $f : \{1, 2, \ldots, 6\} \to \{1, 2, \ldots, 6\}$ with $\tau = 3\ 6\ 2\ 5\ 4\ 1$ we have the grid.

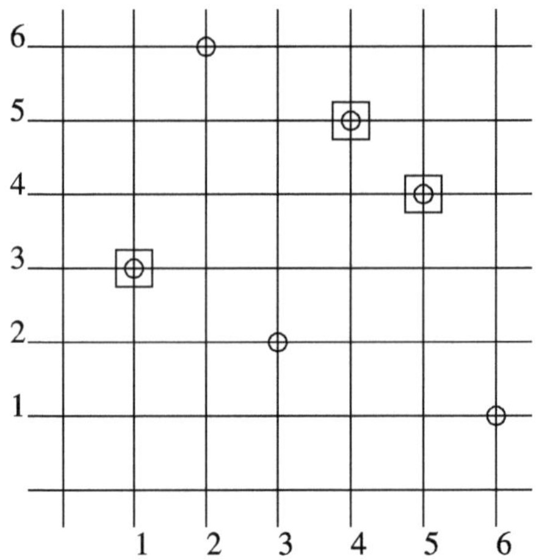

Definition 11.21 Let τ be a permutation in S_n. If $\tau = \tau_1\tau_2\tau_3 \ldots \tau_n$ with no 231-pattern, then τ is a 231-avoiding permutation.

Example 11.22 Some 231-avoiding permutations in S_4 are

$$1\,2\,3\,4, \quad 2\,1\,3\,4, \quad 3\,2\,1\,4, \quad 4\,3\,2\,1, \quad 1\,3\,2\,4.$$

Notation: Let $s_n(231)$ denote the number of permutation of S_n that avoid 231-pattern.

Theorem 11.23 *Let S_n be a set of permutations on $\{1, 2, \ldots, n\}$. Then $s_n(2\ 3\ 1)$ satisfy the following recurrence relation.*

$$s_n(2\ 3\ 1) = \sum_{i=0}^{n-1} s_i(2\ 3\ 1)s_{n-i-1}(2\ 3\ 1)$$

Proof Let τ be a permutation in S_n such that $\tau_i = n$, for $\tau = \tau_1\tau_2 \ldots \tau_n$. The values $\tau_1, \tau_2, \ldots, \tau_{i-1}$ must be smaller in value than those entries $\tau_{i+1}, \tau_{i+2}, \ldots, \tau_n$. If this is not true, then there exists an entry τ_ℓ with $\ell < i$ and τ_k with $k > i$ such that $\tau_k < \tau_\ell$. This means $\tau_k < \tau_\ell < \tau_i$ which is not possible. Hence $\tau_1\tau_2 \ldots \tau_{i-1}$ form a 231-avoiding permutation and $\tau_{i+1}\tau_{i+2} \ldots \tau_n$ also form a 231-avoiding permutation. Hence $s_n(2\ 3\ 1) = \sum_{i=0}^{n-1} s_i(2\ 3\ 1)s_{n-i-1}(2\ 3\ 1)$. □

Remark 11.24 As Catalan number C_n satisfy the recurrence relation $C_n = \sum_{i=0}^{n-1} C_i C_{n-i-1}$ and hence $s_n(2\ 3\ 1) = C_n$,

Definition 11.25 The numbers

$$E_{2\ 3\ 1}(n, k) = \{\tau \in S_n(2\ 3\ 1) \big| |dec(\tau)| = k\}$$

are called as Narayana numbers.

Example 11.26 The Narayana numbers for $n = 1, 2, 3, 4, 5$ with $k = 0, 1, 2, 3$ are given in the following table.

$n \backslash k$	1	2	3	4	5
1	1				
2	1	1			
3	1	3	1		
4	1	6	6	1	
5	1	10	20	10	1

The following is the formula that computes $E_{2\ 3\ 1}(n, k)$

$$E_{2\ 3\ 1}(n, k) = \frac{1}{k+1} C(n, k)C(n-1, k)$$

Note that the set of permutations that avoid 231-pattern can be an interpretation for describing Catalan numbers. We have seen an interpretation of Catalan numbers in terms of lattice paths earlier.

11.5 Schroder Numbers

In combinatorics the Schroder numbers S_n can be used to count several combinatorial objects. For example the nth Schroder number counts the number of distinct ways of subdividing a polygon of $(n+1)$ sides into smaller polygons by drawing diagonals. Also n^{th} Schroder number counts the number of lattice paths from (0, 0) to a point (n, n) such that each such path lie below the diagonal $y = x$. We will define more on Schroder numbers using the lattice paths problem.

For this we need the definitions and details of lattice paths which we recall from Sect. 12.1.

In Sect. 12.1 we had the discussion on rectangular lattice paths that use only 'H' and 'V' steps. In addition to these steps, suppose we consider diagonal steps also, we can name such paths as (H, V, D) lattice paths (Fig. 11.5).

Example 11.27 The string corresponding to the (H, V, D) lattice path is $HVDHVDD$.

Number of lattice paths from (0, 0) to (x_1, x_2) which uses r-H-steps, s-v steps and k-D-steps will be denoted by $L(r, s, k)$. Such paths will use k diagonal steps $x_1 - k$ H steps and $x_2 - k$ 'V' steps. Hence

$$L(r, s, k) = \frac{(x_1 + x_2 - k)!}{(x_1 - k)!(x_2 - k)!k!}$$

Definition 11.28 A large Schroder number S_n is the number of lattice paths from (0, 0) to (n, n) that can use V, H and D steps but none of the paths rise above the line $y = x$.

For $0 \leq n \leq 9$, first 10 large Schroder numbers are

$$1, 2, 6, 22, 90, 394, 1806, 8558, 41586, 206098.$$

Some Schroder paths for $n = 3$ are (Fig. 11.6)

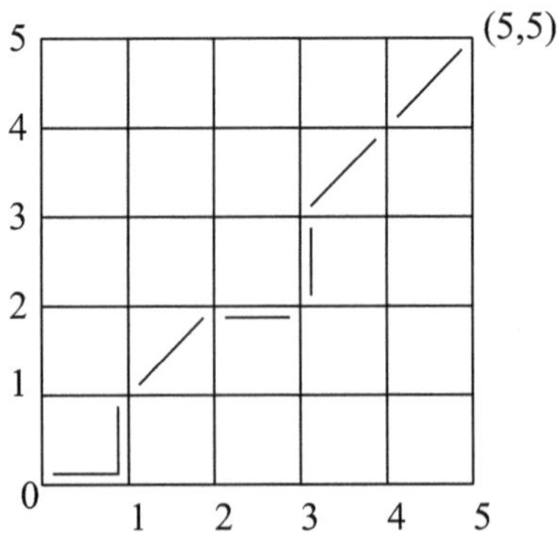

Fig. 11.5 Lattice path

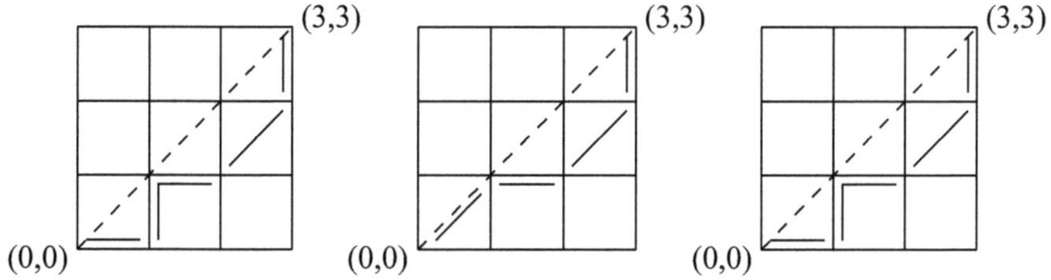

Fig. 11.6 Schroder paths

Fig. 11.7 Small Schroder path

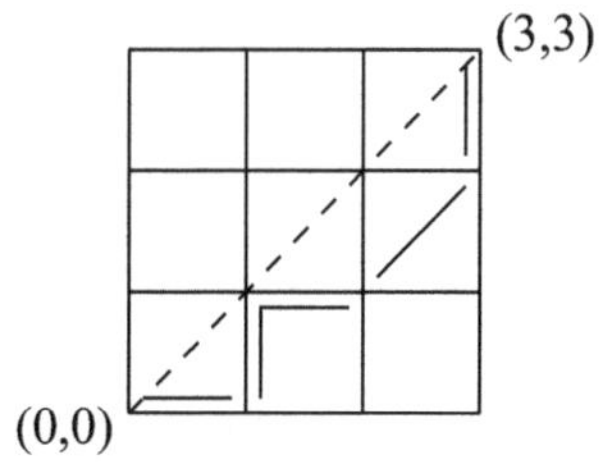

Let $L(n, n)$ denote the number all paths from (0, 0) to (n, n) that use H, V, D steps and that do not cross the line $y = x$.

$$S_n = L(n, n) = \sum_{t=0}^{n} \frac{1}{(n-t+1)} \frac{(2n-t)!}{t!((n-t)!)^2}$$

Here 'S_n' are the large Schroder numbers.

A small Schroder, lattice path from (0, 0) to (n, n) is a path that uses V, H, D steps, do not cross $y = x$ and do not use in 'D' steps to lie on the line $y = x$.

Example 11.29 Consider the following grid (Fig. 11.7).

$HVHDV$ is a small Schroder path.

Definition 11.30 A small Schroder number s_n is the number of lattice paths from (0, 0) to (n, n) that use V, H and D steps but none of the paths rise above the line $y = x$ and none of its 'D' step lie on the line $y = x$.

Some small Schroder numbers are

$$1, 1, 3, 11, 45, 197, 903.$$

Comparing the sequence of small Schroder numbers with large Schroder numbers, one can intutively see that $S_n = 2s_n, n \geq 1$, with $S_0 = 1$. But this need to be proved.

Consider the following Schroder path (Fig. 11.8).

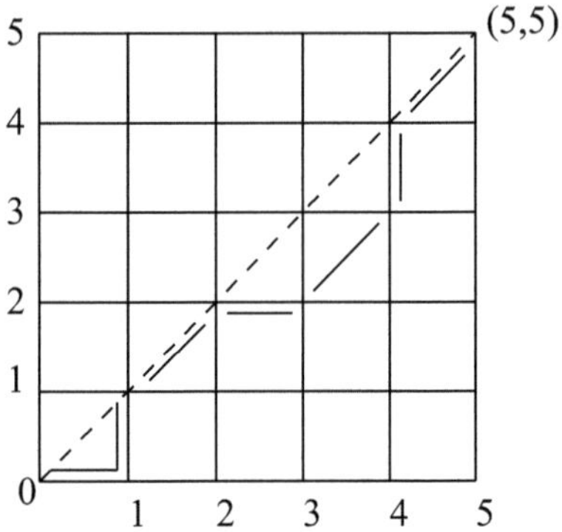

Fig. 11.8 Primitive Schroder path

We can define primitive Schroder path as follows:

1. Every 'D' step on $y = x$ is primitive.
2. Every Schroder path following a 'V' or 'H' step is primitive.

So one can write down the generating function of a primitive Schroder path as $X + Xg(X)$. So the generating function for all paths from (0, 0) to (n, n) will be

$$g(X) = \sum_{t=0}^{\infty}(X + Xg(X))^t = \frac{1}{1 - X - Xg(X)}$$

For small Schroder paths 'D' steps on the diagonal are not allowed. So the generation function will be

$$f(x) = \frac{1}{1 - Xg(X)}.$$

So $g(X) = \sum_{t=0}^{\infty} S_n X^n$ and

$$f(X) = \sum_{t=0}^{\infty} s_n X^n$$

Also

$$g(X) = \frac{1 - X - \sqrt{1 - 6X + X^2}}{2X}$$

$$f(X) = \frac{1 + X - \sqrt{1 - 6X + X^2}}{4X}$$

Exercises

1. If $2n$ students are sitting around a table, in how many ways can they all shake their hands so that there is no crossover of hands?
2. Consider a rectangular grid of size $2 \times m$. Find the number of ways of filling the holes with numbers $\{1, 2, \ldots, 2n\}$ such that the entries are monotonic both row wise and column wise.
3. Consider the recurrence relation on the numbers F_n.

$$F_{n+1} = \sum_{i=0}^{n} F_i F_{n-i}, \quad F_0 = 1$$

   (i) Find the values of F_2, F_3, F_4 and F_5.
   (ii) Prove that Catalan numbers do satisfy the above recurrence relation.
4. Show that there exists a lattice path from $(1, 1)$ to (n, n) that lie below the line $y = x + 1$ corresponding to a lattice path from $(0, 0)$ to (n, n) that completely lie below the diagonal $y = x$.
5. Suppose that a prefix of a binary string contains number of 0's less than number of 1's by 1. If every 0 is changed as 1 and every 1 is changed as 0 in this prefix, what will be ratio of 0's and 1's in the newly formed string.
6. Consider a convex polygon P_{n+1} on $(n + 1)$ sides. Let the sides be denoted as $B, a_1, a_2, \ldots, a_n$ where 'B' is the base of the polygon and the labelling of the sides start with a_1 and go up to the last side a_n in a clockwise manner. By disecting the interior of P_n, we mean that diagonals are drawn inside P_n in a non cross over manner. Note that the interior need not be completely disected. Show that bracketing of the string $a_1a_2 \ldots a_n$ has 1–1 correspondence with disections of P_n.
   Also show that the number of disections of P_n is equal to the small Schroder number s_n.
7. How many different rooted binary trees are there on n-vertices? (Hint: Use Catalan numbers)
8. Find the number of permutations in S_3 that avoid the patterns '132'.
9. Let m, n be natural numbers. Let $S = \{(x, y) | 0 \leq x \leq m, 0 \leq y \leq n\}$ be the set of lattice points in the grid $[0, m] \times [0, n]$. Find the number of lattice paths between $[0, m]$ and $[n, 0]$ that do not meet any point in S.
10. Draw all Dyck paths from $(0, 0)$ to $(6, 0)$ with no peaks (an upstep followed immediately at height 3).
11. Show that $S_m(132, 123) = 2^{m-1}$.

12. Sketch all the lattice paths from (0, 0) to (6, 6) that do not run above the line $y = x$. Use only steps H and V. i.e., $H = (1, 0)$, $V = (0, 1)$.
13. Let $\alpha = a_1 \cdot a_2 \cdot \cdots \cdot a_n$ where $a_i \in \mathbb{R}$. In order to compute α, one can multiply two factors at a time. We can use brackets to indicate such products. For example $t = a_1(a_2a_3)$ is the product of a_1 with a_2a_3. Find the number of ways we can bracket the expression $a_1a_2 \ldots a_n$ to compute α.

Discrete Probability 12

12.1 Activity and Outcomes

Consider the activity or sometimes we may call it as experiment of tossing a coin twice. The outcome of each tossing may turnout to be 'Head' or 'Tail'. We cannot predict the outcome with certainty. We will refer all such activities or experiments as 'random experiments'. Going back to our coin tossing activity, the outcome may be any of the 4 strings.

$$A = \{HH, HT, TH, TT\}$$

where a string HH mean we get in both the tosses 'Head' as the outcome. Similarity we can interpret the strings HT, TH, TT. We also call the set A to be the sample space.

If the number of tosses of the coin increases, the size of the sample space will also increase. The sample space gives all possible outcomes of the experiment. For example, instead of two times, if we toss the coin 4 times, the sample space

$$A = \{HHHH, THHH, HTHH, HHTH, HHHT, TTHH, \\ THTH, THHT, \ldots, TTTT\}.$$

The size of A will be $2^4 = 16$.

One requires to know how big the sample size will be and what subset of the sample space is required.

Example 12.1 Consider the activity of rolling two dice. Then the sample space will be

$$A = \{mn \mid m, n \in \{1, 2, 3, 4, 5, 6\}\}$$

R. Rama, *Topics in Combinatorics and Graph Theory*,
https://doi.org/10.1007/978-3-031-74252-1_12

where in the string 'mn', m denotes the outcome of rolling the first dice and n is the outcome of rolling the second dice. Only in some activities, we can write the sample space explicitly, as in the case of previous example, but look at the following example.

Example 12.2 In a game of bridge, each player is distributed with 13 cards from a deck of 52 cards. The distribution method is not known. In such cases it is not possible to explicitly list the sample space of all possible hands. We can only say how many members will be in the sample space. It is

$$C(52, 13) = \frac{52!}{13!39!}$$

which is a very large number.

Many times, we may not require the full sample space to be explicitly written and it is not possible also. Note that each element of the sample space is distinct. We may need to extract only a subset of the sample space.

Example 12.3 Consider the activity of tossing a coin 4 times whose sample space S is of size $2^4 = 16$. Let B be a subset of S such that each member of B is a string over $\{H, T\}$ of length 4 and has exactly 2 heads will be

$$B = \{HHTT, HTHT, HTTH, THTH, THHT, TTHH\}.$$

The subset 'B' is called an event.

Definition 12.4 Let $\mathcal{H}$ be an activity whose sample space is S. Any subset of S will be called an 'event'.

Example 12.5 Recall the sample space of rolling two dice. The following subset $B = \{22, 33, 44, 55\}$ will be an event.

Example 12.6 Suppose the activity is to distribute three distinguishable balls in three boxes then the sample space will be of size 27. The set

$$B = \{(a, b, c), (a, c, b), (b, a, c), (b, c, a), (c, a, b), (c, b, a)\}$$

is an event that each box has exactly one ball.

Example 12.7 Suppose three indistinguishable balls are to be distributed in 3 urns, then the sample space is simply of size 10. (The event $\{(a, a, a)\}$ has exactly one element. That is the activity is to distribute the balls such that no urn is empty.

Example 12.8 Suppose an elevator starts in the ground floor with 't' passengers. Suppose the building has m floors. The different ways of discharging the passengers is nothing but the number of ways t balls can be distributed in m urns. The sample space is accordingly computed. We do not distinguish each individual person here.

Example 12.9 The activity is to collect a sample of 1000 people who are diabetic for some purpose. Here we are interested only in the number of such persons even though each individual is a distinct person. So we have the sample space as tuple (D, ND) where D stands for diabetic and ND stands for non diabetic. A set of people who are female will be subset of $\{(D, ND)\}$.

The above examples are given with the purpose of understanding that probability theory will deal with real activities whose outcome is not predictable with certainty. Many call such activities as 'random' experiments. What maximum can be done is to find a measure or number that indicates the happening of the event when the activity is carried out large number of times. The sample space may be described in more than one way. The description will vary by varying the conceptions. For example while rolling a 'dice', initially we look for sample space $\{1, 2, 3, 4, 5, 6\}$. But if we include the state of the dice that stand on its edges, as 'e', then the sample space will be $\{1, 2, 3, 4, 5, 6, e\}$. It may so happen that the dice will roll off and lost so now the sample space will be $\{1, 2, 3, 4, 5, 6, e, b\}$.

From the example it is clear that we cannot give rules to the construction of the sample space. It is different for different problems and different situations.

We now have more information on events which we defined earlier to be subsets of sample space. An event is also a description of the activity. When we say an event 'A' has occurred, we mean that the outcome of the activity under consideration is a member of A. Since events are subsets of sample space, all subset operations become meaningful. That is event $A \cup B$ is said to occur when the outcome is either a member of A or a member of B. Similarly $A \cap B$, A^c can be defined in terms of occurrence of the outcome. An usual property of sets which will be used very often will be De-Morgan's law. An impossible event is denoted by the empty set $\emptyset$.

Also two events A and B that cannot occur simultaneously are said to be mutually exclusive events. When the activity is carried out, the outcome is either a member of A or a member of B, but not both. For example, the sample space of the activity of rolling a 'die' will contain either 2 or 4 as outcome. It cannot be both. The situation of defining equality of events also need to be done carefully in terms of outcomes. We can define first an event A being a subset of an event B and then we may proceed to define $A = B$.

Example 12.10 Let a game of tossing coins be played by two persons P_1 and P_2. The tossing is done alternatively. P_1 will receive Rs. 10/- for each tossing when head turns up. Find the sample space for P_1 for tossing the coins 5 times.

Solution: In all the five times, head may not turnup, head may turnup once head may turn up twice and so on. Hence the sample space will be

$$\{0, 10, 20, 30, 40, 50\}.$$

That is, P_1 may receive no money or Rs. 10, or Rs. 20, or Rs. 30 and so on.

12.2 The Definition of Probability Function

In the previous section we have seen what is known as 'sample space' of an experiment. We also saw several mathematical notions of such sets. Now in this section we are ready to describe the mathematical model of the random experiment. A chance or random outcome is having the property that the outcome under the given situation does not guarantee any information in the future outcome. However a random event must guarantee, in a long repeated activity of a randomly selected situations, approaches a stability through outcome.

Let A be a random event. We say that A satisfies a random property if there exist numbers between 0 and 1 which represent the frequency of possible outcomes that happen in a long repetition of the activity. This property is called as 'statistical stability'.

Example 12.11 Suppose a coin is tossed 500 times. The following table is called the frequency table and the associated number between 0 and 1.

Tosses	Number of Heads	Proposition (w.r.to 100)
1–100	54	.54
101–200	53	.53
201–300	51	.51
301–400	51	.51
401–500	52	.52

The statistical stability is the proportion which is near $\frac{1}{2}$.

Now we define what is known as probability function. The probability function is defined from the event sets to a number between 0 and 1. An event A is mapped to a number $p(A)$ which is a number, which can be got only by performing an activity or experiment repeatedly large number of times the members of A. Every time performing an experiment for a large number of times may not be practical. Moreover the probability function should be defined exhaustively for all events of the sample space. Hence the probability function has to be mathematically well defined for simpler events so that we can proceed to look for such functions for all situations. With this background now we formally give the definition of probability functions.

Definition 12.12 A probability function p is a function from the sample space S to a real number between 0 and 1 such that the following axioms are satisfied

(i) $p(A) \geq 0$ for every event A.
(ii) $p(S) = 1$ for the one event S.
(iii) $p(A \cup B) = p(A) + p(B)$, where A and B are events and $A \cap B = \emptyset$.

If n_A is the number of occurrences of the event A, then $p(A) = \frac{n_A}{n}$ where $n = |S|$. So axiom (i) must be true. Also $p(S) = \frac{n}{n} = 1$, so (ii) must be true. Finally if the number of occurrences of A and B be respectively, n_A and n_B, then $p(A \cup B) = p(A) + p(B) = \frac{(n_A+n_B)}{n}$. Hence axiom (iii) is satisfied as $A \cap B = \emptyset$.

So, in order to speak of the probability of an event, all that we need is

(a) A sample space of the activity on which events of which the probability can be defined.
(b) A and B be any two events in S.
The space S that satisfy axioms (i), (ii) and (iii) defined above will be referred to as probability space. The following results can be proved easily using the basic definitions and the definitions of set theory.

Theorem 12.13 *Let S be a probability space. Let A, B be any two events in S.*

(i) $p(\emptyset) = 0$.
(ii) $p(A^c) = 1 - p(A)$.
(iii) $p(A \cup B) = p(A) + p(B) - p(A \cap B)$.
(iv) $p(A \cap B^c) = p(A) - p(A \cap B)$.
(v) $p(A \cap B^c) = p(A) - p(B)$ *if* $A \subset B$.
(vi) $p(A) \leq p(B)$ *if* $A \subset B$.

Example 12.14 Two coins are tossed. Find the sample space. Find the probability of all the events of the sample space.

Solution: The sample space will be $S = \{HH, HT, TH, TT\}$. Then the events are $\emptyset$, $\{HH\}$, $\{HT\}$, $\{TH\}$, $\{TT\}$, $\{HH, HT\}$, $\{HH, TH\}$, $\{HH, TT\}$, $\{HH, HT, TH\}$, $\{HH, TH, TT\}$,{HT, TH, TT},{HT, TH, TT},{HT, $p(\emptyset) = 0$, $p(S) = 1$
$p(\{HH\}) = \frac{1}{4}(= p(\{HT\}), p(\{TH\}), p(\{TT\}))$
$p(\{HH, HT\}) = (p\{HH, TH\}) = p(\{HH, TT\}) = p(\{HT, TH\}) = p(\{HT, TT\}) = p(\{TH, TT\})$
$= p(\{HH\}) + p(\{HT\})$
$= \frac{1}{4} + \frac{1}{4} = \frac{1}{2}$
Similarly other probabilities can be written down.

Definition 12.15 (i) Two events A and B are said to be independent if
$p(A \cap B) = p(A).p(B)$
(ii) If events $A_1, A_2, \ldots, A_m$ are independent, then for all t, $2 \leq t \leq m$, we have

$$p(A_{j_1} \cap A_{j_2} \cap A_{j_3} \cap \cdots \cap A_{j_t}) = p(A_{j_1}).p(A_{j_2}) \ldots p(A_{j_t}),$$
$$1 \leq j_1 < j_2 < j_3 \cdots < j_t \leq m.$$

Remark 12.16 The definition of independent events of an infinite set of events can be defined similarly. That is if $\{A_1, A_2, A_3, \ldots\}$ is an infinite set of events, and if $\{A_1, A_2, \ldots, A_m\}$ are independent for every $m \geq 2$, then $\{A_1, A_2, A_3, \ldots\}$ is said to be independent.

12.3 Finite Probability Space

So far we have seen some basic concepts on sample spaces that lead us to define what is known as probability functions. The probability function need to satisfy the three axioms that are defined on events. We may denote the probability space S in terms of its members $a_1, a_2, \ldots, a_n$ where each $\{a_i\}$ is called an event with single member. In the tossing of a fair coin $S = \{H, T\}$ and $\{H\}$, $\{T\}$ are called singleton events. When two fair coins are tossed, $S = \{HH, HT, TH, TT\}$ and $\{HH\}, \{HT\}, \{TH\}, \{TT\}$ are called singleton events. So a singleton event is an event which contains only one outcome. For example 'HH' is an outcome and $\{HH\}$ is an event. If $A_1, A_2, \ldots, A_t$ are singleton events and $A = A_1 \cup A_2 \cup A_3 \cdots \cup A_t$. Then $p(A) = p(A_1) + p(A_2) + \cdots + p(A_t)$.

A sample space S is said to have equally likely outcomes if the singleton events of S have the same probabilities.

Useful notation

An n-tuple $(t_1, t_2, \ldots, t_n)$ is simply an array of n-symbols $t_1, t_2, \ldots, t_n$. t_1 is called the first element, t_2 is called the second element, and so on, t_n is called the nth element. This is called an ordered n-tuple as the ordering is important.

Example 12.17 (1, 1), (1, 2), (2, 1), (2, 2) are ordered 2-tuples.

Equality of two n-tuples: $(t_1, t_2, \ldots, t_n)$ and $(t'_1, t'_2, \ldots, t'_n)$ are said to be equal iff $t_i = t'_i$, $1 \leq i \leq n$.

Ball-Box Distribution

Suppose there is a box that contains m balls which are numbered from 1 to m. Suppose the activity is to draw a ball from the box, until a collection of n balls is noted.

There are two cases here. We say that the new bag is a sample.

Case (i): The sample is drawn with replacement. That is the sample of size 'n' is made where after each draw of a ball, the drawn ball information is recorded and the ball is returned to the box.

Case (ii): The sample is drawn without replacement. That is the sample of size 'n' is made where after each draw of a ball, the ball drawn is not put back in the box.

In each draw, more than one ball may be drawn.

Example 12.18 Let a box contain 3 distinct balls. If a sample of size 2 is made (i) with replacement (ii) without replacement. Find the sample space.

Solution: Let the 3 balls be numbered as '1, 2, 3'.

(i) With replacement case

The probability sample space will be ordered 2-tuple where the first element indicates the first draw, second element indicates the second draw. The space will be

$$\{(1,1),(1,2),(1,3),(2,1),(2,2),(2,3),(3,1),(3,2),(3,3)\}$$

(ii) Without replacement case

$$\{(1,2),(1,3),(2,1),(2,3),(3,1),(3,2)\}$$

Remark 12.19 (1) We have seen earlier that the number of ways of collecting n balls from a box containing m distinguishable balls is $m(m-1)(m-2)\ldots(m-n+1)$ if the draw is made without replacement, and it will be m^n if it is done with replacement.

(2) The number of events of a sample space S having n elements will be 2^n which includes the impossible event $\emptyset$.

(3) The number of ways in which a sample space having n elements can be partitioned into t ordered (subsets) events in such a way that the first event contains n_1 elements, the second event has n_2 elements and so on will be

$$C(n,n_1).C(n-n_1,n_2)C(n-n_1-n_2,n_3)\cdots C(n-n_1-n_2\cdots-n_{t-1},n_t)=\frac{n!}{n_1!n_2!\ldots n_t!}.$$

We just list the four possible situations arise in any ball-box distribution activity. We can see that we have enumerated them earlier in chapters (combinatorics).

(1) The balls are distinct and the boxes are also distinct. The number of ways of distributing t balls into m boxes where no box is empty will be the number of integer partitions of the integer 't' into 'm' parts where no part is zero.

$$\sum_{\substack{i_1+i_2+\cdots+i_m=t \\ i_j\geq 1}} C(t, i_1, i_2, \ldots, i_m)$$

(2) The balls are distinct and the boxes are identical. The number of ways of distributing t balls into m boxes where no box is empty will be $s(m, t)$, Stirling numbers of second type.

(3) The balls are identical and the boxes are distinct. The number of ways of distributing t balls into m boxes so that no box is empty will be $C(t-1, t-m)$.

(4) The balls are identical and the boxes are also identical. The number of ways of distributing t balls into m boxes where no box is empty will be the number of integer partitions of the integer 't' into 'm' parts where no part is zero.

We defined $p(A \cup B)$ when $A \cap B = \emptyset$. However when A and B are events, they may or may not contain outcomes which are common to both A and B. In combinatorial problems we address '$A \cap B$' to be the part which is counted twice. Hence we have the theorem.

Theorem 12.20 *For any two events A and B we have,*

$$p(A \cup B) = p(A) + p(B) - p(A \cap B)$$

Proof The event $A \cup B$ means that it is the union of outcomes of A and B. An outcome may occur either in A or in B or in both. Hence

$$p(A \cup B) = p(A) + p(B) - p(A \cap B).$$

If A and B are mutually exclusive, then

$$p(A \cup B) = p(A) + p(B).$$

□

Example 12.21 Two fair coins of different sizes are tossed. The probability sample space will be $\{(H, H), (H, T), (T, H), (T, T)\}$.

In the pair (a, b), a denotes the outcome of the smaller coin, b the outcome of the larger coin.

Let $A = \{$outcome such that the smaller coin is Head$\}$.
$B = \{$outcome such that the larger coin is head$\}$.

Then

$$\begin{aligned} A &= \{(H, H), (H, T)\} \\ B &= \{(H, H), (T, H)\} \\ A \cup B &= \{(H, H), (H, T), (T, H)\} \\ p(A \cup B) &= p(A) + p(B) - p(A \cap B) \\ &= \frac{1}{2} + \frac{1}{2} - \frac{1}{4} = \frac{3}{4} \end{aligned}$$

Boole's Inequality

Let $A_1, A_2, \ldots, A_m$ be m events, then

$$p(A_1 \cup A_2 \cup \ldots A_m) \leq p(A_1) + p(A_2) + \cdots + p(A_m).$$

Note 12.22 *If $A_1, A_2, \ldots, A_m$ are mutually exclusive events then*

$$p(A_1 \cup A_2 \cup \cdots \cup A_m) = p(A_1) + p(A_2) + \cdots + p(A_m).$$

Example 12.23 A two digit number is formed using the digits $\{1, 2, 3, 4, 5\}$, where repetitions of the digit is not allowed. Assume that all the outcomes are equally likely. Find the probability that

(i) the number formed is an odd number.
(ii) the number formed is an even number.

Solution: The probability space has 20 elements. The probability each outcome will be $\frac{1}{20}$.

(i) The event that the number will be odd will be numbers
$O = \{13, 15, 23, 25, 35, 45\}$.
$p(0) = \frac{6}{20} = \frac{8}{10}$.
(ii) $E = \{12, 14, 24, 32, 34, 42, 52, 54\}$.
$p(E) = \frac{8}{20} = \frac{2}{5}$.

Conditional Probability

The definition of probability given earlier is based on a direct situation with respect to probability space and events.

Example 12.24 Let a box contain 15 balls out of which 5 are red balls and 10 are white balls. Find the probability of selecting a red ball from the box?

In the above example the solution is approached directly. The number of possible ways a ball is selected is 15 and number of ways a red ball is selected is 5. Hence the probability of selecting a red ball will be $\frac{1}{3}$.

Suppose we have the following situation.

Example 12.25 Let a box contain 15 balls out of which 5 are red balls and 10 are white balls. Two balls are drawn from the box. Find the probability that the second ball drawn is white?

We need to understand the probability space in steps as the above activity has two steps. In the first step one ball is drawn. It may be either white or red. One more ball is drawn in the second step. This may also be white or red. But our event should include only those outcomes which will be white in the second draw. Since the activity is so called 'ball-box' activity, there will be two situations where we look for samples with replacement and samples without replacement. We will write down the solution to this example later.

Definition 12.26 Let S be a probability space. Let A and B be any two events in S. The conditional probability of the event B, given that the event A has occurred is the ratio $\frac{p(A \cap B)}{p(A)}$ and it is denoted by $p[B \mid A]$.

Notation: In the subject probability theory, writing $A \cap B$ as product AB is a standard notation. Here after we will use the product AB to denote $(A \cap B)$.

Example 12.27 Let a box contain 3 distinct balls. If a sample of size 2 is made (i) with replacement (ii) without replacement. Find the sample space.

Solution: Since it is a 'ball-box' activity we discuss the solution in two cases.

Case (i): Let the activity be done without replacement.

Clearly the number of elements in the sample space will be 15×14. Let A be the event that the first is red and the second ball is white. This can be done in 5×10 ways. Let B be the event that the first ball is white and the second ball is also white. This arrangement can be done in 5×4 ways. Hence the probability of second ball being white will be

$$\begin{aligned} p(A) + p(B) &= \frac{5 \times 10}{15 \times 14} + \frac{5 \times 4}{15 \times 14} \\ &= \frac{70}{15 \times 14} = \frac{1}{3}. \end{aligned}$$

The argument for Case (ii) where we consider the activity where a ball drawn is replaced can be argued combinatorially.

Remark 12.28 For such problems it will be easier to use n tuple notation where each of ith component indicates the ith draw where sum of the components will be $\leq n$, the total number of balls in the box. We are computing here any conditional probability as the two draws are independently done.

The case where the balls are replaced can be argued in similar lines.

Now we modify the Example 12.27 further as below.

Example 12.29 Let a box contain 15 balls out of which 5 balls are red in color and rest are white in color. Two balls are drawn from the box without replacement. Find the probability that both the balls drawn are white.

Solution: We understand the situation as before using 2-tuple notation. Let (b_1, b_2) denote the balls drawn where b_1 is the ball drawn in the first draw and b_2 is the ball drawn in the second. We know that the sample space S will be set of all 2-tuples (b_1', b_2') where b_1', b_2' are any two balls in the box. Hence the activity is done without replacement, the cardinality of S will be 15×14. Now the cardinality of (b_1, b_2) will be 10×9. Hence the answer will be $\frac{10 \times 9}{15 \times 14} = \frac{3}{7}$.

In terms of probability space and events, let A be the event where the first draw is white and B be the event where the second draw is white. It is clear that

$$\begin{aligned} p(A) &= \frac{10}{15} = p(B) \\ p(AB) &= \frac{3}{7} \neq p(A).p(B). \end{aligned}$$

In the case of activity with replacement it will be $p(A).p(B) = p(A.B)$. Why?

Example 12.30 In a town T, there are 100 scooters. There are three scooters in red color. There are 20 scooters having registration numbers being odd. Given the probability that the scooter is red in color and the registration number being odd is $\frac{1}{3}$, find the conditional probability that the registration number is odd, given that the scooter is red in color.

Solution: Let A be the event that the scooter is red in color. Hence

$$p(A) = \frac{3}{100} = 0.03$$

Let B be the event that the registration number is odd. Hence

$$p(B) = \frac{20}{100} = \frac{1}{5}$$

Given $p(AB) = \frac{1}{3} = 0.333$.
Hence by formula

$$p(B \mid A) = \frac{p(AB)}{p(A)} = \frac{0.333}{0.03} = 0.0449$$

Next we state a small useful result.

Lemma 12.31 *Let A and B be two events where $p(A) \neq 0$. Then A and B are independent if and only if $p(B \mid A) = p(B)$.*

Proof Let A and B be the two events which are independent. By A, B being independent, it implies $p(AB) = p(A).p(B)$. Hence

$$p(B \mid A) = \frac{p(AB)}{p(A)} = p(B).$$

Conversely if $p(B \mid A) = p(B)$.
Then

$$p(B \mid A) = \frac{p(AB)}{p(A)} = p(B)$$
$$\Rightarrow p(AB) = p(A).p(B)$$

Hence A and B are independent. □

Example 12.32 Let a box contain 10 balls of 5 different sizes. Each ball is either green or red. Find the probability of drawing a smallest ball and that it is colored red.

Solution:
Let the ball sizes be denoted by s_1, s_2, s_3, s_4, s_5, where size $s_i < s_{i+1}, 1 \leq i \leq 4$. The green and red color are denoted by g and r. The object '$s_1 g$' means the ball of size s_1 is green. The sample space will be

$$s_1 g, s_1 r, s_2 g, s_2 r, s_3 g, s_3 r, s_4 g, s_4 r, s_5 g, s_5 r$$

Let A be the event that the randomly drawn ball is in color red. Let B be the event that the ball is the smallest one of size 's_1'. Note that A and B are not independent.

$$\begin{aligned} p(A) &= \frac{5}{10} = \frac{1}{2} \\ p(B) &= \frac{2}{10} = \frac{1}{5} \\ p(AB) &= \frac{1}{2} \cdot \frac{1}{5} = \frac{1}{10} \\ p(B \mid A) &= \frac{\frac{1}{10}}{\frac{1}{2}} = \frac{0.1}{0.5} = \frac{1}{5}. \end{aligned}$$

Extension of Independent Events

So far we have seen two events being independent. Suppose we have three events A, B and C defined on a probability space. What is the meaning of conditional probability of the event A given that B and C have happened? We can very well take the notation as $p[A \mid BC]$. Then the definition of $p[A \mid B, C]$ will be

$$p[A \mid B, C] = p[A \mid BC]$$

Suppose the event A is independent of B and C. For defining this we need to extend the definition of independent events to A, B, C.

Definition 12.33 Three events A, B, C are independent if

$$\begin{aligned} p(AB) &= p(A)p(B) \\ p(AC) &= p(A)p(C) \\ p(BC) &= p(B)p(C) \\ p(ABC) &= p(A)p(B)p(C). \end{aligned}$$

If all the above used probabilites are non zero, then we get the following.

$$\begin{aligned} p(A \mid B, C) &= p(A \mid B) = p(A \mid C) = p(A) \\ p(B \mid A, C) &= p(B \mid A) = p(B \mid C) = p(B) \\ p(C \mid A, B) &= p(C \mid A) = p(C \mid B) = p(C). \end{aligned}$$

Remark 12.34 The conditional probability can be extended to n events $A_1, A_2, \ldots, A_n$.

Total Probability Theorem

Before proceeding further, let us understand more about independent events. Suppose we are given two sets of events say $\mathcal{A}$ and $\mathcal{B}$. The sets of events $\mathcal{A}$ and $\mathcal{B}$ are said to be independent if for any $A \in \mathcal{A}$ and $B \in \mathcal{B}$, A and B are independent, where A and B are events. We call $\mathcal{A}$ and $\mathcal{B}$ to be mutually independent sets of events.

Next we define what are known as 'trials'. Let S be a probability space. Then we know that any member of S may have a tuple description. For example for tossing two fair coins, elements of S will be of the form (ζ_1, ζ_2) where ζ_1 is the outcome on the first coin and ζ_2 is the outcome in the second coin. Then we say that S is madeup of 2 trials. Similarly if S has a n-tuple for n trials. Further suppose in the coin tossing set up mentioned above, one looks for the event A where $\zeta_1 = H$, then we say that A is dependent on the first coin output which must be a 'head'. We see that event 'A' depends on a trial.

Next we give the definition of independent Bernoulli trials. There are problems in probability theory which involve a random activity or experiment which will have exactly two possible outcomes. For example while tossing a coin, there can be only two possible outcomes. The appearance of 'head' or the appearance of a tail will be the possible outcomes. If two coins are tossed, then the outcome will be a tuple (ζ_1, ζ_2) or $\zeta_1\zeta_2$ where each trial ζ_i, $i = 1, 2$ will have only two categories of outcome. Such activities or experiments will be called 'Bernoulli trial' experiments. The general definition can be easily understood as problems in probability theory that have independent trials whose outcomes are only two. Usually such outcomes are called as "success" and 'failure'. For example while tossing a coin, the appearance of 'head' can be categorized as 'success' and the appearance of 'tail' as 'failure'. If 'p' is the probability of the outcome 'success', then $(1 - p)$ will be the probability of the outcome 'failure'.

Example 12.35 A fair coin is tossed 5 times whose probability of the outcome head is 'p' where $0 < p \leq 1$. Find the probability of an event A that the coin shows up 'head' in the first three trials and shows up 'tail' in the last two trials.

Solution: Let S be the sample space whose members are $(\zeta_1, \zeta_2, \zeta_3, \zeta_4, \zeta_5)$ where ζ_i is the outcome of the ith trial which may be 'head (H)' or 'tail (T)'. Hence $|S| = 2^5$. Here $A = \{(H, H, H, T, T)\}$ which is a singleton. Hence the probability will be $p^3(1 - p)^2$.

Binomial Coefficient Law

Consider a problem event that has n independent repeated Bernoulli trials. Let the probability of a 'success' trial be p and the probability of a 'failure' trial be q. Then the probability of such an event will be $C(n, k)p^k q^{n-k}$ where the event has k successes and $(n - k)$ failures.

Clearly

$$p + q = 1$$
$$(p + q)^n = \sum_{k=0}^{n} C(n, k) p^k q^{n-k} = 1.$$

Example 12.36 (*"Odd man out" Problem*) A game is played by a group of people. The game is to choose a ball from a bag that contains two identical balls of different colors say red and green. Each person has to close their eyes while playing. If the outcome is that there is a person 'A' who has picked a different color ball and all others have chosen the same color, then 'A' is the 'odd man out'. Find the probability that any game will have an 'odd man out' situation.

Solution: Let there be n people in the group. Each person picks a ball independently by closing their eyes. If $(\zeta_1, \zeta_2, \ldots, \zeta_n)$ is a member of the sample space, then each $\zeta_i = R$ or G, (Red or Green). The ζ_i's are independent Bernoulli trials. Let the probability of $\zeta_i = R$ be p. Then the probability of $\zeta_i = G$ is $1 - p = q$ (say). Then we need to look for members like $(R, R, R, \ldots, R, G)$, $(R, R, R, \ldots, R, G, R, \ldots, R)$ etc. where exactly one of the ζ_i is G and the rest are 'R' or the other way. Hence the associated probability will be $C(n, 1)p^{n-1}q$ or $C(n, 1)pq^{n-1}$. That is in the n-tuple only one of the trial 'ζ_i' is fixed with different color from the remaining $(n - 1)$ of them. Hence the total probability will be $C(n, 1)p^{n-1}q + C(n - 1)pq^{n-1}$.

Remark 12.37 One define in a probability space whose members are n-tuples what is known as 'dependent' trial. Simply dependent trials are trials which are dependent.

Example 12.38 Let a box contain 10 balls out of which 4 are green balls, the rest of the balls are red in color. Let a collection of 3 balls be made. Find the probability that all the balls are green. Assume the drawing is done without replacement.

Solution: We have taken this example to understand what are known as 'dependent' trials.

Let A_i be the event that ball drawn at the ith instance is green. There are three events A_1, A_2, A_3 under consideration. That is A_1 is the event that the first ball drawn is green and its probability is $p(A_1) = \frac{4}{10} = \frac{2}{5}$.

Then A_2 is the event that the second ball drawn is green. But we want the first ball is green and the second will be green. That

$$p(A_2 \mid A_1) = \frac{p(A_1A_2)}{p(A_1)} = \frac{p(A_1)p(A_2)}{p(A_1)}$$
$$= \frac{3}{9}.$$

Similar $p(A_3 \mid A_1, A_2) = \frac{1}{4}$.

Example 12.39 A coin is tossed three times. What is the probability of an event that

(i) all the three outcomes are tails
(ii) out of the three outcomes, there are exactly two tails
(iii) given the fact that the outcome has atleast one tail, the outcome that there are atleast two tails?

Solution: The tosses of the coin are independent activities. The outcomes will be the tuple $(\zeta_1, \zeta_2, \zeta_3)$ where each trial ζ_i will be either 'head' (H) or 'tail' (T).

(i) The event containing $\{(T, T, T)\}$ has the probability

$$p(T, T, T) = p(T).p(T).p(T), \quad [\because p(H) = \frac{1}{2}] = p(T)$$
$$= \frac{1}{8}$$

(ii) The event for this situation will contain three outcomes $B = \{(T, T, H), (H, T, T), (T, H, T)\}$.
Hence

$$p(B) = p((T, H, T)) + p((H, T, T)) + p((T, T, H)) = \frac{3}{8}$$

(iii) Let A be the event that has atleast one tail. From the method of indirect counting we have

$$A = S - \{(H, H, H)\}$$
$$\#S = 2^3 = 8 \text{ and } p(A) = \frac{7}{8}$$

Let B be the event such that the outcomes contain atleast two tails. By direct counting $\#B = \#\{(H, T, T), (T, H, T), (T, T, H), (T, T, T)\}$.

$$p(B) = \frac{1}{2}.$$

So $p(A_2 \mid A_1) = \frac{p(A_1A_2)}{p(A_1)} = \frac{1}{2}.\frac{8}{7} = \frac{4}{7}$.

The discussions and examples given above are towards what is known as 'total probability' theorem. We state the theorem without proof.

Theorem 12.40 *Let $A_1, A_2, \ldots, A_m$ be mutually non-overlapping events such that $S = A_1 \cup A_2 \cdots \cup A_m$ with $p(A_i) > 0$, for $1 \leq i \leq m$. [By mutually non-overlapping events we mean each outcome of S appears exactly in one A_i]. Then for any event K,*

$$\begin{aligned} p(K) &= p(A_1 \cap K) + p(A_2 \cap K) + \cdots + p(A_m \cap K) \\ &= p(A_1)p(K \mid A_1) + p(A_2)p(K \mid A_2) + \cdots + p(A_m)p(K \mid A_m) \end{aligned}$$

Example 12.41 A math question paper, has three categories of questions. Half the number of questions are tough and the probability of answering any one of them correctly is $\frac{1}{3}$. One fourth of the questions are slightly twisted and the probability of answering any one of them correctly is $\frac{1}{4}$. The remaining questions are such that the probability of answering any one of them correctly is $\frac{1}{2}$. Student M chooses a question. Find the probability that M gets the correct answer.

Solution: Let the questions be divided into three mutually disjoint events A, B, C such that A is the event of selecting a question from the tough set, B is the event of selecting a question from the slightly twisted set, C is the event of selecting a question from the rest. Let K be the event that a question is chosen by M randomly is answered correctly. Then K may be in A or B or C. That is,

$$K = (A \cap K) \cup (B \cap K) \cup (C \cap K)$$

Hence

$$p(K) = p(A \cap K) + p(B \cap K) + p(C \cap K)$$

Now

$$\begin{aligned} p(A \cap K) &= p(A)p(K \mid A) \\ p(B \cap K) &= p(B)p(K \mid B) \\ p(C \cap K) &= p(C)p(K \mid C) \end{aligned}$$

$$p(K \mid A) = \frac{1}{3}, \quad p(K \mid B) = \frac{1}{4}, \quad p(K \mid C) = \frac{1}{2}$$

Hence using total probability theorem,

$$\begin{aligned} p(K) &= p(A).p(K \mid A) + p(B).p(K \mid B) + p(C)p(K \mid C) \\ &= \frac{1}{2}.\frac{1}{3} + \frac{1}{4}.\frac{1}{4} + \frac{1}{4}.\frac{1}{2} \\ &= \frac{1}{6} + \frac{1}{16} + \frac{1}{8} = \frac{17}{48}. \end{aligned}$$

Example 12.42 A box contains three coins. In these coins, two coins are regular coins and one coin has tail printed on both sides. The activity of picking a coin from the bag is randomly performed and tossed. (i) Find the probability that a 'tail' turns up. (ii) By randomly picking a coin and tossing, 'Tail' turns up. Find the probability that the coin is a 'two-tailed' coin (a coin with tail printed on both sides).

Solution: The sample space is the union of two events A and B such that, A is the event that a regular coin is chosen and tossed, and B is the event that the two-tailed coin is chosen and tossed. Hence we use conditional probability rule in two steps.

$$\begin{aligned} p(T \mid A) &= \frac{1}{2} \\ p(T \mid B) &= 1. \end{aligned}$$

Hence, using the law of total probability,

$$\begin{aligned} p(T) &= p(T \mid A)p(A) + p(T \mid B)p(B) \\ &= \frac{1}{2}.\frac{2}{3} + 1.\frac{1}{3} \\ &= \frac{2}{3}. \end{aligned}$$

(ii) In the second part, we need to find the probability that such chosen coin is 'two-tailed'. So by conditional probability condition, and Bayes' theorem, we have

$$\begin{aligned} p(B \mid T) &= \frac{p(T \mid B)p(B)}{p(T)} \\ &= \frac{1.\frac{1}{3}}{\frac{2}{3}} = \frac{1}{2}. \end{aligned}$$

The above problem can be understood as follows. The activity is to select a question from a set of questions which belongs to a larger set of questions. We realize that it is simply an activity of selecting a question from the larger set itself. Bayes' theorem which connects conditional probabilities $p(A \mid B)$ with $p(B \mid A)$ is an interesting consequence of total probability theorem.

Bayes' Theorem: Let $A_1, A_2, \ldots, A_m$ be mutually disjoint events such that $S = A_1 \cup A_2 \cup \cdots \cup A_m$ and $A_i \cap A_j = \emptyset$ for $i \neq j$, $1 \leq i, j \leq m$. Given that $p(A_i) > 0$, for $1 \leq i \leq m$, for any event K of S, we have

$$p(A_i \mid K) = \frac{p(A_i)p(K \mid A_i)}{p(K)}$$

That is

$$p(A_i \mid K) = \frac{p(A_i)p(K \mid A_i)}{p(A_1)p(K \mid A_1) + p(A_2)p(K \mid A_2) + \cdots + p(A_m)p(K \mid A_m)}$$

Verifying the above equality is simple and direct from the definition of conditional probability.

Example 12.43 A box contains 5 white balls, 4 red balls and 3 yellow balls. Four balls are drawn one after the other without replacement. Find the probability of the event that the first and third drawn balls are white, second drawn ball is yellow and the 4th drawn ball is red.

Solution: The sample space will be the tuple $(\zeta_1, \zeta_2, \zeta_3, \zeta_4)$ where each ζ_i is one among 12 balls drawn at the ith draw. We need to find the probability of the event $A = \{(W, Y, W, R)\}$.

$$p(A) = \frac{5}{12}.\frac{3}{11}.\frac{4}{10}.\frac{4}{9}.$$

From Bayes' Theorem we have,

$$p(A) = p(W)p(Y \mid W).p(W \mid WY).p(R \mid WYW)$$

where $P(Y \mid R)$ means probability of drawing a Yellow ball, while a white ball being drawn in the first step. Similarly $p(W \mid WY)$, $p(R \mid WYW)$ can be defined.

Example 12.44 Two cards are drawn successively from a standard deck of 52 cards. Find the probability of

(i) drawing a spade in the second draw given that the first card drawn is a spade.
(ii) that the first card drawn is a space given that the second card drawn is a space.

Solution: Let A be the event of drawing a spade in the first draw and B be the event of drawing a spade in the second draw. Clearly $p(A) = \frac{13}{52}$. Also if $\overline{A}$ is the event that a spade is not drawn first, then $p(\overline{A}) = \frac{39}{52}$. So in the second draw we need to draw a spade irrespective of the first draw. So

(i) $p(B \mid A) = \frac{p(B).p(n)}{p(A)} = \frac{12}{51}$
(ii) $p(A \mid B) = \frac{p(B).p(A)}{p(B)} = \frac{12}{51}$.

Also note that

$$p(B) = p(B \mid A).p(A) + p(B \mid \overline{A}).p(\overline{A})$$
$$= \frac{13}{52} = p(A).$$

12.4 Ball-Box Distribution: Cases and Analysis

We have seen several problems involving ball-box activity where the balls are picked in order and are not replaced. Also more importance was given to the situation in which a distinction is made in the instances of drawing the balls. That is we discussed the ordered n-tuple $(z_1, z_2, \ldots, z_m)$ mean that z_i is the result of ith draw.

Now the discussion will be on instances where the balls are replaced and then a fresh draw is made. Clearly there is nothing like first draw, second draw and so on. Hence the required balls can be extracted all at one time. If 't' balls are to be taken from a box that contains distinct m-balls, then the outcome will be something like a set $\{z_1, z_2, \ldots, z_t\}$ of numbers between 1 and m. We call the set $\{z_1, z_2, \ldots, z_t\}$ as an 'unordered sample'.

Example 12.45 A box contains 3 distinct balls and two balls are drawn from the box. Then we may draw one ball at a time, or pick two balls and then one ball and so on. So let the balls be numbered as $\{1, 2, 3\}$. Hence our unordered sample will be

$$\{1, 1\}, \{2, 2\}, \{3, 3\}$$
$$\{1, 2\}, \{1, 3\}, \{2, 3\}.$$

Remark 12.46 The unordered partition can be written down to situations, where the balls are handled with or without replacement.

Proposition 12.47 *(i) Let the activity be to draw t balls with replacement from a box that contains m balls. Then the size of the sample space will be* $|S| = C(m + t - 1, t)$.
(ii) Let the activity be to draw t balls without replacement from a box that contains m balls. Then the size of the sample space will be $|S| = C(m, t)$.

Proof The proof directly comes from elementary counting methods. □

The following example enumerates all situations.

Example 12.48 Let there be m boxes $B_1, B_2, \ldots, B_t$ and an activity of distributing m balls is to be carried out, where $t < m$. Find the probability that a boxes $B_1, B_2, \ldots, B_t$ will each contain exactly one ball.

Solution: The boxes are distinct. For balls, we have two cases.

(i) Balls are distinct: While distributing the m balls labelled $\{\ell_1, \ell_2, \ldots, \ell_t\}$ a box may be chosen at most one time or any number of times.

(a) The box is chosen at most one time. It can be shown using combinatorial argument that the (probability) sample space will be of size $\frac{m!}{(m-t)!}$.
Let A be the event where each box is distributed with exactly one ball. Then $|A| = t!$. Hence $p(A) = \frac{t!(m-t)!}{m!}$.

(b) The box is chosen any number of times to receive the ball, then the size of the probability space will be m^t and as before $|A| = t!$. So $p(A) = \frac{t!}{m^t}$.

(ii) Balls are not distinct:

(a) The box is chosen atmost one time. Then the size of the probability space S will be $C(m, t)$. Here $|A| = 1$. Hence $p(A) = \frac{1}{C(m,t)}$.

(b) The box is choosen any number of times. Then the size of the sample space S will be $C(m + t - 1, t)$, $|A| = 1$. Hence $p(A) = \frac{1}{C(m+t-1,t)}$.

Remark 12.49 Sometimes we write the probability space as sample space.

Example 12.50 A group of 10 people living in a 5-storied building enters a lift at the basement to reach their homes. At each floor, when the lift stops, some residents get in and some get out. The lift can hold only 10 people at any point of time. Find the probability that the lift is always full.

Solution: Let x_i be the number of people who leave the lift at floor i-clearly x_i people will enter the lift at the same time. Then we have

$$x_1 + x_2 + x_3 + x_4 + x_5 = 10 \text{ where } x_i \in \mathbb{N} \cup \{0\}$$

The size of the sample space will be the number of integral solutions of this equation which will be $C(14, 4)$. The probability is $\frac{1}{C(14,4)}$.

Example 12.51 A box contains 5 white balls, 5 red balls and 6 green balls. Find the probability of choosing 4 balls so that there is atleast one white, one red and one green ball in them.

Solution: First we have to find the total number of ways 4 balls can be chosen from 16 balls. It is $C(16, 4)$. Let A be the event that 4 balls are selected with one ball in each color. It is same as the number of integer solutions of

$$x_1 + x_2 + x_3 = 4 \text{ where } x_i \geq 1$$

$$\Rightarrow x_1 + x_2 + x_3 = 1, x_i \geq 0$$

Hence

$$|A| = C(4, 1)$$

Hence

$$p(A) = \frac{C(4, 1)}{C(16, 4)} = \frac{4}{C(16, 4)}.$$

[The solution is given using the fact that the balls are not distinct and the box may be chosen any number of times].

12.5 Random Variables

We know what a function $f : X \to Y$ means for X and Y being set of numbers. That is $f(t)$ is called the function for a value t in X. The definition of functions are widely used and studied in calculus. How else a function be viewed? Here the input is a single value or parameter 't'. One may also define functions which takes more number of parameters. For example 'area of the rectangle' is a function which can compute the area for a pair (x, y) where x and y are length and breadth of the rectangle. The number of cars that pass a signal in every 30 min is a function that outputs a positive integer. The number of passengers arriving at a railway counter in every 30 min duration is a function. Note that the functions that we deal with in calculus do not have any 'random phenomena' associated with them. But look at the few examples that we have here, that compute area of a rectangle, cars passing a signal, passengers arriving at a railway counter. They have random phenomena associated with them. We now understand the definition of what are known as 'random variables' which associates itself to a number where the so called variable will be function similar to 'passengers arrive at a railway counter', 'area of a rectangle'. Before we proceed for formal definition let us take an example.

Example 12.52 A coin is tossed n number of times. The 'number of times head turns up' is a function that outputs the number for n Bernoulli trials. Let h be the number. Clearly there are 2^n elements in the sample space S and the function 'number of times head turns up' will map every element in S to a number which indicates the 'heads'.

So in one line we write the definition of a random variable as a function that maps a sample space to a real number, that has some random phenomena in it.

Definition 12.53 A random variable x is a function from the probability space S to $\mathbb{R}$, set of real numbers. That is here $x \in \mathbb{R}$ and x is a random phenomena.

(i) If X is finite or countable, then the random variable is said to be discrete.
(ii) If X is continuous, then there exists a function 'f' called the density function such that $\int_a^b f(x)dx \geq 0$, for $a < b$, $p(X) = \int_a^b f(x)dx$ for $a < X < b$ and $\int f(x)dx = 1$.
That is the axioms of probability are satisfied here.

Example 12.54 The following are random variables

(i) Number of 'heads' in n Bernoulli trials.
(ii) Number of combinations of outcome whose sum is 9 while rolling two dice.
(iii) Number of Indians whose earning is less than 2 lakhs per month taken in a particular year.
(iv) Number of odd numbers formed using the numbers $\{1, 2, 3, 4, 5\}$.
(v) The number of combinations of two numbers n and r
(vi) Number of aces in hand in a game of bridge.

Example 12.55 From a college a sample of 100 students is taken and their average weight is recorded. The table will show different averages as the samples differ. Here the 'averaging' is the random function and the average weight is a random variable.

Sometimes the 'random function' is referred to as probability function of a random variable X. X may assume different values $x_1, x_2, \ldots$ (something like range). These together assume a real value $f(x_i)$, for $i = 1, 2, \ldots$. That is, we write it as

$$p(X = x_i) = f(x_i).$$

Sometimes we may write it as $pX(x_i) = f(x_i)$. It is called the probability mass function.

Example 12.56 Let a fair coin be tossed two times. Let 'X' be the random variable that gives the number of heads in the tosses. Then we can see that the number of times head turns will be 0, 1, 2. So,

$$p(X = 0) = \frac{1}{4} \text{ ('0' head)}$$

$$p(X = 1) = \frac{1}{2} \text{ ('1' head)}$$

$$p(X = 2) = \frac{1}{4} \text{ ('2' heads)}$$

Sometimes we write a situation

$$p(X = t) = 0 \text{ otherwise}$$

Clearly $\sum_{i=0}^{2} p(X = i) = 1$ which is the total probability condition.

Definition 12.57 Let X be a random variable. Clearly for the values $x_1, x_2, \ldots$ that the variable X may assume, we have,

$$p(X = x_i) > 0 \text{ for } i = 1, 2, \ldots$$
$$p(X = x_i) = 0 \text{ otherwise. Also}$$
$$p(X = x_1, x_2, \ldots) = \sum_{i=1}^{\infty} p(x_i) = 1$$

Next we look into two basic random variables. They are (i) The Bernoulli random variable (ii) The Poisson random variable.

The Bernoulli Random Variable: We have already seen what a Bernoulli trial means. Now we just understand Bernoulli random variable is the variable that takes only two values p or q. That is

$$p(X = x) = \begin{cases} p & \text{if } x = \ell \\ q = 1 - p & \text{if } x = \neg\ell \end{cases}$$

For example if a coin is tossed, the outcome will be either 'head' or 'tail'. Here the random variable

$$X = \begin{cases} 1 & \text{if 'head'} \\ 0 & \text{if 'tail'} \end{cases}$$

So

$$p(X = 1) = p$$
$$p(X = 0) = q = 1 - p$$
$$p(X = t) = 0, \text{ otherwise}$$

will be the probability mass function. Bernoulli random variables are very simple and found to be more useful.

From Bernoulli random variable, we define what is popularly known as 'Binomial Random Variable'. Suppose a fair coin is tossed as stated above. The same description is borrowed to define the Binomial random variable, where the trials are only with two probabilities. Let H_n be the random variable with n-tosses of a fair coin showing 'heads'. Then H_n is a binomial random variable with parameters (n, p). The probability mass function of this variable will be

$$p(H_n = i) = C(n, i)p^i(1-p)^{n-i}, i = 0, 1, 2, \ldots, n$$
$$= 0 \text{ otherwise}$$

Clearly $\sum_{i=0}^{n} p(H_n = i) = 1$.

Example 12.58 Let three fair coins be tossed. Find the probability that the outcome has two heads and one tail.

Solution: Let H_3 be the random variable and is equal to the number of 'heads' in the three tossed coins. Clearly H_3 is a binomial random variable.

$$p(H_3 = 2) = C(3, 2)\left(\frac{1}{2}\right)^2\left(1 - \frac{1}{2}\right)$$
$$= 3.\frac{1}{8} = \frac{3}{8}.$$

Example 12.59 Let a die be rolled 6 times. The probability that an odd number turns up on a die is 0.3. Let X be the random variable that indicates the number of odd numbers that turns up in the rolling activity. Find

(i) $p(X = 3)$
(ii) $p(X = 2)$
(iii) $p(1 < X \leq 3)$

Solution:
(i) $p(X = 3)$
Here $n = 6$, $p = 0.3$, $q = 0.7$
$p(X = 3) = C(6, 3)(0.3)^3(0.7)^3$
$p(X = 2) = C(6, 2)(0.3)^2(0.7)^4$
$p(1 < X \leq 3) = p(X = 2) + p(X = 3)$

Example 12.60 In an examination, the examiner sets a question paper in which any student passing the test is with probability 0.3. If 3 students appear for the test, find the probability that at most 1 student will pass the test.

Solution: Let X be the random variable which indicates the number of students who pass the test. The possibilities are only two, 'pass' or 'fail'. Hence X is a binomial random variable. Hence

$$\begin{aligned} p(X=0,1) &= p(X=0)+p(X=1) \\ &= C(3,0)(0.3)^0(0.7)^3 + C(3,1)(0.3)^1(0.7)^2 \end{aligned}$$

Notation:
(i) Let X and Y be random variables on the same sample space S. Then $X+Y, kX$ and XY are functions defined on S such that

$$\left.\begin{aligned} (X+Y)(x) &= X(x)+Y(x) \\ (kX)(x) &= kX(x) \\ (XY)(x) &= X(x).Y(x) \end{aligned}\right\} \text{ where } x \in S$$

(ii) $p(t_1 \leq X \leq t_2) = p(\{x \mid t_1 \leq X(x) \leq t_2\})$.

Notation: Let X be a Binomial random variable with parameters (m, p) or (k, m, p) where random variable is associated with the number $X = k$. Here m is the total number of trials, p is the probability of 'success', for any $k \geq 0$ and X is discrete. Here (m, p) or (k, m, p) are called the Binomial distributions.

Geometric Random Variable

Consider an activity consisting of independent Bernoulli trials until a 'success' is reached. For example two coins are tossed simultaneous until both the coins turn up 'head' is a Bernoulli trial repeated until a condition of 'success' is reached. Let X be the Bernoulli random variable. Then clearly

$$\text{range of } X = \{0, 1, 2, 3, 4, \ldots\}.$$

That is,

$$p(X=t) = p(\text{first } t \text{trials are failures and } (t+1)th \text{ trial is a success}).$$

Then the probability computation will be

$$p(X=t) = (1-p)^t.p$$

The distribution must satisfy the condition that $\sum_{i=0}^{\infty} p(1-p)^i = 1$.

Example 12.61 Two fair coins are tossed until two 'heads' turn up. Let the probability that two heads turn up be 0.2. Let X be the random variable that denote the number of trials made prior to a success trial 'HH'. What is the probability that the fifth trial is a success?

Solution: The random variable can take values as 0, 1, 2, 3, 4, ...

$$
\begin{aligned}
p(X = 4) &= p(\text{first 4 trials are failures}).p \\
&= (0.8)^4(0.2)
\end{aligned}
$$

Example 12.62 In a university network, a person can log in successfully with a probability 0.9. Several log-in attempts are made independently and each attempt has the probability of success. Mr. K logs in repeatedly until Mr. K succeeds. Find the number of attempts made by Mr. K prior to his successful attempt.

Solution: Let X be the random variable that indicates the trials made before a successful log in is made. Hence X can take values 0, 1, 2, 3, ...
Now

$$
\begin{aligned}
p(X = t) &= p(1 - p)^t \\
&= 0.9(1 - 0.9)^t \\
&= 0.9(0.1)^t \\
&= \frac{0.9}{10^t} = \frac{9}{10^{t+1}}
\end{aligned}
$$

Hence

$$
\begin{aligned}
\frac{9}{10} + \frac{9}{10^2} + \frac{9}{10^3} + \cdots + \frac{9}{10^m} &= 1 \\
\frac{9}{10}\left(1 + \frac{1}{10} + \frac{1}{10^2} + \cdots + \frac{1}{10^{m+1}}\right) &= 1 \\
\frac{9}{10}\left(\frac{0.1}{1 - \left(\frac{1}{10}\right)^m}\right) &= 1 \\
\frac{9}{10^2} \cdot \frac{10^m}{10^m - 1^m} &= 1 \\
9.10^{m-2} &= 10^m - 1 \\
10^{m-2}(100 - 9) &= 1 \\
10^{m-2} &= \frac{1}{91} \\
(m - 2) &= \log \frac{1}{91}
\end{aligned}
$$

The Poisson Random Variable

The random variable that is defined below is called Poisson random variable. As before let X count the number of events that take place in a given space X is said to be Poisson random variable with parameter 'λ' if

$$p(X = m) = e^{-\lambda}\frac{\lambda^m}{m!} \quad m = 0, 1, 2, \ldots$$

Here $\lambda > 0$. Clearly

$$\sum_{t=0}^{\infty} p(t) = e^{-\lambda}\sum_{t=0}^{\infty}\frac{\lambda^t}{t!} = e^{-\lambda}.e^{\lambda} = 1.$$

Hence the defined 'p' is a probability mass function.

The Poisson probability mass function and Binomial probability mass functions are related as below:

$$p(X = k) = e^{-\lambda}.\frac{\lambda^k}{k!} \cong C(n, k)p^k(1 - p)^{n-k}, k = 0, 1, 2, \ldots, n.$$

for $\lambda = np$, n being very large and p is very small.

Hence depending on the problem situations, that depending on p and n values, a suitable random variable may be chosen.

Example 12.63 Suppose in a hospital child birth occurs randomly at an average rate of 2 per hour. Find the probability of observing 6 births in a given hour.

Solution: Let X be the Poisson random variable which is the number of child births in a given hour. Here $\lambda = 2$, which is the mean rate of birth.

$$p(X = 6) = e^{-2}.\frac{2^6}{6!}$$

Example 12.64 Suppose in a railway booking counter passengers arrive at the counter at the rate of 3 per minute. Find the probability of 4 persons arrive at a given minute.

Solution: Identical to the previous question.

12.6 Joint Distributions

So far we have seen activities handled using one discrete random variable. Suppose there are two random variables defined on some sample spaces S_1 and S_2 respectively. The joint probability mass function is denoted by $p(X = x_1, Y = y_1)$ and is defined as $p(x_1, y_1) = p(X = x_1, Y = y_1)$. It satisfies the 3 axioms

(i) $p(x_1, y_1) \geq 0$.
(ii) $\sum_{y_1 \in S_2} \sum_{x_1 \in S_1} p(x_1, y_1) = 1$.
(iii) $p((X, Y) \in A) = \sum\sum_{\substack{x \in X \\ y \in Y \\ (x,y) \in A}} p(x, y)$
where A is a subset of (S_1, S_2).

Remark 12.65 (i) Many standard books on this topic refer S_1 as support of X, S_2 as support of Y.
(ii) The joint probability function can be denoted as $f(x, y)$ instead of $p(x, y)$ for better clarity.

Next we define the independence of the two random variables X and Y.

Definition 12.66 Two random variables X and Y are said to be independent if and only if

$$p(X = x, Y = y) = p(X = x) \times p(Y = y)$$

for every pair $(x, y) \in S_1 \times S_2$ where S_1 is the sample space of X and S_2 is the sample space of Y. Otherwise, they are dependent.

Example 12.67 Suppose two fair coins of different sizes are tossed. Find the probability that both coins turn up 'head'.

Solution: Here clearly the tossing is an independent activity.

Let X be the random variable of associated with the appearance of head or tail by the smaller coin.

Similarly Y can be taken. The probability space of X is $\{0, 1\}$, 0 indicates 'tail' appearance, 1 indicates 'head' appearance. The probability space of Y is also $\{0, 1\}$. The joint probability space of X and Y will be

$$S_1 \times S_2 = \{(0, 0), (0, 1), (1, 0), (1, 1)\}$$

Hence $p(X = 1, Y = 1) = \frac{1}{4}$.

Table 12.1 Probability mass function

		Small coin		
	$f(x, y)$	0	1	$f(X = 0, Y = y)$
Large coin	0	$\frac{1}{4}$	$\frac{1}{4}$	$\frac{2}{4}$
	1	$\frac{1}{4}$	$\frac{1}{4}$	$\frac{2}{4}$
	f(X = x, Y = 0)	$\frac{2}{4}$	$\frac{2}{4}$	1

If $f(x, y)$ is the joint probability mass function, then we have the following table w.r.to f and X, Y.
Clearly,

$$p(X = 1, Y = 1) = p(X = 1) \times p(Y = 1)$$
$$= \frac{1}{2} \cdot \frac{1}{2} = \frac{1}{4}$$

Remark 12.68 In the above Table 12.1, a calculation of the probability distribution values of X and Y are given. It is seen that the sum of the probabilities of $X(Y)$ is 1. These entries are called as 'marginal distributions'.

Example 12.69 Two balls are to be distributed in two boxes. Construct the tables for the joint distribution where X_i denote the number of balls in the box i.

Solution: Let X_i be the random variable as given in the problem. Here we assume the boxes are distinct. Total number of ways to distribute the two balls will be

Box 1	Box 2		Box 1	Box 2		Box 1	Box 2
0 0				0 0		0	0
(i)			(ii)			(iii)	

Let Y be the random variable which takes 1 for (i) & (ii) and 2 for (iii). Clearly

$$p(Y = 1) = \frac{2}{3}$$
$$p(Y = 2) = \frac{1}{3}$$

One can construct joint distribution table for (Y, X_1) and (X_1, X_2). The table for (Y, X_1) is as below

		X_1			
	$f(x, y)$	0	1	2	$p(Y = 1, Y = 2)$
Y	1	$\frac{1}{3}$	0	$\frac{1}{3}$	$\frac{2}{3}$
	2	0	$\frac{1}{3}$	0	$\frac{1}{3}$
	$p(X_1 = x)$	$\frac{1}{3}$	$\frac{1}{3}$	$\frac{1}{3}$	1

Similarly the table for (X_1, X_2) can be constructed.

12.7 Expectations

The probability mass function of a random variable X describes the distribution of values of the random variable that they may take. There is an elaborate way of describing the distributed values. It is desirable to describe, if possible a few representative values of the distribution for simplicity in using them. Look at the following example.

Example 12.70 A die is rolled m times. The probability that the numbers 1 to 6 turn up be $p_1, p_2, \ldots, p_6$. If the sum of the numbers in the m rolling is even, then you win the game. What will be the number of rollings you expect to be made so that you win the game.

In this example we are seeing for the first time the word 'expect' in the question. The interpretation will be as follows. If in the m rolls, let 1 occur t_1 times, 2-occur t_2-times 3 occur t_3 times and so on.

So, $t_1 + 2t_2 + 3t_3 + 4t_4 + 5t_5 + 6t_6$ is even and

$$\frac{t_1 + 2t_2 + \cdots + 6t_6}{m} = n$$

will be the average. Also p_i is approximately equal to $\frac{t_i}{m}$, $i = 1, 2, \ldots, 6$. So n will be approximately equal to $t_1 p_1 + t_2 p_2 + \cdots + t_6 p_6$ for any large m.

Now we formally define expectation.

Definition 12.71 The expected value or the mean value of a random variables X with a probability mass function $p(x)$ is defined as $\sum xp(x)$ provided $p(x) > 0$ and the series converges absolutely. It is denoted by $E[X]$.

Remark 12.72 If $\sum xp(x)$ diverges, then X is said to have no expectation.

Example 12.73 The activity is to roll a six sided die. If X is the random variable which indicates the number that turn up, find $E[X]$.

Solution: Clearly $p(X = 1) = p(X = 2) = \cdots = p(X = 6) = \frac{1}{6}$.
So

$$E[X] = 1.\frac{1}{6} + 2.\frac{1}{6} + 3.\frac{1}{6} + 4.\frac{1}{6} + 5.\frac{1}{6} + 6.\frac{1}{6}$$
$$= \frac{7}{2}$$

Example 12.74 Let two independent coins be tossed. The probability that a head turns up in each coin is $\frac{1}{2}$. Let X be the random variable to indicate that a head is obtained. Find $E[X]$.

Solution: First let us write the probability mass function as below.

$$p(X = 0, X = 0) = \frac{1}{2}.\frac{1}{2} = \frac{1}{4}$$
$$[\text{Two coins being tossed } = p[X = 0] \times p[X = 0]$$
$$p(X = 0, X = 1) = p(X = 0).p(X = 1)$$
$$= 2.\frac{1}{4}.\frac{3}{4} = \frac{3}{8}$$
$$p(X = 1, X = 1) = (p(X = 1))^2 = \left(\frac{3}{4}\right)^2 = \frac{9}{16}$$

Hence

$$E[X] = 0.\left(\frac{1}{4}\right)^2 + 1.\frac{3}{8} + 2\left(\frac{9}{16}\right) = \frac{3}{2}$$

If m random variables are defined on the same sample space S_1, then their sum will be a new random variable. What is the expected value of the sum? We state the following theorems without proof. The proof of (i) can be given using mathematical induction technique.

Theorem 12.75 *(i) Let $X_1, X_2, \ldots, X_m$ be random variables defined on S and let $E[X_i]$ be the expectation, $1 \leq i \leq m$. Then*

$$E[X_1 + X_2 + \cdots + X_m] = E[X_1] + \cdots + E[X_m].$$

(ii) If X and Y are two mutually independent random variables with $E[X]$ and $E[Y]$ being finite, then $E[XY] = E[X].E[Y]$.

Example 12.76 Let X_m be the number of 'successes' in m Bernoulli trials with the success probability being p. Then we know

$$\begin{aligned}
p(X_m = k) &= C(m,k)p^k(1-p)^{n-k} \\
E[X_m] &= \sum kC(m,k)p^k(1-p)^{m-k} \\
&= mp \text{ (verify)}
\end{aligned}$$

Example 12.77 Suppose the random variable X has the following probability mass function,

$$p(X=0) = \frac{1}{4}, \quad p(X=1) = \frac{1}{2}, \quad p(X=2) = \frac{1}{4}.$$

Calculate $E[X^2]$.

Solution: Here the new random variable Y be such that $Y = X^2$. The corresponding probability mass function will be,

$$\begin{aligned}
p(Y = 0^2) &= \frac{1}{4} \\
p(Y = 1^2) &= \frac{1}{2} \\
p(Y = 2^2) &= \frac{1}{4}
\end{aligned}$$

Hence

$$\begin{aligned}
E[X^2] &= 0.\frac{1}{4} + 1.\frac{1}{2} + 4.\frac{1}{4} \\
&= \frac{3}{2}
\end{aligned}$$

Note

$$\begin{aligned}
(E[X])^2 &= \left(0.\frac{1}{4} + 1.\frac{1}{2} + 2.\frac{1}{4}\right)^2 \\
&= 1 \neq E[X^2].
\end{aligned}$$

Example 12.78 A box contains 10 balls numbered from 1 to 10. Balls are drawn. After noticing the numbers, they are put back in the box. Let X be the random variable which indicates the largest numbered ball that is drawn in m drawings. Find $E[X]$.

Solution: We first find the probability mass function p. $p(X \leq k)$ means that each of the numbers on the balls drawn in m trials are less than or equal to k in value. So $p(X \leq k) = \left(\frac{k}{10}\right)^m$. So

$$\begin{aligned} p(X = k) &= p(X \leq k) - p(X \leq \overline{k - 1}) \\ &= \left(\frac{k}{10}\right)^m - \frac{(k-1)^m}{10^m} \\ &= \frac{k^m - (k-1)^m}{10^m} \end{aligned}$$

Therefore,

$$\begin{aligned} E[X] &= \sum_{t=1}^{10} t.p(X = t) \\ &= \frac{1}{10^m}\left[10^{m+1} - \sum_{t=1}^{10}(t-1)^m\right] \end{aligned}$$

12.8 Mean, Variance and Standard Deviation

There are three important quantities that are very useful. They are mean, variance and standard deviation.

Definition 12.79 Let X be a random variable. Then

(i) Mean: $\mu = E[X]$
(ii) Variance: $E[X^2] - \mu^2$
(iii) Standard deviation: $\sqrt{E(X^2) - \mu^2}$.

Definition 12.80 Let X and Y be two random variables. The covariance of X and Y is denoted by $cov(X, Y)$ and is defined by

$$cov(X, Y) = E[XY] - E[X].E[Y].$$

Remark 12.81 If X and Y are independent random variables, then we know that $E[XY] = E[X].E[Y]$ and hence $cov(X, Y) = 0$.

There are some properties of covariance of two random variables. We state three of them without proof.

Property 12.82 $cov(X, X) = var(X)$.

Property 12.83 $cov(X, Y) = cov(Y, X)$.

Property 12.84 *$cov(kX, Y) = k\, cov(X, Y)$, k being a scalar.*

Property 12.85 *$cov(X, Y + Z) = cov(X, Y) + cov(X, Z)$.*

Proof Consider

$$\begin{aligned} cov(X, Y + Z) &= E[X(Y + Z)] - E[X]E[Y + Z] \\ &\quad \text{(By definition)} \\ &= E[XY] + E[XZ] - E[X]E[Y] - E[X]E[Z] \\ &= COV(X, Y) + COV(X, Z). \end{aligned}$$

□

Property 12.86 *Let $X_1, X_2, \ldots, X_m$ be independent and have the same type of probability distribution with mean μ and variance σ^2. Let $\overline{X} = \frac{X_1+X_2+\cdots+X_m}{m}$. Then*

(i) $E[\overline{X}] = \mu(mean)$.
(ii) $V(\overline{X}) = \frac{\sigma^2}{m}$.
(iii) $cov(\overline{X}, X_j - \overline{X}) = 0$ *for* $j = 1, 2, \ldots, m$.

Remark 12.87 The random variable $X - \mu$ of a random variable X is called the deviation of X from the mean μ. In general if X is a random variable with probability mass function $p(X = x_j)$ and let $r \geq 0$ be an integer, then

$$E[X^r] = \sum x_j^r p(X = x_j).$$

This is called the rth moment of X. First moment is the 'mean'.

Example 12.88 Find the variance of X if X is the random variable that indicates the number turns up when a die is rolled.

Solution: We have seen earlier $E[X] = \frac{7}{2}$.

$$\begin{aligned} E[X^2] &= 1.\frac{1}{6} + 2^2.\left(\frac{1}{6}\right) + 3^2.\left(\frac{1}{6}\right) + 4^2.\frac{1}{6} + 5^2.\frac{1}{6} + 6^2.\frac{1}{6} \\ &= \frac{1}{6}\left[1 + 2^2 + 3^2 + 4^2 + 5^2 + 6^2\right] = \frac{91}{6} \end{aligned}$$

$$\text{Variance of } X = \frac{91}{6} - \left(\frac{7}{2}\right)^2 = \frac{35}{12}$$

Notation: Let us denote variance of X as $Var(X)$.

Example 12.89 Let X be a random variable which has probability mass function defined as below.

$$p(X = x) = \begin{cases} \frac{1}{6} & \text{for } x = -3, -2, -1, 0, 1, 2, 3 \\ 0 & \text{otherwise} \end{cases}$$

Find $E[X]$ and $Var(X)$.

Solution:

$$E[X] = -3.\frac{1}{6} + -2.\frac{1}{6} + -1.\frac{1}{6} + 0.\frac{1}{6} + 1.\frac{1}{6} + 2.\frac{1}{6} + 3\frac{1}{6} = 0$$

$$E[X^2] = 2\left(9.\frac{1}{6} + 4.\frac{1}{6} + 1.\frac{1}{6}\right) = \frac{14}{3}$$

$$Var(X) = \frac{14}{3} - 0^2 = \frac{14}{3}$$

12.9 Law of Large Numbers

So far we have seen that from the probability distribution or probability law, one can determine mean and variance which are very important measures. However from the known mean and mean of a distribution it is not possible to determine the probability law or mass function. Also small variance indicates that nothing about the large deviations from the mean. If it is possible to have some information on the probability distribution like it is a binomial distribution, with parameter m, k and Poisson distribution with parameter λ then one can relate the known parameters the type of distribution mean and variance. Otherwise the mean and variance makes a rough estimate of the probability distribution (mass) function. The following important theorem is very useful for the above mentioned purpose.

Chebyshev Inequality

Let X be a random variable with mean μ and variance σ^2. Then for any value $k > 0$,

$$p(|X - \mu| \geq k\sigma) \leq \frac{1}{k^2}.$$

Proof We know that $E(X) = \mu$, $Var(X) = \sigma^2$.

$$\sigma^2 = \sum_i (x_i - \mu)^2 p(X = x_i) \tag{12.1}$$

where x_i's are in the domain of X. Consider the x_i's in the domain of $(X - \mu)$ such that $|X - \mu| \geq k\sigma$. That is $(x_i - \mu)^2 \geq k^2\sigma^2$.

So

$$\begin{aligned}\sum_i (x_i - \mu)^2 p(X = x_i) &\geq \sum_{|X-\mu|\geq k\sigma} (x_i - \mu)^2 p(X = x_i) \\ &\geq \sum k^2\sigma^2 p(X = x_i) \\ &= k^2\sigma^2 \sum_{|X-\mu|\geq k\sigma} p(X = x_i) \\ &= k^2\sigma^2 p(|X - \mu| \geq k\sigma)\end{aligned}$$

Hence $p(|X - \mu| \geq k\sigma) \leq \frac{1}{k^2}$. □

Example 12.90 Let X be a random variable with mean $\mu = 42$ and standard deviation $\sigma = 7$. Using Chebyshev inequality comment on the probability mass function near the mean for $k = 2$ and $k = 4$.

Solution: Here $\mu = 42$, $\sigma = 7$.
For $k = 2$.

$$\begin{aligned}&|X - \mu| \geq k\sigma \text{ means} \\ &\mu + k\sigma \geq X \geq \mu - k\sigma \\ &\mu - k\sigma = 42 - 2.7 = 28 \\ &\mu + k\sigma = 42 + 14 = 56\end{aligned}$$

Hence $p(28 \leq X \leq 56) \leq 1 - \frac{1}{4} = \frac{3}{4}$.
That is the probability that the values of X that lies between 28 and 56 will be atleast $\frac{3}{4}$.
Similarly discuss for $k = 4$.

We state the following theorem without proof which is called as Chernoff Bounds.

Theorem 12.91 *Let a sequence of Bernoulli trials be conducted n times with the usual notation of 'success' probability p. Let the trials be independently conducted. Let X be a random variable that counts the total number of 'successes' in the n trials. Then*

$$p(X \geq 6\mu) \leq 2^{-6\mu}.$$

Example 12.92 Let an unbiased coin be tossed 100 times consecutively and the 'success' be the appearance of 'head'. Let the 'success' probability be given as $\frac{2}{5}$ in each toss. Find the upper bound on the probability that there will be atleast 80 heads appear in the total tossing.

Solution:

$$\mu = E[X] = n.p(\text{we have seen earlier})$$
$$= 100.\frac{2}{5} = 40.$$

By Chernoff bounds,

$$p(X \geq 80) = p(X \geq 6\mu) \leq 2^{-6.\mu}$$
$$= 2^{-6.40} = 2^{-240}.$$

Finally we now see what is known as 'law of averages' which is also called as 'law of large numbers' of discrete random variables.

Let $X_1, X_2, \ldots, X_m$ be a set of identically distributed random variables with mean, $\mu_1, \mu_2, \ldots, \mu_m$ and variance $\sigma_1^2, \sigma_2^2, \ldots, \sigma_m^2$. Let

$$Y = X_1 + X_2 + \cdots + X_m$$
$$\mu = E[Y] = \mu_1 + \mu_2 + \cdots + \mu_m$$
$$s^2 = Var(Y) = \frac{\sigma_1^2 + \sigma_2^2 + \cdots + \sigma_m^2}{m}$$

Using Chebyshev inequality we obtain

$$p(|Y - \mu| \geq \epsilon) \leq \frac{s^2}{m.\epsilon^2} \quad \text{for any } \epsilon > 0.$$

Clearly as $m \to \infty$, for a fixed ϵ, $\frac{s^2}{m.\epsilon^2} \to 0$. Thus we have,

The weak law of large numbers

Let $X_1, X_2, \ldots, X_m$ be independent random variables that have identical probability distribution, with mean μ as computed above. Then for every $\epsilon > 0$, we have the following:

$$p(|Y - \mu| \geq \epsilon) = p\left(\left|\frac{X_1 + X_2 + \cdots + X_m}{m} - \mu\right| \geq \epsilon\right)$$
$$\to 0 \text{ as } m \to \infty.$$

Remark 12.93 As m becomes larger and larger, the distribution of Y is near μ, the mean.

Example 12.94 Let a fair coin be tossed n times. Let X be the random variable that denotes the number of times 'head' turns up. Then $\frac{S_n}{n}$ denotes the number of times 'head' turns up which will be a fraction between 0 and 1. The law of large numbers tells that the outcomes of the random variable X will be having the probability $\frac{1}{2}$ when $n \to \infty$.

Example 12.95 Let $X_1, X_2, \ldots, X_n$ be the random variables which follow Bernoulli distribution with probability 0.3 for 'success' and 0.7 for 'failure'. Here $X_j = 1$ for success and $X_j = 0$ for failure. Define $A_n = \frac{X_1+X_2+\cdots+X_n}{n}$, find $p(|A_{100} - \frac{1}{3}| \geq 0.1)$, $p(|A_{1000} - \frac{1}{3}| \geq 0.1)$.

Solution: Clearly

$$p(|A_n - \frac{1}{3}| > \epsilon) \leq \frac{s^2}{n.\epsilon^2} = \frac{Var(A_n)}{n.\epsilon^2}$$
$$Var(A_n) = \frac{.21}{n}$$

$n = 100$

$$p(|A_{100} - 0.3| \geq 0.1) \leq \frac{0.21}{100 \times 0.1} = 0.21$$

$n = 1000$

$$p(|A_{1000} - 0.3| \geq 0.1) \leq 0.21$$

Exercise

1. For each of the following events draw a Venn diagram and shade the area corresponding to the events.
   (i) $A \cup B^c$ (ii) $A^c \cap B^c$ (iii) $(A \cup B)^c$ (iv) $A^c \cap B$.
2. Let $S = \{1, 2, 4, 6, 8, 10, 12, 14, 16\}$, $A = \{1, 2, 10, 14\}$, $B = \{10, 16\}$. Find the members of the event $A \cup B$, $A \cap B$, $A \cup B^c$, $A^c \cap B$.
3. An activity of rolling two dice is carried out. Find the sample space. List the members of the events
   $A = \{\text{sum of numbers shown on dice} = 8\}$
   $B = \{\text{sum of numbers shown on dice} \leq 5\}$
   Are A and B disjoint? Justify.
4. Let the sample space of an activity be $S = \{1, 2, 3, 4\}$. Find the set of all possible events of S.
5. A coin is tossed until a head turns up. What is the sample space?
6. A four sided die is rolled until a 3 comes up. Find the same space. Find the members of the event of getting a 2 in the first three rolls.
7. A student of a college may or may not choose general science. Even if the student chooses general science, the student may or may not choose Mathematics. Even if the student chooses Mathematics, he may or may not take the elective 'commutative algebra'. Consider the activity of selecting a student and determine whether the student is a science student and the student has not chosen 'commutative algebra'.

8. Let A, B, C be three events. Show that

$$p(A \cup B \cup C) \leq p(A) + p(B) + P(C)$$

whatever A, B, C may be.

9. Extend the above inequality to m events and prove it using mathematical induction on m.
10. A box contains 9 balls of different sizes. A ball is chosen at random.

    (i) Find the probability that the smallest ball is chosen.
    (ii) Find the probability that the largest ball is chosen.
    (iii) Find the probability of choosing any two balls.

11. Let a fair coin be tossed 5 times. List the elements of the event that contains 'four tails and one head' in the total of 5 tosses. Interpret the event and find its probability.
12. Let there be m persons in a group and assume that no body is born on 29th February. Find the probability sample space. What is probability that two or more persons have the same birthday?
13. Let A and B be two events. Define $(B - A)$ as the event that contains all those elements which are in B, but not in A. Write $(B - A)$ in terms of S and A^c.
14. If A and B are two independent events. What can you say about $(B - A)$?
15. Three players p, q, r play the game of tennis (singles matches) as per rules. In the beginning p and q play a match. The winner of this match is to play with player r. The winner of the second match has to play with the loser of the first match. The matches continue, until a player wins two successive matches. Write down the infinite probability sample space.
16. A box contains 100 balls. Suppose the balls are numbered as either 1 or 2. Two balls are drawn from the box (i) with replacement (ii) without replacement. Find the probability of

    (i) both the drawn balls are numbered '1'
    (ii) both the drawn balls are of different numbers
    (iii) atleast one ball drawn is numbered '1'
    (iv) exactly one ball drawn is numbered '1'.

17. A fair coin is tossed m times. Find the probability that the

    (i) all the m coins show 'head'
    (ii) all the m coins show 'tail'
    (iii) there is exactly one head shown up.

18. Consider that there are three boxes B_1, B_2, B_3. Box B_1 contains 2 green and 4 blue balls, Box B_2 contains 3 green and 2 blue balls, Box B_3 contains 2 green and 2 blue balls. Two balls are drawn from the boxes. Find the probability that the 2 drawn balls are green.
19. Eight persons are arranged in a circular manner. Let A and B are two persons in the arrangement. Find the probability that A and B are adjacent to each other.
20. Let a three digit number be formed using $\{1, 2, 3, 4, 5\}$. A number is to be assigned to a number plate of a scooter. Suppose the scooter owner is asked to choose a number and the number is an odd number. Find its probability.
21. A fair coin is tossed three times in sequence. Let A be the event that more tails turn up than heads. Let B be the event that in the first toss, head turns up. Find $p(A \mid B)$.
22. A competition is conducted between two teams to suggest a model for a car. One team has more experienced engineers whose success probability is 2/3. The other team's success probability is 1/2. Also the models suggested by one of the teams is always successful and its probability is 3/4. If there is a situation where both the teams' car models are successful and in which case, then the model suggested by the experienced engineers is taken. If there is a situation that there is exactly one successful design model. What is the probability that it was designed by the team of experienced engineers?
23. Assume the multiplication principle

$$p(A_1 \cap A_2 \cap \cdots \cap A_n) = p(A_1)p(A_2 \mid A_1) \ldots p(A_n \mid A_1 \cap A_2 \cdots \cap A_{n-1}) \quad (12.2)$$

Solve the following problem.
From a deck of 52 cards 4 cards are drawn. Find the probability that none of the cards is a spade.
24. An elective class consists of 15 undergraduate students and 5 graduate students. The whole class is divided in to 5 groups to make a group project. What is the probability that each group has a graduate student?
25. A fair coin is tossed 4 times

    (i) Find the probability of the event that has outcomes that have atleast one 'head'.
    (ii) Find the probability of the event that has outcomes that have all the four heads.
    (iii) Given that the first three tosses are having the outcome tail, find the probability of such an event that the last toss gives a 'head'.

26. I know that a friend of mine has two kids. The friend brought one child to my family function and that child is a girl. What is the probability that the other kid is also a girl?
27. A three digit number is formed using the numbers $\{1, 2, 3, 4, 5\}$. Find the probability that the number is odd number, given the information that the first decimal place has

an even number and the second decimal place contains an odd number. Use direct combinatorial argument to verify your computation.

28. Let S be a probability space and A and B are events on S. Given $p(A) = \frac{1}{12}$, $p(B) = \frac{1}{3}$, $p(AB) = \frac{1}{52}$. Find (i) $p(A \mid B)$ (ii) $p(B \mid A)$.
29. Box B_1 contains 4 white balls and 4 black balls. Box B_2 contains 6 white balls and 8 black balls. A ball is transferred from box B_1 to box B_2. From box B_2 a ball is drawn. Find the probability that the ball is white in color?
30. Three dice are rolled on a table. The first die shows 5 and the second die shows 4. One die falls under the table. What is the probability that the number of the die rolled under the table is 3?
31. A book shelf in a library has 8 maths books, 6 physics books and 7 general knowledge books. If M selects two books at random, find the probability that both the selected books are of different subjects.
32. Two dice are rolled. Find the probability that the sum of the two dice will be greater than 7 given that one die shows 4.
33. M has to clear two qualifying exams for a medical seat in a medical university. The probability for M to clear the first qualifying test is $\frac{4}{5}$. The probability for M to get a seat is $\frac{2}{3}$. Find the probability of M clearing the second qualifier given that she has cleared the first test.
34. Two dice are rolled. Let A be the event that the outcome of the first die is $\{1, 2, 3\}$. Let B be the event that the outcome of the second die is $\{3, 4, 5\}$. Let C be the event that the sum of the two dice is 8. Find $p(A \cap B \cap C)$ and show that it is not equal to $p(A).p(B).p(C)$.
35. 6 dice are thrown simultaneously. Find the probability that each number appears exactly twice.
36. A deck of 52 cards are distributed to 4 people. Find the probability that a hand consists of four spades, three hearts, five diamonds and one club?
37. An elevator starts at floor 1 with 9 passengers and stops at all 10 floors. What is the probability that no two passengers leave at the same floor?
38. A die is rolled 4 times. Find the probability of getting a sum '6' atleast in one roll?
39. 10 balls are to be distributed in 5 boxes. Find the probability that

    (i) a box may be empty
    (ii) atleast one ball in each box
    (iii) each box contains exactly two balls

    Discuss the solution by assuming all the 4 cases, where boxes distinct and boxes identical, balls distinct and balls identical.
40. Using three a's, four b's and six c's, a word of length 5 is formed. What is the probability that the word may contain

(i) all four b's?
(ii) all c's?
(iii) atleast one letter from each category?

41. Let X be a discrete random variable with the following probability distribution function

$$p(X = x) = \begin{cases} 0.2 & \text{for } x = 0 \\ 0.3 & \text{for } x = 1 \\ 0.4 & \text{for } x = 2 \\ 0.1 & \text{for } x = 3 \\ 0 & \text{otherwise.} \end{cases}$$

(i) Write down the range of the variable X.
(ii) Find $p(X < 3)$.
(iii) Find $p(1 \leq X < 3)$.

42. Let two dice be rolled. Let X and Y be the numbers that turn up.

(i) Find range of X, range of Y, probability mass function of X and Y.
(ii) Let $Z = XY$. Find the range and probability mass function of XY.
(iii) Find $p(X = 3 \mid Z = 12)$.

43. A fair coin is tossed repeatedly until two consecutive tails turn up. If X is the total number of times the coin is tossed, find $p(X = k)$ for $k = 0, 1, 2, \ldots$.
44. A box contains 6 green balls, 4 yellow balls and 10 white balls. Two balls are drawn randomly from the box.

(i) Find the sample space of the activity.
(ii) If X is a random variable that indicates number of white balls selected. Find the range of X.
(iii) Find $p(X = 0)$ and $p(X = 2)$.

45. A students writes an examination that consists of 20 multiple-choice questions. The number of options for each question is 4. Assume that the student knows answers to 50% of the questions and the answers to the remaining questions are picked randomly. Each correct answer is given 1 mark. Let X denote the possible 'marks' scored by any individual (i) Find the range of X and the probability mass function (ii) Find $p(X > 12)$.
46. Let the number of cars crossing a 't'-junction be a Poisson variable. On an average 20 cars pass the t-junction in one hour duration. Find the probability that the number of cars crossing the 't'-junction at a peak hour is more than 20 but less than 30.

47. A box contains three distinctly by numbered balls. Two balls are drawn at random without replacement. Let X be the random variable that indicates the number on the first ball and Y be the random variable that indicates the number on the second ball. Then construct the joint distribution table indicating the marginal distributions.
48. Five balls are distributed at random in five boxes. Let X_i denote the number of the box that contain exactly i balls. Construct joint probability distribution tables of (X_1, X_2), (X_2, X_3) indicating the marginal distributions.
49. Find $E[X]$ where X is binomially distributed random variable with parameters n, k and p.
50. Find $E[X]$ where X is a Poisson random variable with parameter λ.
51. Suppose n people go for a night party wearing their hats. The leave their hats in a basket that is kept outside. Sudden power failure occurs. They come out and each person picks a hat before leaving. Let X be the random variable that indicates the number of persons who takes their hats. Find $E[X]$.
52. Suppose a box contains 10 red balls and 15 green balls. A random sample of 8 balls is drawn without replacement. Let X_t be the random variable that indicates by assigning value 0 or 1 according as the tth ball drawn is black or not for $t \leq 8$. Find the mean and variance of X_t.
53. Let a box contain three balls numbered 1, 2, 3. Two balls are drawn at random from the box without replacement. Let X be the random variable that indicates the number on the first ball and Y be the random variable that indicates the number on the second ball. Find $cov(X, Y)$.
54. Let a box contain balls of K colors. A ball is drawn at random and such a ball can have any of the K colors. Find the expected number of balls to be drawn so that the selection has atleast a ball from each color (Assume the draw being done without replacement).
55. A box originally contains 't' green balls. A multiple step activity is conducted as below: At each step, a ball is randomly taken from the box. If the ball taken is a green one, then a yellow ball is put back in the box replacing the green ball. If the ball taken is yellow one, then it is returned to the box.
    Find the probability of drawing 't' green balls in n trials. If $p(n, t)$ denote this answer, find its connection with Stirling numbers of second kind?
56. A teacher conducts a surprise test to a class of n students. After the test is over, the teacher decides to distribute the answer papers to the same set of students for correction so that no student is get to receive his/her own answer paper. Find the probability of this occurrance? Make a remark on the probability thus optained related to the size of the class?
57. 15 balls are distributed into 5 boxes.

    (i) Find the probability that the first box will contain 2 balls.
    (ii) Find the probability that each box contains three balls.

58. A box contains 5 green balls and 4 yellow balls. The activity is to draw 3 balls at random from the box without replacement.

   (i) Find the sample space.
   (ii) Find the probability that the first ball drawn being green.
   (iii) Find the probability that atleast one out of the three balls is yellow.
   (iv) Find the conditional probability that the last ball drawn is a green one given that the first two balls drawn being yellow.

59. A fair die of six sides is rolled 8 times and the outcomes are recoved in order. Find the probability that the outcome is '4' atleast two times.

Problems and Solutions—Combinatorics

1. Prove that

$$\sum_{t=0}^{m} t^2 = \frac{m(m+1)(2m+1)}{6}$$

for all natural numbers m.

Solution:
The equality in the question is established using mathematical induction. For $m = 0, 1$ the equality is true. Assume the equality to be true for $m = n$. That is

$$\sum_{t=0}^{n} t^2 = \frac{n(n+1)(2n+1)}{6} \tag{12.3}$$

We now show that it is true for $m = n + 1$.
Consider the summation

$$\begin{aligned}
\sum_{t=0}^{n+1} t^2 &= 0^2 + 1^2 + 2^2 + \cdots + n^2 + (n+1)^2 \\
&= \sum_{t=0}^{n} t^2 + (n+1)^2 \\
&= \frac{n(n+1)(2n+1)}{6} + (n+1)^2, \text{ using (12.3)} \\
&= \frac{1}{6}(n+1)(2n^2 + 7n + 6) \\
&= \frac{1}{6}(n+1)(n+2)(2n+3)
\end{aligned}$$

2. Are the following true? Justify your answer.

(i) $n^2 > n + 1, \forall n \in \mathbb{N}$.
(ii) $n! > n^2, \forall n \geq 4$.

Solution:
Both (i) and (ii) can be proved using mathematical induction.
(i) No.
For $n = 2$, $2^2 > 2 + 1$.
Assume the inequality to be true for $n = m$. That is

$$m^2 > (m + 1). \tag{12.4}$$

Consider

$$\begin{aligned}(m + 1)^2 &= m^2 + 2m + 1 \\ &> (m + 1) + (2m + 1), \text{ using (12.4)} \\ &= 3m + 2 > m + 1 \text{ for } m \geq 1.\end{aligned}$$

(ii) Left as an exercise.

3. Are the following sets countable? Justify your answer.

(i) $X = \{(e, o)|e \in \{0, 2, 4, 6, 8, \ldots\}, o \in \{1, 3, 5, 7, \ldots\}\}$
(ii) $Y = \{\frac{1}{n^2}|n \in \mathbb{N}\}$
(iii) $T = (0, 0.5]$.

Solution:
(i) X is countable as X being the cartesian product of two countable sets.
(ii) Y is countable as there exists a one-to-one correspondence from $\mathbb{N}$ to Y.
(iii) T is uncountable as it is in the form of semi open interval.

4. If $S = \{(x, y)|x^2 + y^2 = 16\}$ and $T = \{(x, y)|x^2 + 9y^2 = 81\}$, then find $|S \cap T|$.

Solution:
The set S describes the set of points (x, y) lying on a circle with center 'O' and radius 4. The set T describes the set of points lying on an ellipse $\frac{x^2}{9^2} + \frac{y^2}{3^2} = 1$. Clearly these two curves intersect at 4-points. Hence $|S \cap T| = 4$.

5. Consider the function $f(x) = \frac{x^4-1}{x^4+1}$ for $x \in \mathbb{R}$. Find the minimum value of $f(x)$.

Solution:
Consider

$$\begin{aligned} f(x) &= \frac{x^4 - 1}{x^4 + 1} \\ &= 1 - \frac{2}{x^4 + 1} \end{aligned}$$

Here

$$\begin{aligned} \frac{2}{x^4 + 1} &< 2 \text{ as } x^4 + 1 > 1 \\ \Rightarrow 1 - \frac{2}{x^4 + 1} &\geq 1 - 2 \\ \Rightarrow f(x) \geq -1 &\text{ and } f(x) < 1. \end{aligned}$$

So, -1 is the minimum value of $f(x)$.

6. Let S be a set which is countable and infinite. Then show that there exists a bijection between S and the set of natural numbers.

Solution:
Define a function $f : S \to \mathbb{N}$, as $f(s) = n$, such that s is the nth smallest element of S. As there are only finitely many natural numbers less than 's', f is well defined. Also for $s_1, s_2 \in S$, $f(s_1) = f(s_2) = n$ is impossible. So f is clearly a one-to-one mapping. f is also onto as S is infinite.

7. Let $f(x) = x^2 - 2x$. Find

(i) $f(-4)$
(ii) $f(t - 1)$

Solution:
(i) $f(-4) = (-4)^2 - 2(-4)$
$= 24.$
(ii) $f(t - 1) = (t - 1)^2 - 2(t - 1)$
$= t^2 - 4t + 3.$

8. Let $f(x) = x + 1$ and $g(x) = x^2 + 2x - 1$. Find $(f - g)(x)$, $f \circ g(x)$ and $\left(\frac{f}{g}\right)(x)$

Solution:

$$\begin{aligned}(f - g)(x) &= x + 1 - x^2 - 2x + 1 \\ &= -x^2 - x + 2\end{aligned}$$

$$\begin{aligned}f \circ g(x) &= f(x^2 + 2x - 1) \\ &= x^2 + 2x - 1 + 1 \\ &= x^2 + 2x\end{aligned}$$

$$\frac{f}{g}(x) = \frac{f(x)}{g(x)} = \frac{x + 1}{x^2 + 2x - 1}$$

9. Let $f(x) = \sqrt{x + 1}$ and $g(x) = \frac{1}{x}$. Find the domain of $(g - f)(x)$.

Solution:

$$\begin{aligned}(g - f)(x) &= g(x) - f(x) \\ &= \frac{1}{x} - \sqrt{x + 1}\end{aligned}$$

The domain is $(0, \infty) \cup (-1, 0)$.

10. Let f be a function defined on an interval $[0, 1]$ such that

$$\begin{aligned}f(0) = f(1) = 1 \text{ and} \\ |f(x) - f(y)| < |x - y| \; \forall \, x \neq y, x, y \in [0, 1].\end{aligned}$$

Show that $|f(x) - f(y)| < \frac{1}{2}$.

Solution:

$$\begin{aligned}|f(x) - f(y)| &= |f(x) - f(1) + f(0) - f(y)| \\ &\leq |f(x) - f(1)| + |f(0) - f(y)| \\ &\leq |x - 1| + |0 - y| \end{aligned} \tag{12.5}$$

If $|x - y| \leq \frac{1}{2}$, $|f(x) - f(y)| < \frac{1}{2}$.
If $|x - y| > \frac{1}{2}$, using (12.5), $|f(x) - f(y)| < \frac{1}{2}$
So $|f(x) - f(y)| < \frac{1}{2}$.

11. Let n be a positive integer. Let $\pi(n)$ denote the product of non zero digits of n. Suppose 'n' is a single digit number, then $\pi(n) = n$. If

$$S = \pi(1) + \pi(2) + \cdots + \pi(99),$$

then what is the largest prime factor of S?

Solution: The positive integers less than 100 are

$$1, 2, 3, 4, \ldots, 9, 10, 11, 12, \ldots, 99.$$

These are written as

$$01, 02, 03, 04, 05, 06, \ldots, 09, 10, 11, 12, \ldots, 99.$$

Now the sum of the product of the digits of these numbers will be

$$\begin{aligned}&(0 \cdot 0 + 0 \cdot 1 + 0 \cdot 2 + 0 \cdot 3 + \cdots 9 \cdot 9) - 0 \cdot 0\\ &= (0 + 1 + 2 + \cdots + 9)^2 - 0 \cdot 0 \qquad (12.6)\end{aligned}$$

Now by definition of $\pi(n)$, the sum of these digits can be found by replacing '0' by 1 in the expression (12.6) above as we need to ignore '0's. This is equivalent to considering '1's in the products. Note that $0 \cdot 0$ in the above expression is contributing '1' to the count as it is changed as $1 \cdot 1$. So,

$$\begin{aligned}\mathbb{Z} &= (1 + (1 + 2 + \cdots 9))^2 - 1\\ &= 46^2 - 1 = 3 \cdot 3 \cdot 5 \cdot 47\end{aligned}$$

So 47 is the largest prime in $\mathbb{Z}$.

12. Let $f(n)$ and $g(n)$ be asymptotic positive functions. Prove or disprove the following equations.

(i) $f(n) = O(g(n)) \Rightarrow g(n) = O(g(n))$
(ii) $f(n) = O(g(n)) \Rightarrow g(n) = \Omega(f(n))$
(iii) $\theta(g(n)) = o(f(n)) + f(n)$

Solution:
(i) Suppose $f(n) = 2n + 3$ and $g(n) = n^3$.
Given $2n + 3 \in O(n^3)$ or $2n + 3 = O(n^3)$.
But $n^3 \notin O(n)$.
So (i) is false.

(ii) By definition, $f(n) = O(g(n))$ means $0 \leq f(n) \leq kg(n)$ for all $n \geq n_0$, k being constant, $n_0 > 0$. This inequality can be rewritten as

$$\begin{aligned} 0 &\leq \frac{1}{k} \cdot f(n) \leq g(n) \\ &\Rightarrow g(n) = \Omega(f(n)) \end{aligned}$$

Hence (ii) is true.
(iii) Let $g(n) = o(f(n))$.
This means there exists a positive constant k and $n_0 > 0$, such that

$$\begin{aligned} &0 \leq g(n) < kf(n) \; \forall \, n \geq n_0. \\ &\Rightarrow f(n) \leq f(n) + g(n) < (1+k) f(n). \\ &\Rightarrow f(n) \leq f(n) + o(f(n)) < (1+k) f(n). \end{aligned}$$

This means $f(n) + o(f(n)) \in \theta(f(n))$, and hence (iii) is true.
13. Using mathematical induction show the following equality

$$\left(1 - \frac{1}{2^2}\right)\left(1 - \frac{1}{3^2}\right) \cdots \left(1 - \frac{1}{n^2}\right) = \frac{n+1}{2n}.$$

Solution:
For $n = 2$, the equality is true. Assume the equality to be true for $n = m$. That is

$$\left(1 - \frac{1}{2^2}\right)\left(1 - \frac{1}{3^2}\right) \cdots \left(1 - \frac{1}{m^2}\right) = \frac{m+1}{2m}. \tag{12.7}$$

Consider the product for $n = m + 1$

$$\begin{aligned} &\left(1 - \frac{1}{2^2}\right)\left(1 - \frac{1}{3^2}\right) \cdots \left(1 - \frac{1}{m^2}\right)\left(1 - \frac{1}{(m+1)^2}\right) \\ &\quad = \frac{m+1}{2m}\left(1 - \frac{1}{(m+1)^2}\right) \quad \text{by using (12.7)} \\ &\quad = \frac{(m+1)}{2m}\left[\frac{(m+1)^2 - 1}{(m+1)^2}\right] \\ &\quad = \frac{m^2 + 2m}{2m(m+1)} = \frac{m+2}{2(m+1)} \end{aligned}$$

Hence the equality is true for $n = m + 1$ and hence for all n.

14. Show by mathematical induction that $(m^3 - 7m + 3)$ is divisible by 3 for integers $m \in \mathbb{Z}$, $m \geq 0$.

Solution:
For $m = 0$, the statement is true. Assume the statement to be true for $n = m$
That is $m^3 - 7m + 3$ is divisible by 3.
Consider the expression for $n = m + 1$.

$$\begin{aligned}(m+1)^3 - 7(m+1) + 3 &= (m^3 - 7m + 3) + (3m^2 + 3m - 6) \\ &= A + B \text{ (say)}\end{aligned}$$

$A = m^3 - 7m + 3$ is divisible by 3 by assumption. $3m^2 + 3m - 6$ is also divisible by 3. So the expression $(m^3 - 7m + 3)$ is divisible by 3 for all m.

15. Let r be a real number with $r > -1$. Prove that

$$(1+r)^n \geq 1 + nr \;\forall\, n \in \mathbb{N}.$$

Solution:
The inequality is true for $n = 1$. Assume the inequality to be true for $n = m$. That is

$$(1+r)^m \geq 1 + mr \tag{12.8}$$

Consider

$$\begin{aligned}(1+r)^{m+1} &= (1+r)^m(1+r) \\ &\geq (1+mr)(1+r) \text{ by (12.8)} \\ &= 1 + (m+1)r + mr^2 \\ &\geq 1 + (m+1)r.\end{aligned}$$

So the inequality is true for $m + 1$ and hence for all n.

16. A tennis tournament has 20 participants. It is a singles tournament and any two players play one game against each other. Show that at time 't' there are two players who played against same number of participants.

Solution:
Each player can finish either no game or one game or two games and so on. Upto $(n - 1)$ games which totals to n possibilities. If there is a player who finished $(n - 1)$ games, then there cannot be a player who finished zero game. This means '0' and '$n - 1$' cannot occur in the list of played games. Hence, by direct argument there must be 2 players in the n players who at some time 't' played the same number of matches. Note that PH - principle cannot be used as there are n players and n possibilities.

17. There are 80 sports bikes in a shop. Each bike has no more than 60 spokes in its wheels. Show that there are atleast two bikes that have the same number of spokes in their wheels.

Solution:
Let a bike be put in room 't' if it has 't' spokes. So as per the statement of the problem there must be 60 rooms into which the bikes are put. The shop actually has 80 bikes. So by PH - principle there must be a room containing atleast 2-bikes. Hence the solution follows.

18. Show that among any 5 numbers, it is always possible to find two numbers whose difference is divisible by 5.

Solution:
Let n_1, n_2, n_3, n_4, n_5 be the five numbers when each n_i is divided by 5, they leave a remainder which is either 0, 1, 2, 3 or 4. Hence by PH - principle, out of the 5 numbers, two of them will give the same remainder when divided by 5. Let these numbers be n_i and n_j, $1 \leq i, j \leq 5$. That is

$$\begin{aligned} n_i &= 5t_1 + r \\ n_j &= 5t_2 + r, \ t_1, t_2 \in \mathbb{N} \\ \Rightarrow \ (n_i - n_j) &= 5(t_1 - t_2) \end{aligned}$$

Hence $n_i - n_j$ is divisible by 5.
19. Let $n_1, n_2, \ldots, n_t$ be positive integers. Show that if

$$n_1 + n_2 + n_3 + \cdots + n_t - t + 1$$

balls are distributed into 't' boxes, then either the first box contains atleast n_1 balls or 2nd box contains n_2 ball and so on, the tth box contains atleast n_t balls.

Solution:
Suppose that the 'then' claim is not true. This means that when

$$n_1 + n_2 + \cdots + n_t - t + 1$$

balls are distributed in t boxes, then box 1 contains $(n_1 - 1)$ balls, box 2 contains $(n_2 - 1)$ balls and so on, box t contains $(n_t - 1)$ balls. Then the total number of balls in the t-boxes will be $n_1 + n_2 + \cdots + n_t - t$. But this count is not correct. Hence the conclusion follows.

20. If $n_1, n_2, \ldots, n_{t^2+1}$ is a sequence of real numbers, then show that this sequence contains a subsequence of length $(t + 1)$ which is either an increasing or a decreasing one.

Solution:
Let there be no decreasing subsequence of length $t+1$. Then we need to show that there exists a subsequence of length $t+1$ which is increasing.

Let 's_f' be the length of the longest subsequence which is decreasing and let the subsequence start with the number 'n_f' where $1 \leq f \leq t^2+1$. By our assumption on decreasing subsequences we have, $1 \leq s_f < t+1$ for all 'f'. By the previous problem, we have that the (t^2+1) integers must contain $(t+1)$ integers such that

$$s_{f1} = s_{f2} = s_{f3} = \cdots = s_{f(t+1)},$$

where $1 \leq f_1 \leq f_2 \leq \cdots < f_{(t+1)} \leq t^2+1$. If there is one f_i such that $n_{f_{i+1}} > n_{f_i}$, then any decreasing subsequence of length $s_{f_{i+1}}$ that begins with $n_{f_{i+1}}$ will lead to a subsequence of length $s_{f_{i+1}}$, which can starts with the term n_{f_i} by appending n_{f_i} in the beginning. Hence we have $s_{f_i} > s_{f_{i+1}}$ which is a contradiction as $s_{f_i} = s_{f_{i+1}}$. Hence we have

$$n_{f1} \geq n_{f2} \geq \cdots \geq n_{f_{t+1}}$$

which is increasing subsequence of length $(t+1)$.

21. In the month of April, a maths instructor conducts atleast one surprise test but plans to conduct no more than 46 such tests. Show that there must be a period of some 't' number of consecutive days during which the teacher conducts 14 surprise tests.

Solution:
Let t_i denote the number of tests conducted on day 'i' in April. So the total number of tests conducted starting from day 's' and ending on day 'f' will be

$$M = t_s + t_{s+1} + \cdots + t_f.$$

Now we look for partial sums as follows. That is

$$\begin{aligned}
n_1 &= t_1 \\
n_2 &= t_1 + t_2 \\
n_3 &= t_1 + t_2 + t_3 \\
&\cdots \\
n_{30} &= t_1 + t_2 + \cdots + t_{30}.
\end{aligned}$$

Clearly $M = n_f - n_{s-1}$.
Now to show $M = 14$ for some f and s, where $1 \leq f, s \leq 30$. Clearly $n_i \leq 46\ \forall\ i$. Also the instructor conducts atleast one test on each day. Hence $n_1, n_2, \ldots, n_{30}$ are distinct integer from $\{1, 2, \ldots, 46\}$.

Now we treat the distinct integers n_i's to be the number of pigeons. The aim is to fix the pigeonholes satisfying the condition that $M = 14$. The possibilities are

$$(1, 15), (2, 16) \ldots (14, 28)$$
$$(29, 43), (30, 44), (31, 45), (32, 46)$$
$$\{33\}\{34\} \ldots \{42\}.$$

So the total number of holes will be 28. Hence two n_i's must be in the same hole. It means we have n_i, n_j such that $n_j - n_j = 14$.

22. Let n, t be nonnegative integers. Show that

$$\sum_{t=1}^{n} tC(n,t) = n \cdot 2^{n-1}$$

Solution:
We can see that the summation $\sum_{t=1}^{n} C(n,t)$ will occur as coefficients of $(1+x)^n$ with $x = 1$.

That is

$$(1+x)^n = \sum_{t=1}^{n} C(n,t)x^t$$
$$\frac{d}{dx}(1+x)^n = \frac{d}{dx}\left(\sum_{t=1}^{n} C(n,t)x^t\right)$$
$$n(1+x)^{n-1} = \sum_{t=1}^{n} tC(n,t)x^{t-1}$$

For $x = 1$, we get,

$$\sum_{t=1}^{n} tC(n,t) = n2^{n-1}.$$

23. Show that

$$(x+y+z)^4 = 1(x^4+y^4+z^4) + 4(x^3y + x^3z + xy^3 + y^3z + xz^3 + yz^3) + 6(x^2y^2 + x^2z^2 + y^2z^2) + 12(x^2yz + xy^2z + xyz^2)$$

Solution:
Coefficient of x^4, y^4, z^4 is 1 when x, y, z from each term in the product $(x+y+z)(x+y+z)(x+y+z)(x+y+z)$ is taken. Hence it is 1.

Coefficient of x^3y is 4. This is because one has to take three terms in the product $(x + y + z)(x + y + z)(x + y + z)(x + y + z)$ to form x^3 and y from the single term. The argument will be identical for the terms $x^3z, xy^3, y^3z, zx^3, yz^3$ as well.

Coefficient of x^2y^2 is 6. This is because one has to take two terms in the product $(x + y + z)^4$ to form x^2 and y^2 from the remaining two terms.

Coefficient of x^2yz being 12 can be identically seen as above.

24. Find the expansion of $(1 + t)^k, k = 3, 4, 5, 6$.

Solution:
The coefficients of the terms in the expansion of $(1 + t)^3$ can be seen to occur in the 3rd row of Pascal triangle. So

$$(1 + t)^3 = 1 + 3t^2 + 3t + t^3$$

Similarly, by referring the Pascal triangle, we have

$$(1 + t)^4 = 1 + 4t + 6t^2 + 4t^3 + t^4$$
$$(1 + t)^5 = 1 + 5t + 10t^2 + 10t^3 + 5t^4 + t^5$$

25. Prove the following

(i) $\sum_{t=0}^{n} C(n, k) = 2^n$
(ii) The sum of the entries in column 'i' ($i \geq 0$) of Pascal's triangle starting from row '0' upto row 'n', equals the $((n + 1), (i + 1))^{th}$ entry. That is

$$\sum_{t=0}^{n} C(t, i) = C(n + 1, i + 1)$$

Solution:
(i) Consider

$$(1 + x)^n = \sum_{t=0}^{n} C(n, t)x^n$$

Substitute $x = 1$, we get

$$2^n = \sum_{t=0}^{n} C(n, t)$$

(ii) We prove this by using induction on the 'row number' m.

The sum entry below row '0' will be 1 in column $i = 0$. The entries from row '0' upto row m in any other column will be '0'.

So,

$$C(1, i+1) = \begin{cases} 1 & \text{if } i = 0 \\ 0 & \text{otherwise.} \end{cases}$$

Hence the result is true for $m = 0$. Assume the result to be true for some $m = n$. That is for $n \geq 1$,

$$\sum_{t=0}^{n-1} C(t, i) = C(n, i+1)$$

Now consider

$$\begin{aligned} \sum_{t=0}^{n} C(t, i) &= \sum_{t=0}^{n-1} C(t, i) + C(n, i) \\ &= C(n, i+1) + C(n, i) \\ &= C(n+1, i+1). \end{aligned}$$

26. Prove the following

(i) $\sum_{t=0}^{n} C(n, t)2^t = 3^n$.

(ii) $\sum_{t=1}^{n} tC(n, t) = n \cdot 2^{n-1}$.

Solution:
(i) Let $X = \{1, 2, \ldots, n\}$.
A subset A of X of size 't' can be chosen in $C(n, t)$ ways.
A subset $\overline{A}$ of A can be chosen in 2^t ways.

Hence there are $\sum_{t=0}^{n} C(n, t)2^t$ ways are the number of pairs $(A, \overline{A})$ with $A \subseteq X$ which is the left hand side expression.

For the right hand side count, we observe the following.

For every element $x \in \{1, 2, \ldots, n\}$ the following are the possibilites for forming the pair $(A, \overline{A})$, $A \subseteq X$.

(i) The element $x \in A$ and $\overline{A}$.
(ii) The element $x \in A$ but not in $\overline{A}$.
(iii) The element $x \notin A$ and hence not in $\overline{A}$.

Hence the total of pairs $(A, \overline{A})$ will be 3^n.
(ii) This problem can also be seen as a selection and combination problem.

For the term $tC(n, t)$, we can say that from a set of n elements, we choose a t-element subset A and then choose an element $\overline{x}$ from A as a representative. For all possible 't' values, we have $\sum_{t=1}^{n} C(n, t)$ to give the count of the pairs $(A, \overline{x})$ in every possible way.

The right hand side counts the same pair by choosing a representative $\overline{x}$ from an 'n' element set and then pick a subset $\overline{A}$ of size $(t-1)$. Then add $\overline{x}$ to $\overline{A}$ to form a 't' element subset. Hence this can be done in $n \cdot 2^{n-1}$ ways.

27. Show that

$$\sqrt{7} = \sqrt{6}\left[1 + \frac{1}{1!} \cdot \frac{1}{2} \cdot \frac{1}{6} - \frac{1}{2!} \cdot \frac{1}{4}\left(\frac{1}{6}\right)^2 + \frac{1}{3!} \cdot \frac{1}{8} \cdot \left(\frac{1}{6}\right)^3 + \cdots\right]$$

Solution:
Consider

$$\begin{aligned}
\sqrt{7} &= (6+1)^{1/2} \\
&= \sqrt{6}\sum_{t=0}^{\infty} C(\frac{1}{2}, t)\left(\frac{1}{6}\right)^t \\
&= \sqrt{6}\left[1 + \frac{1}{1!} \cdot \frac{1}{2} \cdot \frac{1}{6} - \frac{1}{2!} \cdot \frac{1}{4}\left(\frac{1}{6}\right)^2 + \cdots\right]
\end{aligned}$$

28. Find the binomial expansion of $(1+x)^{-1}$. Is the expanded series convergent? Justify your answer.

Solution:

$$\begin{aligned}
(1+x)^{-1} &= \sum_{t=0}^{\infty} C(-1, t)(x)^t \\
&= \sum_{t=0}^{\infty} \frac{(-1)^t 1 \cdot 2 \cdot 3 \cdots t}{t!} \cdot x^t \\
&= 1 - x + x^2 - \cdots x^n + \cdots
\end{aligned}$$

This series converges $\forall\ x$ such that $|x| < 1$.

29. Let n and t be non negative integers. Show that

$$C(n, t) + C(n-1, t) + C(n-2, t) + \cdots + C(t, t) = C(n+1, t+1).$$

Solution:
From $C(n+1, t+1)$, we can see that we count the number of subsets of $(n+1)$ having $(t+1)$ elements.

This can be done via a system which selects 't' elements from $t+i$ elements $t+i \leq n$ such that a $(t+1)$ element subset can be formed by taking a 'leading' element of the set being $t+i+1$. Note $t+i+1$ is like a 'leader' of the group of $(t+i)$ elements. We have the left hand side of the equation for $0 \leq i \leq n$. Note that this left-hand side summation can be seen to be sum of the diagonal elements whose first position being the left most entry in the row corresponding to 't'. The summation leads to the right hand side, an entry in the same diagonal where we started at.

30. Expand $\sqrt{1-9x}$ as s power series.

Solution:

$$\sqrt{1-9x} = (1-9x)^{\frac{1}{2}}.$$

Using Newton's Binomial theorem we have

$$(1-9x)^{\frac{1}{2}} = \sum_{n=0}^{\infty} C(\frac{1}{2}, n)(-9x)^n$$

Here

$$C(\frac{1}{2}, n) = \frac{\frac{1}{2} \cdot \frac{-1}{2} \cdot \frac{-3}{2} \cdots \frac{-2n+3}{2}}{n!}$$
$$= (-1)^{n-1} \frac{1 \cdot 3 \cdot 5 \cdot 7 \cdots (2n-3)}{n! \cdot 2^n}$$

So

$$(1-9x)^{\frac{1}{2}} = \sum_{n=0}^{\infty} 3^n \cdot \frac{1 \cdot 3 \cdot 5 \cdot 7 \cdots (2n-3)}{n! \cdot 2^n} \cdot x^n$$

31. A sequence of numbers $a_0, a_1, \ldots, a_n$ is said to be unimodal if there is an integer 't' with $0 \leq t \leq n$ such that

$$a_0 \leq a_1 \leq \cdots \leq a_t, a_t \geq a_{t+1} \geq a_{t+2} \geq \cdots \geq a_n.$$

Such an 'a_t' need not be unique. Consider the sequence $C(n, 0), C(n, 1), \ldots, C(n, n)$, the binomial coefficients for 'n' being a positive integer. Show that this sequence is unimodal.

Solution:
We know

$$C(n, t) = \frac{n!}{t!(n-t)!}$$

The ratio

$$\frac{C(n,t)}{C(n,t-1)} = \frac{n!}{t!(n-t)!} \times \frac{(t-1)!(n-t+1)!}{n!}$$
$$= \frac{n-t+1}{t} \tag{12.9}$$

We have,

$$C(n,t-1) < C(n,t), C(n,t-1) = C(n,t)$$
$$\text{or } C(n,t-1) > C(n,t).$$

This means, by (12.9)

$$t < n-t+1, t = n-t+1 \text{ or } t > n-t+1$$

If $t < n-t+1$, then $t < \frac{n+1}{2}$ and if $t < \frac{n+1}{2} \Rightarrow t < n-t+1$.
If n is even, we have $t \le \frac{n}{2}$.
If n is odd, we have $t \le \frac{n-1}{2}$.
Hence the binomial coefficients increase as seen below for n even.

$$C(n,0) < C(n,1) < C(n,2) < \cdots < C(n,\frac{n}{2})$$
$$C(n,\frac{n}{2}) > C(n,\frac{n}{2}+1) > \cdots > C(n,n).$$

For 'n' being odd we have,

$$C(n,0) < C(n,1) < \cdots < C(n,\frac{n+1}{2})$$
$$C(n,\frac{n+1}{2}) > C(n,\frac{n+1}{2}+1) > \cdots > C(n,n).$$

For $t = n-t+1$, imply $2t = n+1$. If n is even, we can see that the two consecutive coefficients at the middle being equal. For n odd also we can see that $C(n,\frac{n-1}{2})$ and $C(n,\frac{n+1}{2})$ being the same.

32. Show that the number of surjective functions f from a set $X = \{1,2,\ldots,n\}$ to a set $\overline{X} = \{1,2,3,\ldots,t\}$ is $t!S'(n,t)$.

Solution:
From the definition of f, we can see that the elements of X are partitioned to 't' blocks. The blocks are labelled and there are exactly t of such blocks. If we place the elements of $\overline{X}$ into the 't' blocks in $S'(n,t)$ ways, there are $t!$ ways to form t member subsets of X. Hence the number of surjective functions will be $t!S'(n,t)$.

33. Show that

$$\sum_{n\geq t} \frac{S'(n,t)x^n}{n!} = \frac{(e^x-1)^t}{t!}$$

Solution:
We show the equality using mathematical induction on 't'.
For $t=1$, $S'(n,1)=1$ and hence the equality $\sum_{n\geq 1}\frac{x^n}{n!}=(e^x-1)$ is true.
Assume the equality to be true for $t=m-1$. Then,
Consider

$$\frac{(e^x-1)^m}{m!} = \frac{1}{m}(e^x-1)\cdot\frac{(e^x-1)^{m-1}}{(m-1)!}$$
$$= \frac{1}{m}\left(\sum_{n\geq 1}\frac{x^n}{n!}\right)\left(\sum_{n\geq(m-1)}\frac{S'(n,m-1)}{n!}x^n\right)$$

Consider $S'(n,t)=0$ for all $n<t$, we have

$$\frac{(e^x-1)^m}{m!} = \frac{1}{m}\left(\sum_{n\geq 1}\frac{x^n}{n!}\right)\left(\sum_{n\geq 1}\frac{S'(n,m-1)}{n!}x^n\right)$$

Coefficient of $\frac{x^n}{n!}$ will be

$$\frac{n!}{m}\sum_{j=1}^{n-1}\frac{S'(n-j,m-1)}{(n-j)!j!} = \frac{1}{m}\sum_{j=1}^{n-1}C(n,j)S'(n-j,m-1) \tag{12.10}$$

Now

$$S'(n,t) = C(n-1,0)S'(n-1,t-1)+C(n-1,1)S'(n-2,t-1)$$
$$+\cdots+C(n-1,n-2)S'(1,t-1)$$
$$i.e., S'(n,t) = \sum_{j=1}^{n-1}C(n-1,j-1)S'(n-j,t-1) \tag{12.11}$$

$S'(n,t)$ means the number of partitions of $X=\{1,2,\ldots,n\}$ into 't' arbitrary sets. The part 'j' that contains 'n' of an arbitrary partition means that there are $C(n-1,t-1)$ choices of the remaining elements in that part. Note that $S'(n-j,t-1)$ gives the number partitions of the remaining $(n-j)$ elements into $(t-1)$ parts. Hence we have (12.11).

From (12.11) we can see that the rhs of (12.10),

$$\frac{1}{m}\sum_{j=1}^{n-1} C(n, j)S'(n-j, m-1) = \frac{1}{m}[S'(n+1, m) + S'(n, m-1)]$$
$$= S'(n, t).$$

34. How many 4 letter words can be formed from the first 7 letters of English alphabet that always includes 'a' in it (The words need not be meaningful).

Solution:
We need to include always 'a' in the word. The letter a can be in any of the 4 positions.

$$a - - -, -a - -, - - a -, - - - a.$$

There are $P(4, 1)$ ways = 4. The remaining 3 positions will use the letters $\{b, c, d, e, f, g\}$. This can be done in $P(6, 3)$ ways. Hence the total number of words will be

$$P(4, 1) \times P(6, 3) = 4 \times 120 = 480.$$

35. How many 5 letter words can be formed using the letters in the word 'S A U S E L I T O' that contains atleast one vowel.

Solution:
The number of vowels in 'S A U S E L I T O' is 5. The total number of symbols in the given word is 9. Hence totally there are $P(9, 5)$ words of length 5 can be formed. That is 15120 words of length 5 are there and every such word must use a vowel as there are only 4 non-vowels in the given word.

36.

(i) A bag contains 6 wooden balls. A ball is taken out randomly and painted with any one of 6 distinct colors. Find the number of ways the 6 balls be selected.
(ii) There are 9 flowers plants placed in 9 different pots in a garden. Find the number of ways can 4-plants be placed side-by-side near the entrance of the house.

Solution:

(i) Once a ball is taken out it is painted. So there are $P(6, 6)$ ways to pick a ball. Hence there are 720 ways to do so.
(ii) The problem is a permutation problem where 4 plants to be selected from 9 plants to be placed side-by-side. Hence the answer will be $P(9, 4) = 9 \times 8 \times 7 \times 6 = 3024$.

37. A mathematics course has 20 students. The course instructor conducts a surprise quiz. Four students who did not prepare want to leave the class on the quiz day. They leave the class one after another telling some health reasons to the course instuctor. Find the number of ways can this happen.

Solution:
There are 20 students who are in the class while at the start of the quiz.

Any one among the 4-students will leave the class first by saysing a lie to the instructor. After this time, the class will have 19 students and the next student will leave the class. This continues till all the 4 students leave the class. Hence there are $P(20, 4)$ ways that this can happen.

38.

(i) Find the number of 4-digit numbers that can be formed using only the numbers $\{1, 2, 3, 4, 5, 6, 7, 8, 9\}$.
(ii) Find the number of 4-digit odd numbers that can be formed using the digits $\{1, 2, 3, 4, 5, 6, 7, 8, 9\}$.
(iii) Find the number of 4-digit even numbers that can be formed using the digits $\{1, 2, 3, 4, 5, 6, 7, 8, 9\}$.
(iv) Find the number of 4-digit numbers that can be formed using the digits $\{1, 2, 3, 4, 5, 6, 7, 8, 9\}$ and always end with either 5 or 6.

Solution:

(i) $P(9, 4) = 3024$.
(ii) If the four digit number '$a_1a_2a_3a_4$' is odd, then the last digit $a_4 = \{1, 3, 5, 7, 9\}$. In this case $a_1a_2a_3$ can be any number from the set $\{1, 2, 3, 4, 5, 6, 7, 8, 9\}$. Hence there are $5 \times P(8, 3)$ such numbers.
(iii) $4P(8, 3)$.
(iv) This is a subcase of (ii) and (iii). The count is $2P(8, 3)$.

39. Show that the following statements are equivalent.

(i) Number of ways of distributing 12 balls in three identical boxes.
(ii) Number of non-negative integer solutions (x, y, z) of the equation $x + y + z = 12$.

Solution:

(i) Let 'x' be the number of balls put in box 1, y be the number of balls put in box 2 and z be the number of balls put in box 3. So we have to find out the number of tuples (x, y, z)

satisfying the equation $x + y + z = 12$ which is nothing but statement (ii). Here $x \geq 0$, $y \geq 0$, $z \geq 0$. The answer will be $C(14, 12) = 91$.

(ii) implying (i) can be identically proved.

40. A worker has to work in a company for 3 d in March. However the worker is not permitted to work on consecutive days in the company. Find the number of choices for the worker.

Solution:

The worker must make sure that his choice does not include two consecutive days. Let x_1, x_2 and x_3 be the three days choosen by him, in the order that he will work on day x_1, following day x_2 and then day x_3. Clearly $1 \leq x_i \leq 31$, $i = 1, 2, 3$. Also $x_2 \neq x_1 + 1$, $x_3 \neq x_2 + 1$. So we have $1 \leq x_1 < x_2 - 1 < x_3 - 2 \leq 29$. That is the choice must be such that $1 \leq x_1 < x_2 - 1 < x_3 - 2 \leq 29$. That is simply choosing the 3 d of $\{1, 2, 3, \ldots, 29\}$ which is $C(29, 3)$.

41. Let $r_1, r_2, \ldots, r_k$ be positive integers with $\sum_{i=1}^{k} r_i < n$. Then show that

$$r_1!r_2!\ldots r_k! < n!$$

Solution:

Let us suppose $\sum_{i=1}^{k} r_i < n$. Then

$$D = \frac{n!}{r_1!r_2!\ldots r_k!}$$

is nothing but arrangement of n elements in a row so that the object 'i' is repeated r_i-times. Hence D is a positive integer. Hence $n! \geq r_1!r_2!\ldots r_k!$. Hence if $\sum_{i=1}^{k} r_i < n$, then $r_1!r_2!\ldots r_k! < n!$.

42. Find the coefficient of x^{96} in the generating function $F(x)$ given below.

(i) $F(x) = \frac{1}{1+2x}$
(ii) $F(x) = \frac{1}{(1+3x)^2}$

Solution:

(i)

$$\begin{aligned} F(x) &= (1 + 2x)^{-1} \\ &= 1 - 2x + 2^2x^2 - 2^3x^3 + 2^4x^4 - \cdots . \end{aligned}$$

Coefficient of x^{96} is $+2^{96}$.

(ii)

$$\begin{aligned} F(x) &= (1+3x)^{-2} \\ &= (1+3x)^{-1}(1+3x)^{-1} \\ &= (1-3x+3^2x^2-3^3x^3+\cdots) \\ &\quad (1-3x+3^2x^2-3^3x^3+\cdots). \end{aligned} \tag{12.12}$$

x^{96} can be seen to be obtained from the product (12.12) as

$$tx^{96} = t_0 1x^{96} + t_1 x x^{95} + t_2 x^2 x^{94} + t_3 x^3 x^{93} + \cdots + t_{95} x^{95} x + t_{96} x^{96} 1$$

Here,

$$\begin{aligned} t_0 &= (-3)^{96} \\ t_1 &= (-3)(-3)^{95} \\ t_2 &= (-3)^2(-3)^{94} \\ &\cdots \\ t_{95} &= (-3)^{95}(-3) \\ t_{96} &= (-3)^{96} \end{aligned}$$

$$t = t_0 + t_1 + t_2 + \cdots + t_{96} = 97 \cdot 3^{96}.$$

43. Find the number of integer solutions of $x_1 + x_2 + x_3 = 7$, $1 \le x_1 \le 2$, $1 \le x_2, x_3 \le 4$.

Solution:
The generating function for the condition $1 \le x_1 \le 2$ will be $f_1(x) = x + x^2$.
Similarly the generating functions for x_2, x_3 will be,

$$\begin{aligned} f_2(x) &= x + x^2 + x^3 + x^4 \\ f_3(x) &= x + x^2 + x^3 + x^4 \end{aligned}$$

Hence the product generating function $F(x)$ will be

$$\begin{aligned} F(x) &= f_1(x) f_2(x) f_3(x) \\ &= (x+x^2)(x+x^2+x^3+x^4)^2 \\ &= x^3(1+x)(1+x+x^2+x^3)^2 \end{aligned}$$

Coefficient of x^7 will be coefficient of x^4 in the product $(1+x)(1+x+x^2+x^3)^2$. It will be 7.

44. Find the closed form expression for the generating functions of the following sequences.

(i) $2, 2, 2, 2, 2, \ldots$
(ii) $7, 0, 7, 0, 7, 0, \ldots$
(iii) $1, 6, 19, 40, \ldots, 1 + n + 4n^2, \ldots$.

Solution:
(i) The generating function for $2, 2, 2, \ldots$ is

$$\begin{aligned}
&2 + 2x + 2x^2 + 2x^3 + \cdots \\
&= 2(1 + x + x^2 + \cdots) \\
&= \frac{2}{1-x}.
\end{aligned}$$

(ii) The generating function for $7, 0, 7, 0, 7, \ldots$ is

$$\begin{aligned}
&7 + 0x + 7x^2 + 0x^3 + 7x^4 + \cdots \\
&= 7(1 + x^2 + x^4 + x^6 + \cdots) \\
&= \frac{7}{1-x^2}.
\end{aligned}$$

(iii) nth term of this sequence is $1 + n + 4n^2$. So the generating function is $\sum_{t=0}^{\infty} x^t + \sum_{t=0}^{\infty} tx^t + 4\sum_{t=0}^{\infty} t^2x^t$.

We now try to write $1 + n + 4n^2$ as sum of Binomial coefficients. That is

$$1 + n + 4n^2 = a + bC(n+1, 1) + dC(n+2, 2)$$

For $n = 0$,

$$a + b + d = 1 \tag{12.13}$$

For $n = 1$,

$$a + 2b + 3d = 6 \tag{12.14}$$

For $n = 2$

$$a + 3b + 6d = 19 \tag{12.15}$$

$a = 4, b = -11, c = 8$

$$\sum_{t=0}^{\infty}(1+n+4n^2)x^t = 4\sum_{t=0}^{\infty}x^t - 11\sum_{t=0}^{\infty}C(n+1,1)x^t + 8\sum_{t=0}^{\infty}C(n+2,2)x^t$$
$$= \frac{4}{(1-x)} - \frac{11}{(1-x)^2} + \frac{8}{(1-x)^3}$$

45. Find the number of ways of placing n balls into 2 boxes such that no box contains exactly 3 balls.

Solution:
Number of ways distributing n balls into 2 boxes such that no box contain exactly 3 balls means that the generating function associated will be

$$(1+x+x^2+x^4+\cdots)^2$$
$$= [1+x+x^2+x^4(1+x+x^2+\cdots)]^2$$
$$= [1+x+x^2+x^4(1-x)^{-1}]^2$$

The coefficient of x^n will be the number of ways of placing the n balls in 2 boxes satisfying the given condition.

46. A gardener arranges n flower pots in a row. The owner of the garden splits the arranged flower pots at different positions on the row, there by forming a smaller non empty group of pots for using them in turn for decoration. Find the number of ways the owner can do this.

Solution:
Let a_n be the number of ways the gardener can arrange the pots in a row.

If $a_n = 1$, for $n \geq 1$, there is one way which is the original arrangement itself. For the original arrangement, if the number of splits make 't' groups, then the owner chooses one of the 't' groups for decoration. Let b_n be the number of ways to choose the subset of pots for decoration by the owner and let $b_0 = 1$. Clearly $b_n = 2^n$ as the owner simply makes a choice on the set of groups. Hence

$$F(x) = x + x^2 + x^3 + \cdots = \frac{x}{1-x}$$
$$G(x) = 1 + 2x + 2^2x^2 + \cdots = \frac{1}{1-2x}.$$

This means $F(x)$ is the generating function for the arrangements of the gardener and $G(x)$ is the generating function for the group being chosen by the owner from the arrangement of the gardener. So choice of the owner is dependent on the arrangements by the gardener. Hence the generating function $K(x)$ corresponding to the final choice of the owner will be composition of F and G. That is

$$\begin{aligned}K(x) = G(F(x)) &= G\left(\frac{x}{1-x}\right)\\ &= \frac{1}{1-\frac{2x}{1-x}}\\ &= \frac{1-x}{1-3x} = \frac{1}{1-3x} - \frac{x}{1-3x}\\ &= \sum_{k=0}^{\infty} 3^k x^k - \sum_{k=1}^{\infty} 3^{k-1} x^k\\ &= 1 + \sum_{k=1}^{\infty} 2 \cdot 3^{k-1} x^k.\end{aligned}$$

So there are $2 \cdot 3^{n-1}$ ways for the owner to his choice.

47. Find a direct expansion for a_t if $a_0 = 0$ and $a_{t+1} = a_t + 2^t$, $t \geq 1$ using generating function method.

Solution:

Let $F(x) = \sum_{t=0}^{\infty} a_t x^t$ be the generating function for the sequence $\{a_t\}_{t=0}^{\infty}$.

Consider

$$a_{t+1} = a_t + 2^t \tag{12.16}$$

Multiply (12.16) throughout x^{t+1} and sum over from 0 to ∞.

$$\begin{aligned}\sum_{t=0}^{\infty} a_{t+1} x^{t+1} &= \sum_{t=0}^{\infty} a_t x^{t+1} + \sum_{t=0}^{\infty} 2^t x^{t+1}\\ &= x \sum_{t=0}^{\infty} a_t x^t + x \sum_{t=0}^{\infty} (2x)^t\end{aligned}$$

$$\text{So } F(x) = xF(x) + \frac{x}{1-2x}$$

$$\begin{aligned}i.e.,\ F(x) &= \frac{x}{(1-x)(1-2x)}\\ &= x(1 + x + x^2)(1 + 2x + 2^2x^2 + \cdots)\end{aligned}$$

Coefficient of x^t is

$$a_t = \sum_{j=0}^{t-1} 2^j = 2^t - 1.$$

48. Solve the following recurrence relation using generating functions.

$$a_n + a_{n-1} - 6a_{n-2} = 0, n \geq 2, a_0 = 1, a_1 = 3$$

Solution:

Let $F(x) = \sum_{k=0}^{\infty} a_k x^k$ be the generating function of the sequence $\{a_k\}_{k=0}^{\infty}$.

Multiply throughout x^{k+2} and sum the recurrence relation from $0 \to \infty$.

$$\sum_{k=0}^{\infty} a_{k+2}x^{k+2} + \sum_{k=0}^{\infty} a_{k+1}x^{k+2} - 6\sum_{k=0}^{\infty} a_k x^{k+2} = 0$$

$$i.e., (F(x) - a_1 x - a_0) + x(F(x) - a_0) - 6x^2 F(x) = 0$$

$$(1 + x - 6x^2)F(x) - x(a_1 + a_0) - a_0 = 0$$

Substitute $a_0 = 1, a_1 = 3$, we get

$$F(x) = \frac{1+4x}{1+x-6x^2}$$

$$= \frac{1+4x}{(1-2x)(1+3x)} = \frac{6}{5(1-2x)} - \frac{1}{5(1+3x)}$$

$$= \frac{6}{5}\sum_{k=0}^{\infty} 2^k x^k - \frac{1}{5}\sum_{k=0}^{\infty}(-3)^k x^k$$

Hence

$$a_k = \frac{6}{5}2^k - \frac{1}{5}(-3)^k$$

49. Solve the following recurrence relation using the generating function $\sum_{n=0}^{\infty} \frac{a_n x^n}{n!}$ with $a_0 = 2$.

$$a_{n+1} - a_n - n a_{n-1} = 0, n \geq 2.$$

Solution:

For $n = 0$, the recurrence relation reduces as $a_1 - a_0 = 0 \Rightarrow a_1 = a_0$.

Now multiply throughout the recurrence relation by $\frac{x^n}{n!}$ we get,

$$a_{n+1}\frac{x^n}{n!} - a_n\frac{x^n}{n!} - n\frac{a_{n-1}x^n}{n!} = 0$$

Taking summation throughout from $n = 0$ to ∞ we get

$$\sum_{n=0}^{\infty} a_{n+1}\frac{x^n}{n!} - \sum_{n=0}^{\infty} a_n\frac{x^n}{n!} - x\sum_{n=1}^{\infty} a_{n-1}\frac{x^n}{(n-1)!} = 0$$

$$F'(x) - F(x) - xF(x) = 0$$

$$\Rightarrow \frac{F'(x)}{F(x)} = (1+x)$$

Integrating throughout and using the value $F(0) = 0$ we get

$$F(x) = e^{\left(x+\frac{x^2}{2}\right)}$$

$$= 1 + \left(x + \frac{x^2}{2}\right) + \frac{\left(x+\frac{x^2}{2}\right)^2}{2!} + \cdots$$

Coefficient of $\frac{x^n}{n!}$ will give a_n.

50. Consider the recurrence relation

$$a_{n+1} = (n+1)a_n + 2, \ a_0 = 1.$$

Using the generating function $G(x) = \sum_{n=0}^{\infty} a_n \frac{x^n}{n!}$, find an explicit formula for the generating function $G(x)$.

Solution:

$$a_{n+1} = (n+1)a_n + 2$$

$$\sum_{n=0}^{\infty} a_{n+1}\frac{x^{n+1}}{(n+1)!} = \sum_{n=0}^{\infty}(n+1)a_n\frac{x^{n+1}}{(n+1)!} + 2\sum_{n=0}^{\infty}\frac{x^{n+1}}{(n+1)!} \tag{12.17}$$

Clearly $\sum_{n=0}^{\infty} a_{n+1}\frac{x^{n+1}}{(n+1)!} = G(x) - 1$ as $a_0 = 1$. Hence (12.17) can be written as

$$G(x) - 1 = x\sum_{n=0}^{\infty} a_n\frac{x^n}{n!} + 2\left[\sum_{n=0}^{\infty}\frac{x^n}{n!} - 1\right]$$

$$= xG(x) + 2(e^x - 1) + 1$$

Hence

$$G(x) = \frac{2(e^x - 1) + 1}{1 - x} = \frac{2e^x}{1 - x} - \frac{1}{1 - x}$$
$$= 2\left(\sum_{n=0}^{\infty} \frac{x^n}{n!}\right)\left(\sum_{n=0}^{\infty} x^n\right) - \left(\sum_{n=0}^{\infty} x^n\right)$$
$$= 2\left(\sum_{n=0}^{\infty} \frac{x^n}{n!}\right)\left(\sum_{n=0}^{\infty} n!\left(\frac{x^n}{n!}\right)\right) - \left(\sum_{n=0}^{\infty} n!\frac{x^n}{n!}\right)$$
$$= 2\left(\sum_{n=0}^{\infty}\left(\sum_{t=0}^{n} C(n,t)t!\right)\frac{x^n}{n!}\right) - \sum_{n=0}^{\infty} n!\frac{x^n}{n!}$$
$$= \left(\sum_{n=0}^{\infty} 2\left(\sum_{t=0}^{n} C(n,t)t!\right) - n!\right)\frac{x^n}{n!}$$
$$= \sum_{n=0}^{\infty} n!\left(\sum_{t=0}^{n} \frac{2}{(n-t)!} - 1\right)\frac{x^n}{n!}$$

Therefore

$$a_n = n!\left(\sum_{t=0}^{n} \frac{2}{(n-t)!} - 1\right).$$

51. Solve for a_n explicity if

$$a_{n+1} = a_n + 5^n, n \geq 0, a_0 = 13.$$

Solution:
Here

$$a_{n+1} = a_n + 5^n$$

By repeated substitution, we get

$$a_n = a_{n-1} + 5^{n-1}$$
$$= a_{n-2} + 5^{n-2} + 5^{n-1}$$
$$\cdots\cdots$$
$$= a_0 + 5^0 + 5^1 + \cdots + 5^{n-1}$$
$$= 13 + \frac{5^n - 1}{5 - 1} = \frac{1}{4}[5^n - 51].$$

52. (Tower of Brahma) It is given that there are three rod stands, and n number of disks of different diameter can slide on to any rods. Initially one rod contains all the n disks in the descending order of their diameter from bottom to top. The problem is to find the number of moves required to move the present arrangement of disks on to a new rod with the following conditions.

(i) Only one disk can be moved at a time.
(ii) Only the topmost disk can be moved.
(iii) Always, an arrangement of any of the three rods be such that a disk of larger diameter being on the top of a disk of a smaller diameter is not permitted.

Solution:
The conditions imply that at any intermediate stage of moves, all the three rods have proper arrangement as per conditions. Let the disks be numbered as $1, 2, 3, \ldots, n$ such that number 1 is given to a disk of the smallest diameter and the disks are numbered in the increasing order of their diameter. Number 'n' is given to the disk of largest diameter. Let the rods be named as R_1, R_2 and R_3. The problem is to move the disks from R_1 to R_3 say. Let a_n denote the number of moves required to do this. First we move $(n-1)$ disks from R_1 to R_2 which are properly arranged in a_{n-1} moves and then move the nth disk to R_3. Now using the vacant R_1, move the disks in R_2 to R_3 which needs a_{n-1} moves. The recurrence relation will be

$$a_n = 2a_{n-1} + 1.$$

Solving by repeated substitution method we have

$$\begin{aligned} a_n &= 2a_{n-1} + 1, \ n \geq 1, a_1 = 1 \\ &= 2(2a_{n-2} + 1) + 1 \\ &\ldots\ldots \\ &= 2^{n-1}a_1 + (1 + 2 + \cdots + 2^{n-2}) \\ &= 2^n - 1. \end{aligned}$$

53. Find the number of ways of selecting 5 cards so that the selection has atleast one card from each category.

Solution:
The total number of ways selecting 5 cards is $C(52, 5)$. Let N_i denote the set of all possible 5 cards being selected from the deck that will not contain a card from category i, $1 \leq i \leq 4$. So,

$$\begin{aligned} \#N_i &= C(39, 5), 1 \leq i \leq 4 \\ \#N_i \cap N_j &= C(26, 5), 1 \leq i, j \leq 4, i \neq j \\ \#(N_i \cap N_j \cap N_k) &= C(13, 5), 1 \leq i, j, k \leq 4, i \neq j \neq k \\ \#(N_i \cap N_j \cap N_k \cap N_t) &= 0 \end{aligned}$$

Using PIE (Principle of Inclusion, Exclusion) we have,

$$\begin{aligned}
\#\overline{N} &= C(52,5) - \sum_{i=1}^{4} \#N_i + \sum_{1\le i,j\le 4} \#(N_i \cap N_j) \\
&\quad - \sum_{1\le i,j,k\le 4} \#(N_i \cap N_j \cap N_k) + 0 \\
&= C(52,5) - 4\,C(39,5) + 6\,C(26,5) - 4\,C(13,5).
\end{aligned}$$

54. How many integers are there in the set $\{1, 2, 3, \ldots, 180\}$ that are not relatively prime to 180.

Solution:
The prime factors of 180 are 2, 3, 5. Let A_2, A_3 and A_5 denote the numbers that have prime factors 2, 3 and 5 respectively and they are less than 180. We need to compute $|A_2 \cup A_3 \cup A_5|$. By PIE,

$$\begin{aligned}
|A_2 \cup A_3 \cup A_5| &= |A_2| + |A_3| - A_3 - |A_2 \cap A_3| - |A_2 \cap A_5| \\
&\quad - |A_3 \cap A_5| + |A_2 \cap A_3 \cap A_5| \\
&= \frac{180}{2} + \frac{180}{3} + \frac{180}{5} - \left\lfloor \frac{180}{2\times 3} \right\rfloor - \left\lfloor \frac{180}{2\times 5} \right\rfloor \\
&\quad - \left\lfloor \frac{180}{3\times 5} \right\rfloor + \frac{180}{2\times 3\times 5}
\end{aligned}$$

55. In a fine arts class, it is known that out of 50 students

30 students know to sing
18 students know to dance
26 students know to paint
9 students know to sing and dance
16 students know to sing and paint
8 students know to dance and paint
47 students know atleast one of the art.

(i) Find the number of students who know none of the three arts.
(ii) Find the number of students who know all the three.

Solution:
(i) 3 students do not know any of the art.
(ii) Let A, B, C denote the three sets of students of the class who know singing, dancing and painting respectively.

$$|A \cup B \cup C| = |A| + |B| + |C| - |A \cup B| - |B \cap C| - |A \cap C| + |A \cap B \cap C|$$
$$47 = 30 + 18 + 26 - 9 - 8 - 16 + |A \cap B \cap C|$$
$$\Rightarrow |A \cap B \cap C| = 6.$$

So there are 6 students who know all the three arts.

56. Consider the following Hasse diagram of the set $S = \{1, 2, 3, 4, 5, 6, 7, 8, 9\}$. Find the

(i) upper and lower bounds of $\{4, 5, 6\}$, $\{1, 3\}$, $\{2, 3\}$, $\{7, 9\}$.
(ii) glb and lub of the same sets given in (i) (Fig. 12.1).

Solution:
(i) Lower bound of $\{4, 5, 6\}$, $\{1, 3\}$, $\{2, 3\}$ is $\emptyset$.
Upper bound of $\{4, 5, 6\}$, $\{7, 9\}$ is also $\emptyset$.
Upper bound of $\{1, 3\}$ is 8.
Upper bound of $\{2, 4\}$ will be $\{4, 7\}$.
Lower bound of $\{7, 9\}$ is 5.
The least upper bound does not exist for $\{4, 5, 6\}$, $\{7, 9\}$.
The greatest lower bound of $\{4, 5, 6\}$ and $\{1, 4\}$ does not exist.
The lub of $\{1, 3\}$ is 8.
The lub of $\{2, 4\}$ is 4.
The glb of $\{2, 4\}$ is 2.
The glb of $\{7, 9\}$ is 5.

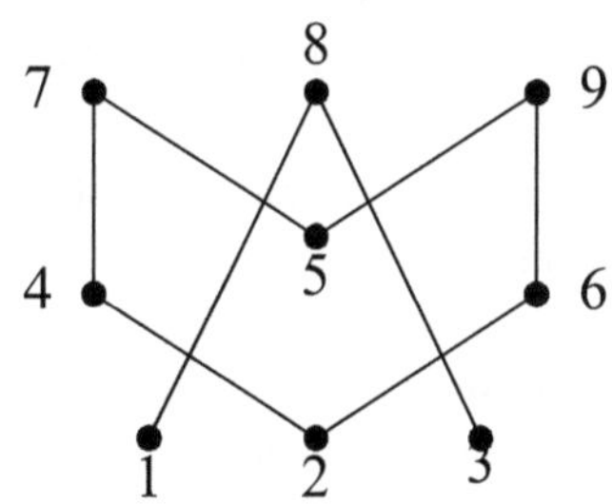

Fig. 12.1 A Hasse diagram

57. Check whether the following Hasse diagrams are Lattices or not. Justify your answer (Figs. 12.2 and 12.3).

(i)

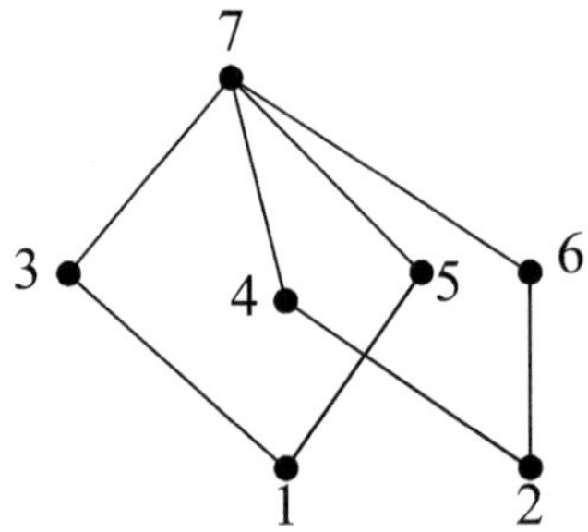

Fig. 12.2 Hasse diagram for (i)

(ii)

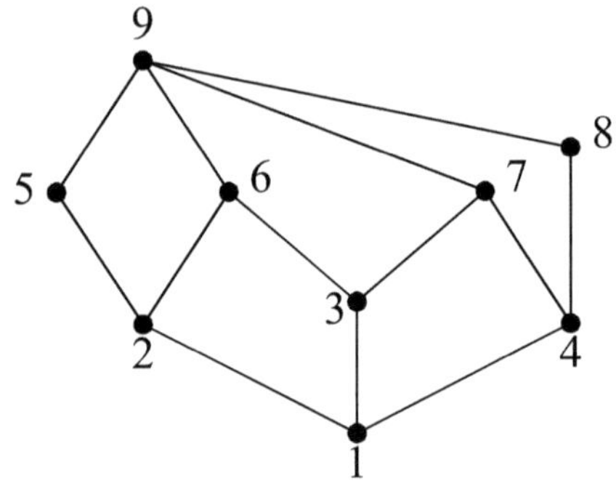

Fig. 12.3 Hasse diagram for (ii)

Solution:

(i) is not a Lattice as $\{1, 2\}$ does not have a lub.
(ii) Yes as every pair $\{i, j\}$ has both lub & glb for $1 \leq i, j \leq 9$.

58. Let $(\mathcal{P}, \leq)$ be a finite lattice. Then is it always true that it has a minimum and a maximum element?

Solution:
We prove by method of contradiction by assuming that $\mathcal{P}$ does not have minimum element. Then there must be two minimal elements u, v in $\mathcal{P}$ such that $u \neq v$. By definition $u \wedge v$ has to be either less than u, v or $u \wedge v = u, u \wedge v = v$. Since both u, v being minimal elements

$u \wedge v$ must be equal to both u, v forcing $u = v$ which is not true. Hence $\mathcal{P}$ has a minimum element. Identically one can show that $\mathcal{P}$ has a maximum element.

59. Let X, Y be two sets such that $X \subseteq Y$. Show that

$$\sum_{X \subseteq A \subseteq Y} (-1)^{|A|-|X|} = \begin{cases} 1, & X = Y \\ 0, & X \neq Y \end{cases}$$

Solution:
When $X = Y$, the result is true as the set X is the only subset between X and X.

So consider $X \subset Y$ and let A be a subset of Y containing X and let $|A| - |X| = t$. Here t can be between 0 and $|Y| - |X| = m$. Number of such subsets A will be $C(m, t)$. Hence the following summation on the 't' such subsets of Y containing X will be

$$\sum_{X \subseteq A \subseteq B} (-1)^{|A|-|X|} = \sum_{t=0}^{m} (-1)^t C(m, t) \tag{12.18}$$

Here the RHS of (12.18) is nothing but a binomial expansion of $(-1+1)^m$ which is '0' for $m \geq 1$.

60. Show that the relation '$\leq$' on $\mathbb{N}$ given by $x \leq y$ iff $y = 3^t x$ for some integer $t \geq 0$ is a partial order.

Solution:
First, for any natural number $n \in \mathbb{N}$ one can see $n = 3^0 n$. Hence for any natural n, '$n \leq n$'. So $\leq$ is reflexive.

Next, we consider $n, m \in \mathbb{N}$ with $n \leq m$ and $m \leq n$. This means $n = 3^{t_1} m$ and $m = 3^{t_2} n$. This means $n = 3^{t_1+t_2} n$ and hence $t_1 = t_2 = 0$. So we get $m = n$. This means '$\leq$' is antisymmetric.

Finally for $n, m, q \in \mathbb{N}$ with $n = 3^{t_1} m$, $m = e^{t_2} q$ imply $n = 3^{t_1+t_2} q$. So $n \leq q$. So '$\leq$' is a partial order relation.

61. Find the number of non-isomorphic graphs on 4 vertices.

Solution:
We know that there are totally 2^4 graphs on 4 vertices. But this count does not exclude isomorphic graphs. So we can understand the graph G on 4-vertices with some edges present and some edges absent between a pair of vertices. That is consider the graph K_4 and color its edges either with black or with white. The graph G is the graph with only black colored edges in it. The white colored edges are left out.

There will be six edges in K_4 and we can understand that the edges of K_4 are induced by a permutation of vertices of the graph. This permutation group S_4 will contain (4!=) 24 elements. The induced permutation group on edges of K_4 be $\overline{S}_4$. It will has its elements edges of K_4.

$$\begin{aligned}S_4 = \{&(1)(2)(3)(4), (1\ 3)(2\ 4), (1\ 4)(2\ 3), (1\ 2)(3\ 4), (1)(2\ 3\ 4), (2)(1\ 3\ 4),\\ &(3)(1\ 2\ 4), (4)(1\ 2\ 3), (1)(2\ 4\ 3), (2)(1\ 4\ 3), (3)(1\ 4\ 2), (4)(1\ 3\ 2),\\ &(1)(2\ 3)(4), (1)(2)(3\ 4), (1\ 2)(3)(4), (2)(3)(1\ 4), (1)(2\ 4)(3),\\ &(1\ 2\ 3\ 4), (1\ 3\ 2\ 4), (1\ 4\ 2\ 3), (1\ 2\ 4\ 3), (1\ 3\ 4\ 2), (1\ 4\ 3\ 2),\\ &(4\ 3\ 2\ 1)\}\end{aligned}$$

Here $\overline{S}_4$ will have 61 elements which are edges $(i\ j)$ of G with $1 \leq i, j \leq 4$. Here the element $\psi(ij)$ will be mapped to the pair $(\pi(i), \pi(j))$ where ψ is the permutation in $\overline{S}_4$ and π is a permutation in S_4. Hence we count the number of cycles in each permutation ψ in $\overline{S}_4$ to know the presence of an edge.

Case (i): The identity permutation on vertices induces an identity permutation in $\overline{S}_4$. This identity permutation has all one cycles. Hence every element in $\overline{S}_4$ can get either of the two colors. So it will be 2^6 possible ways of coloring.

Case (ii): For a permutation $\pi \in S_4$ that is a 4-cycle, $\overline{S}_4$ will have an induced permutation comprising of a product of a 4-cycle and a two cycle. There are totally 6 four cycles in S_4. Hence there will be 6.2^2 possible ways of coloring edges.

Case (iii): For permutations $\pi \in S_4$ that are product of a 3-cycle and one 1-cycle will have a induced permutation in $\overline{S}_4$ with two 3-cycles. Hence there are $4.2 = 8$ different permutations of vertices leading to 8.2^2 possible coloring.

Case (iv): For permutations in S_4 that are product of two 2-cycles. The induced permutations are one with product of two 2-cycles and two 1-cycles. Hence there are 3.2^4 possible colorings.

Case (v): Finally a permutation with two 1-cycles and a 2-cycle induces a permutation with two 1-cycles and two 2-cycles. The possible coloring will be 6.2^4.

So the number of distinct colorings of G will be the sum of all possible colorings discussed in the above 5 cases. It will be

$$\frac{1}{24}(2^6 + 6.2^2 + 8.2^2 + 3.2^4 + 6.2^4) = \frac{264}{24} = 11$$

Hence using Burnside's Lemma there are 11 non-isomorphic graphs on 4-vertices.

62. Find the number of distinct colorings of the vertices of a regular pentagon modulo D_5 so that there is exactly one vertex in white color, two in black color and two in yellow color?

Solution:
Here D_5 is the group acting on the coloring to fix the equivalent colorings. D_5 is the dihedral group. Using Polya enumeration theorem it will be

$$\frac{1}{|D_5|}\sum_{r\in D_5} p_{\mu(r)}(x_1, x_2, \ldots, x_k) = \frac{1}{|D_5|}\sum_{r\in D_5} |F(r)|$$

where $F(r)$ is the permutation r that fixes one white, two black and two yellow colors to the vertices of the pentagon.

By identical argument above we can show that there are 4 different colorings of the vertices with the given condition.

63. Let $\sigma = \begin{pmatrix} 1\,2\,3\,4 \\ 2\,3\,4\,1 \end{pmatrix}$ and $\tau = \begin{pmatrix} 1\,2\,3\,4 \\ 2\,1\,4\,3 \end{pmatrix}$. Find $\sigma \circ \tau$ and $\tau \circ \sigma$. Here '$\circ$' is the composition operation.

Solution:
One need to understand the composition operation '$\circ$' in a systematic manner. We write

$$\sigma \circ \tau = \sigma(\tau(x)) \text{ for } x \in S = \{1, 2, 3, 4\}.$$

That is

$$\begin{aligned} \sigma \circ \tau &= \begin{pmatrix} 1\,2\,3\,4 \\ 3\,2\,1\,4 \end{pmatrix} \\ \tau \circ \sigma &= \tau(\sigma(x)) \\ &= \begin{pmatrix} 1\,2\,3\,4 \\ 1\,4\,3\,2 \end{pmatrix} \end{aligned}$$

It is not wrong if one considers $\sigma \circ \tau$ directly as $\tau(\sigma(x))$.

64. Find the inverse σ^{-1} of $\sigma = \begin{pmatrix} 1\,2\,3\,4\,5 \\ 2\,1\,4\,5\,3 \end{pmatrix}$.

Solution:
Note $\sigma \circ \sigma^{-1} = \begin{pmatrix} 1\,2\,3\,4\,5 \\ 1\,2\,3\,4\,5 \end{pmatrix}$ the identity permutation. So σ^{-1} can be computed by observing the entries of σ in the upside down manner. That is,

$$\sigma^{-1} = \begin{pmatrix} 1\,2\,3\,4\,5 \\ 2\,1\,5\,3\,4 \end{pmatrix}$$

65. Find the cycle representations of the following permutations

(i) $\begin{pmatrix} 1 & 2 & 3 & 4 \\ 4 & 1 & 2 & 3 \end{pmatrix}$

(ii) $\begin{pmatrix} 1 & 2 & 3 & 4 & 5 \\ 2 & 1 & 4 & 3 & 5 \end{pmatrix}$

(iii) $\begin{pmatrix} 1 & 2 & 3 & 4 & 5 & 6 \\ 4 & 5 & 1 & 2 & 6 & 3 \end{pmatrix}$

Solution:

(i) $\begin{pmatrix} 1 & 2 & 3 & 4 \\ 4 & 1 & 2 & 3 \end{pmatrix} = (4\ 3\ 2\ 1)$

(ii) $\begin{pmatrix} 1 & 2 & 3 & 4 & 5 \\ 2 & 1 & 4 & 3 & 5 \end{pmatrix} = (2\ 1)(4\ 3)(5)$

(iii) $\begin{pmatrix} 1 & 2 & 3 & 4 & 5 & 6 \\ 4 & 5 & 1 & 2 & 6 & 3 \end{pmatrix} = (4\ 2\ 5\ 6\ 3\ 1)$

66. Let

$$G = \{(1)(2)(3)(4)(5)(6)(7), (1, 7)(2, 6)(3, 5)(4)\}$$

be a subgroup of S_7. If '$\sim$' is an equivalence relation induced by G, then
(i) is $1 \sim 3$?
(ii) is $3 \sim 5$?
(iii) is $4 \sim 7$?

Solution:
(i) No. $1 \nsim 3$
(ii) Yes
(iii) No.

67. Find the equivalence classes in G of the previous question.

Solution:
$E = \{\{1, 7\}, \{2, 6\}, \{3, 5\}, \{4\}\}$.

68. For the set $S = \{1, 2, 3, 4, 5, 6, 7\}$ and the question 66 above, find the orbit of 2 and fix of the permutations of G. Also find the stabilizer of 2.

Solution:
$O(2) = \{2, 6\}$
$F(i) = S$ is the identity permutation.

For $\tau = (1, 7)(2, 6)(3, 5)(4)$
$F(\tau) = \{4\}$
$T(2) = i$, the identity permutation.

69. Find all the lattice paths from (0, 0) to (2, 2).

Solution:

70. Let $d_3(n)$ denote the number of permutations $\tau \in S_n$ with no decreasing subsequence of length 3. Find $d_3(3)$.

Solution:
The lattice paths that avoid 321 pattern in any permutation of S_3 can be drawn as below (Figs. 12.4 and 12.5).

Here $\tau : \begin{pmatrix} 1 & 2 & 3 \\ 2 & 3 & 1 \end{pmatrix}$.

Other lattice paths are

Hence $d_3(3) = 5$.

Fig. 12.4 Lattice paths

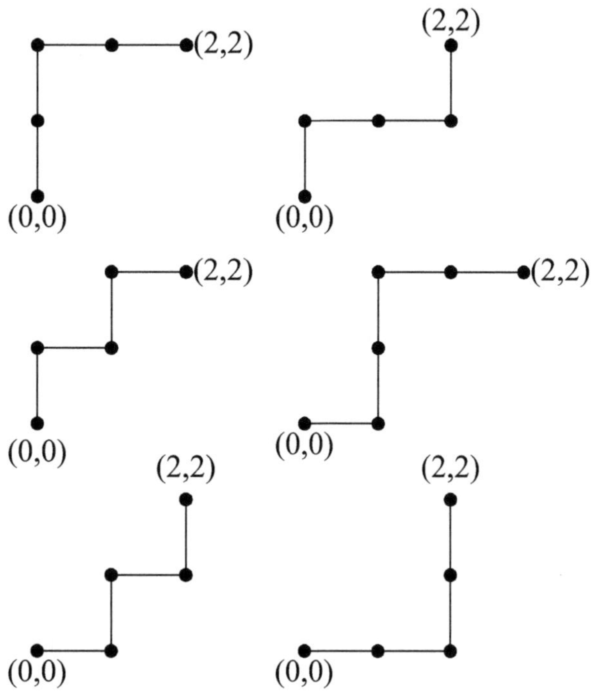

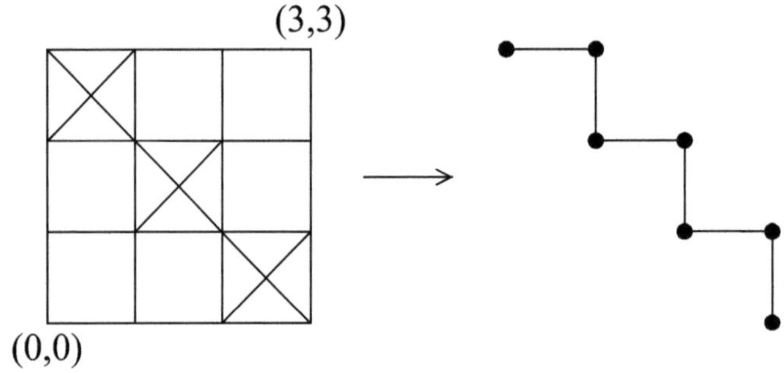

Fig. 12.5 321 pattern avoiding path

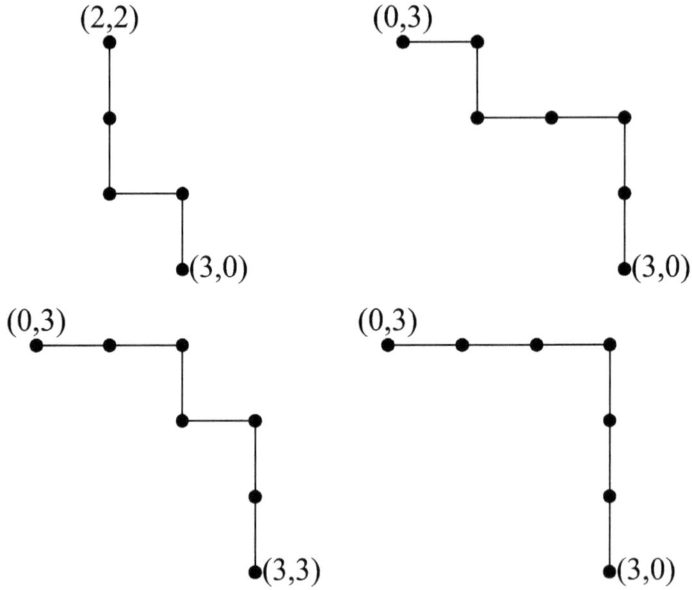

Fig. 12.6 Other lattice paths

71. Consider the cartisian coordinate points (0, 0), (4, 0). Using the steps (1, 1), (1, −1) (the diagonal steps) draw all the lattice paths from (0, 0) to (4, 0) that do not fall below x-axis. Such paths are called as 'Dyck' paths.

Solution:

72. Let $S_n(132, 123)$ denote all permutations of S_n that avoid both 132 and 123. Show that $|S_n(132, 123)| = 2^{n-1}$ (Figs. 12.6 and 12.7).

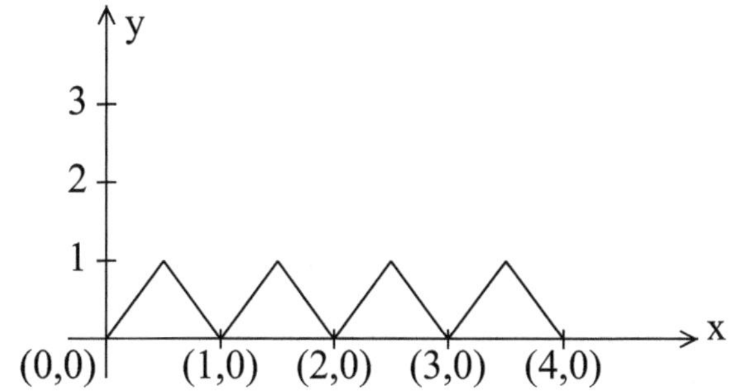

Fig. 12.7 Dyck paths

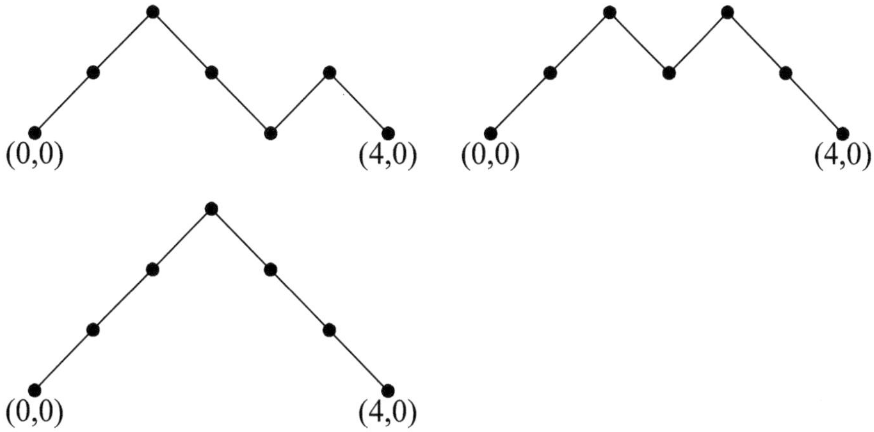

Fig. 12.8 Other Dyck paths

Solution:
The proof is by method of induction on n.

For $n = 3$, the permutations

$$\begin{pmatrix}1\,2\,3\\1\,2\,3\end{pmatrix}, \begin{pmatrix}1\,2\,3\\1\,3\,2\end{pmatrix}, \begin{pmatrix}1\,2\,3\\3\,2\,1\end{pmatrix}, \begin{pmatrix}1\,2\,3\\2\,1\,3\end{pmatrix}, \begin{pmatrix}1\,2\,3\\2\,3\,1\end{pmatrix}, \begin{pmatrix}1\,2\,3\\3\,1\,2\end{pmatrix}$$

There are 4 here which avoid both 132 and 123 patterns.

Let $|S_n(132, 123)| = 2^{n-1}$ be true for $n = m - 1$. That is $|S_{m-1}(132, 123)| = 2^{m-2}$ (Fig. 12.8).

Consider any one permutation of S_{m-1} that avoid both these patterns. Now new m-permutations are formed as below:

For

$$\tau = \begin{pmatrix}1 & 2 & 3 & \ldots & m-1\\a_1 & a_2 & a_3 & \ldots & a_{m-1}\end{pmatrix} \in S_{m-1}$$

that avoid both the above patterns new permutations on m-element are formed as below.

$$\overline{\tau} = \begin{pmatrix} 1 & 2 & 3 & \dots & m-1 & m \\ a_1+1 & a_2+1 & a_3+1 & \dots & a_{m-1}+1 & 1 \end{pmatrix}$$

$$\overline{\overline{\tau}} = \begin{pmatrix} 1 & 2 & 3 & \dots & m-1 & m \\ a_1+1 & a_2+1 & a_3+1 & \dots & 1 & a_{m-1}+1 \end{pmatrix}$$

Clearly both $\overline{\tau}$ and $\overline{\overline{\tau}}$ are in $S_m(132, 123)$. This procedure gives all permutations in $S_m(132, 123)$. The insertion of '1' in the last and last-but-one position guarantees the count of $S_m(132, 123)$ to be $2^{m-1}.2 = 2^m$. '1' inserted anywhere else will give a permutation of S_m having either 132 or 123 pattern. Hence the solution.

73. Find the decents and ascents of the permutations.
(i) 3412567
(ii) 534216

Solution:
(i) For $\tau = 3412567$
$dec(\tau) = \{2\}$
$asc(\tau) = \{1, 3, 4, 5, 6\}$
(ii) $\sigma = 534216$
$dec(\sigma) = \{1, 3, 4\}$
$asc(\sigma) = \{2, 5\}$

Note: By writing a one-line notation a permutation $\pi = p_1 p_2 \dots p_n \in S$, we mean $\pi(j) = p_j$ for $1 \le j \le n$.

74. Definition: Let $P = p_1 p_2 \dots p_n \in S_n$ and let P have k decents. P can be written as the union of $(k+1)$ increasing subsequences of consecutive entries. Such subsequences are called as the "runs" or ascending runs of P.

Express the following permutations as the union of ascending runs.
(i) 2415367
(ii) 342176

Solution:
(i) Let $\tau = 2415367$
$dec(\tau) = \{2, 4\}$
The ascending runs are
24, 15, 367
$\tau = \{24, 15, 367\}$.

(ii) $\sigma = 342176$
$dec(\sigma) = \{2, 3, 5\}$
The ascending runs are
34, 2, 17, 6
Then $\sigma = \{34, 2, 17, 6\}$.

75. Definition: A strict ballot sequence α of length $(2n + 1)$ is a sequence of $+1$ and -1 such that 'α' contains $(n + 1)$ 1's and the remaining 'n' as -1's with the condition that every non-empty partial sum being positive.

Definition: Given a sequence $\alpha = a_1 a_2 \dots a_k$ we define a cyclic shift or conjugate of 'α' as a sequence

$$a_s a_{s+1} \dots a_k a_1 \dots a_{s-1}$$

There can be k such shifts of α. All such sequences need not be distinct.

Using the above two definitions show the following.

Let $\alpha = a_1 a_2 \dots a_{2n+1}$ be a sequence with $(n + 1)$ 1's and remaining as -1's. Then all the $(2n + 1)$ cyclic shifts of α are distinct and exactly one of them is a strict ballot sequence.

Solution:
For example if

$$\alpha = 1\,1\,1\,\,-1\,\,-1\,1\,\,-1\,\,-1\,1, n = 4, n + 1 = 5$$

some cyclic shifts are $1\,1\,1\,1\,\,-1\,\,-1\,1\,\,-1\,\,-1, -1\,1\,1\,1\,1\,\,-1\,\,-1\,1\,\,-1, -1\,\,-1\,1\,1\,1\,1\,\,-1\,\,-1\,1$. Here $1\,1\,1\,1\,\,-1\,\,-1\,1\,\,-1\,\,-1$ is one with every partial sum being positive.

Now we prove the result by mathematical induction on 'n'. Note that $n, n + 1$ are always relatively prime to each other.

For $n = 0$, $\alpha = 1$. Hence it is a strict ballot sequence.

Assume the result to be true for 'n', that is, if 'α' is a sequence of length $(2n + 1)$ having $(n + 1)$ 1's and the remaining as -1, the result is true.

Consider $\alpha = a_1 a_2 \dots a_{2n+3}$ which has $(n + 2)$ 1's and the remaining as -1's. Clearly there will be a position 't' such that $a_t = 1$ and $a_{t+1} = -1$, $t < 2n + 3$.

For $t = 2n + 3$, consider $t \mod (2n + 3)$. Now consider a new sequence $\overline{\alpha} = b_1 b_2 \dots b_{2n+1}$ obtained from 'α' by deleting a_t and a_{t+1} from α. Now for 'β' being of length $(2n + 1)$, the induction assumption is true. That is β has a unique cyclic shift say $\overline{\beta} = b_s b_{s+1} \dots b_{2n+1} b_1 \dots b_{s-1}$ which is a strict ballot sequence. If $b_s = a_f$, then in $a_f a_{f+1} \dots a_{f-1}$ every partial sum is positive as this subsequence matches with $\overline{\beta}$ upto a_t. Now add $a_t = 1$ and $a_{t+1} = -1$, in $\overline{\alpha}$, the partial sums remain positive. Such an $\overline{\alpha}$ is unique as the cyclic shift beginning with a_t will have partial sum to be 0 and the cyclic shift beginning with a_{t+1} will have partial sum -1. Making another cyclic shift of $\overline{\alpha}$ at some b_j $(j \neq t)$

will have its partial sums that include all the partial sums of the cyclic shift $b_j b_{j+1} \dots b_{j-1}$ of $\overline{\alpha}$. By induction, one of the partial sums calculated must be negative. Hence the solution.

76. Definition: A ballot sequence of length $2n$ is a sequence equal number of 1's and -1's such that every partial sum is non negative.

Show that the Catalan number C_n is the count of ballot sequences of length $2n$.

77. Let a box contain, one orange ball. A ball is drawn at random (as a first trial there will be only one chance) then the ball drawn is put back in the box along with a green ball. The procedure is repeated n times. A successful trial is one when an orange ball is drawn. Find the probability of t-successes in n trials.

Solution:
Let A_s be the event of picking an orange ball in trial s. So, $p(A_1) = 1$ as an assumption. Now our aim is to compute $p(A_s)$. We know when we start sth trial, the box has one orange ball and $(s-1)$ green balls.
Hence

$$p(A_s) = \frac{1}{s}.$$

$$p(\overline{A_s}) = 1 - \frac{1}{s} = \frac{s-1}{s}.$$

The 'n' trials are independent and we need t-successes in these n trials. Hence the total probability will be

$$T_s = \sum p(A_{s_1})p(A_{s_2}) \dots p(A_{s_t})p(\overline{A}_{s_{t+1}}) \dots p(\overline{A}_{s_n})$$

Here the summation will be taken over the collection K of all possible $(n-t)$ subsets $\{s_{t+1}, s_{t+2}, \dots s_n\}$ of the set $\{2, 3, 4, \dots, n\}$. First trial is always successful. So

$$T_s = \frac{1}{n!} \sum_K (s_{t+1} - 1)(s_{t+2} - 1) \dots (s_n - 1)$$

[Here the sum is Stirling number of first kind without sign]

78. Find the probability that an arrangement of the symbols in the word 'TIRUCHIRAPALLI' has
(i) 6 consecutive vowels?
(ii) no consecutive vowels?

Solution:
(i) Total length of the given string 'TIRUCHIRAPALLI' is 14. In this 6 vowels must occur as a single block. The word has 5 vowels. The remaining symbols are T, R, C, H, R, P, L, L and there are two R's and two L's. For a block of vowels of length 6, we have $\frac{6!}{3!2!} = 60$ arrangements.

For the two positions of R's we have $C(8, 2)$ ways and for two L's we have $C(6, 2)$ ways to fix positions for R, R, L, L.

For T, C, H, P, we have 4! arrangements. So the total number of arrangements will be

$$60 \times C(8, 2) \times C(6, 2) \times 4!$$

Total number of arrangements will be

$$C(14, 2) \cdot C(12, 2) \cdot 10!$$

So, the probability of 6 consecutive vowels in the arrangement of 'TIRUCHIRAPALLI' is

$$\frac{60 \times C(8, 2) \times C(6, 2) \times 4!}{C(14, 2)C(12, 2) \times 10!}$$

(ii) The 6 vowels can be arranged in 6! ways. The 8 non vowel symbols should be inserted in between any two vowels and at the beginning as well as at the end of the string. Hence there are 8! ways to insert non vowels. Hence the total arrangements will be $6! \times 8!$. The associated probability will be

$$\frac{6! \times 8!}{C(14, 2) \times C(12, 2) \times 10!}.$$

79. An unbiased coin is tossed 20 times. Find the probability that
(i) the first 'head' occurs after exactly 15 tails
(ii) the 4th head occurs after exactly 10 tails?

Solution:
(i) When the coin is tossed 20 times, there are 2^{20} possible out comes. The appearance of 15 tails successively followed by the appearance of a head has 2^{20-16} ways to finish the activity. Hence the probability will be 2^{-16}.
(ii) The 4th head appearance happens after tossing the coin 13 times. In this 13 tosses there must be 10 tails and 3 heads which can happen in $C(13, 3)$ ways. The 14th toss out come is fixed with an 'head' appearance. The remaining 6 tosses are unrestricted and they can be done in 2^6 ways. Hence the probability will be $\frac{C(13,3).2^6}{2^{20}} = C(13, 3).2^{-4}$.

80. Definition: The probability generating function associated with any discrete distribution is a polynomial whose coefficients are the probabilities $p_X(y)$. Here $X = 0, 1, 2, \ldots,$ is the random variable. That is

$$PG_X(y) = P(X = 0)y^0 + P(X = 1)y^1 + \cdots + P(X = n)y^n$$

$P(X = r)$ is the coefficient of y^r in $PG_X(y)$.
(i) If the random variable $X(r)$ denote the number of appearances of heads in a fair coin toss of n-times, find $PG_X(y)$.
(ii) Discuss about $PG_X(y)$ if the coin tossed is not a fair one.

Solution:
For a coin toss, denote the probability of head appearance as 'p' and of tail as $q = 1 - p$.
For $p(X = k) = C(n, k)p^k q^{n-k}$
(i) Here $p = \frac{1}{2}, q = \frac{1}{2}$.

$$\begin{aligned}
PG_X(y) &= C(n, 0)p(X = 0) + C(n, 1)p(X = 1)y + C(n, 2)p(X = 3)y^2 + \cdots + C(n, n)p(X = n)y^n \\
&= C(n, 0)q^n + C(n, 1)pq^{n-1}y + \cdots + C(n, n)p^n y^n \\
&= C(n, 0)\left(\frac{1}{2}\right)^n + C(n, 1)\left(\frac{1}{2}\right)^n y + \cdots + C(n, n)\left(\frac{1}{2}\right)^n y^n \\
&= \left(\frac{1}{2}\right)^n \left[C(n, 0) + C(n, 1)y + \cdots + C(n, n)y^n\right] \\
&= \left(\frac{1}{2}\right)^n (1 + y)^n.
\end{aligned}$$

(ii)

$$\begin{aligned}
PG_X(y) &= C(n, 0)q^n + C(n, 1)pq^{n-1}y + \cdots + C(n, n)p^n y^n \\
&\quad \text{[Here } p \neq q \text{ as the coin tossed is not a fair coin]} \\
&= (q + py)^n.
\end{aligned}$$

81. Suppose 11 indistinguishable balls are distributed in 3 indistinguishable boxes. The distribution is an independent event. Let p_i be the probability of the picked ball being put in box i, $i = 1, 2, 3$ with $p_1 + p_2 + p_3 = 1$.
(i) Find the expected number of boxes that are empty.
(ii) Find the expected number of boxes that have exactly one ball.

Solution:
(i) Here p_i is the probability of the ball being placed in the ith box. So the probability that it is not put in box i will be $1 - p_i$. Hence the probability generating function for 't' balls being put in box i will be

$$PG_X(y) = C(11, 0)p_i^0(1 - p_i)^{11-0} + C(11, 1)p_i^1(1 - p_i)^{1-0} + \cdots + C(11, 11)p_i^{11}(1 - p_i)^0$$

The box being empty will be when $x = 0$

$$\begin{aligned} C(11, 0)p_i^0(1 - p_i)^{11} &= C(11, 0)(1 - p_i)^{11} \\ E(\text{box empty}) &= \sum xp(x) \\ &= \sum_{i=1}^{3} 1.(1 - p_i)^{11} \end{aligned}$$

(ii)

$$\sum_{i=1}^{3} C(10, 1)(p_i)^1(1 - p_i)^{10}$$

82. A box contains 8 green balls and 5 yellow balls.
(i) Find the probability of picking 5 balls in combination of 3 green balls and 2 yellow balls.
(ii) Find the probability of (i) if one replaces the picked ball at each step of its draw.
(iii) Let the desired combination of balls drawn be that it is 2 green balls from the box. Discuss the likelyhood of this with and without replacement.

Solution:
(i) The answer will be direct and simple for this case. It will be

$$\frac{C(8, 3)C(5, 2)}{C(13, 5)} = \frac{560}{1287} \approx 0.435$$

(ii) In case of replacement the probability will be

$$\begin{aligned} C(5, 3)\left(\frac{8}{13}\right)^3\left(\frac{5}{13}\right)^2 &= 10 \times \frac{512}{2197} \times \frac{25}{169} \\ &\approx 0.3447 \end{aligned}$$

(iii) Without replacement the probability will be

$$\frac{C(8, 2)C(5, 1)}{C(13, 3)} = \frac{28 \times 5}{286} \approx 0.490$$

With replacement, the probability will be

$$C(3, 1)\left(\frac{8}{13}\right)^2\left(\frac{5}{13}\right) \approx 0.437$$

Hence the probability without replacement shows that it is more likely to get the desired combination.

83. Suppose X and Y play a game by tossing two fair coins. If both the coins turn as 'head (H)', or as 'tail (T)', then X pays one rupee to Y. Otherwise Y has to pay 2 rupees to X.
(i) Find the expected value of the game for X.
(ii) Find the expected value of the game for Y.
(iii) Discuss the fairness of the game.

Solution:
The possible outcome of the two coin toss can be drawn as a table below.

	H	T
H	HH	HT
T	TH	TT

(i) From the above table, one can see that two out of 4 outcomes produce HH, TT. Hence its probability will be $\frac{1}{2}$. So Y gets one rupee in this case. With the same probability X will get two rupees. Hence the expected value of the game for X is

$$\frac{1}{2}(1) + \frac{1}{2}(-2) = -\frac{1}{2}.$$

So -50 paise X loses in each play or toss.
(ii) The money X loses, will be the money that Y gains. Hence Y is expected to win 50 paise on each game the plays.
(iii) The expected value of the game being not zero, it is not a fair game.

Part II
Graph Theory

Basic Concepts 13

13.1 Introduction

We start with some examples where the discussion has graph theory base.

Example 13.1 1. Suppose a hospital has spread over 7 buildings. They are OP-unit, Administration block, Pharmacy, ICU-unit, Canteen unit, General ward and Operation wing. The problem is to lay roads between the units so that the distance to be covered between them is minimum.

2. Konigsberg bridge problem.
   Suppose river Pragel runs through cities. The river encloses two islands say A and B. The banks are named as say C and D. There are seven bridges as shown below. $b_1, b_2, b_3, b_4, b_5, b_6, b_7$ are the bridges. The question is: Is it possible to start from any one of land area, use the bridges to walk through A, B, C, D and come back to the starting point such that only once each bridge is used to cross across? This is known as Konigsberg Bridge problem and the problem is settled with a solution by Euler using graph theory idea (Fig. 13.1).

Now we proceed for formal presentations of graphs and one can at a later point of reading, understand the solutions to the above problems using graph theory.

Definition 13.2 A graph G is an ordered pair (V, E) where V and E are finite sets and $V \cap E = \emptyset$. Here V is called the vertex set of G, E is called the edge set of G. The set $E = \{\{x_1, x_2\}/x_1, x_2 \in V$ and x_1 and x_2 are related or joined$\}$. Here x_1, x_2 need not be distinct. One can see that $E \subseteq V \times V$ except that the pair $\{u, v\}$ is understand to be same as $\{v, u\}$.

R. Rama, *Topics in Combinatorics and Graph Theory*,
https://doi.org/10.1007/978-3-031-74252-1_13

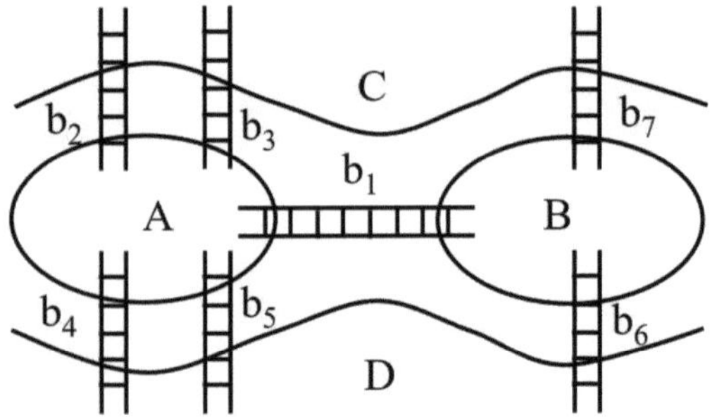

Fig. 13.1 Bridge problem

So if for $e \in E$ and $e = \{x_1, x_2\}$ we can say that x_1 and e are incident, x_2 and e are incident with each other. One can also say that x_1 and x_2 are adjacent vertices. The above definition describes an undirected graph.

Example 13.3 Let $G = (V, E)$ where $V = \{1, 2, 3, 4, 5\}$, $E = \{e_1, e_2, e_3, e_4\}$. The following table describes the relation between the elements of V and the elements of E.

E	e_1	e_2	e_3	e_4
I_G	$\{1, 3\}$	$\{1, 5\}$	$\{2, 4\}$	$\{4, 5\}$

Here I_G is the incidence function. e_1 and vertex 1, e_1 and vertex 3 are incident. Similarly other entries of the table can be understood.

Notation. Let $|G| = |V|$ denote the order of G and $\|G\| = |E|$ denote the size of G. Also the relation $\{u, v\}$ can be written simply as uv.

13.2 Pictorial Representation of a Graph

A graph can be pictorially represented as follows. The vertices are distinctly drawn on a place and edges are indicated by curves or lines between the points.

For example if $G = (V, E)$, $V = \{1, 2, 3, 4\}$, $E = \{13, 14, 23, 24\}$, the graph is shown below (Fig. 13.2).

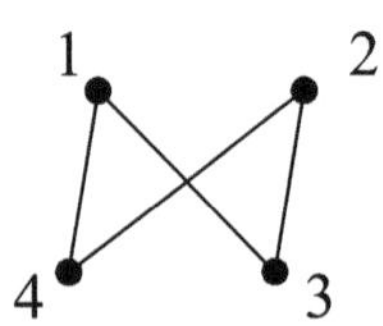

Fig. 13.2 A graph

Fig. 13.3 Non simple graph

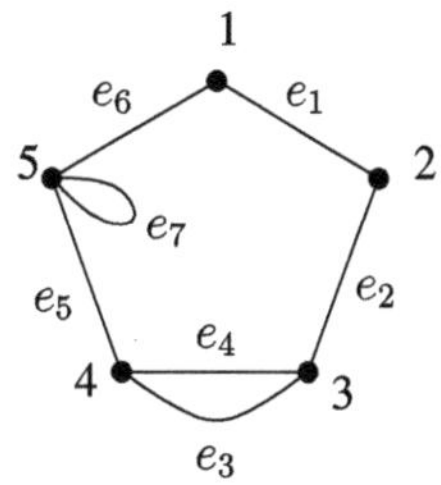

Example 13.4 Consider the pictorial representation given below (Fig. 13.3). Here $V = \{1, 2, 3, 4, 5\}$

$E = \{12, 23, 34, 43, 45, 15, 55\}$.

The edges e_3 and e_4 are called parallel edges and edge e_7 is called a loop.

Definition 13.5 A graph without loops and multiedges is called a simple graph.

Remark 13.6 Mostly the topics and concepts here in this book are developed for simple graphs.

Example 13.7 The following are the pictorial representations of simple graphs (Fig. 13.4).

Definition 13.8 Given a graph $G = (V, E)$, the number of edges incident with a vertex v is called its degree is denoted as $d(v)$or $d_G(v)$.

Example 13.9 Consider the following graph (Fig. 13.5).
Here $V = \{1, 2, 3, 4, 5, 6, 7, 8\}$

$E = \{12, 13, 14, 15, 16, 17, 34, 45, 56, 57, 67\}$.

$d(1) = 6, d(2) = 1, d(3) = 2, d(4) = 3,$
$d(5) = 4, d(6) = 3, d(7) = 3, d(8) = 0.$

The following theorem is known as the first theorem of graph theory.

Fig. 13.4 Simple graphs

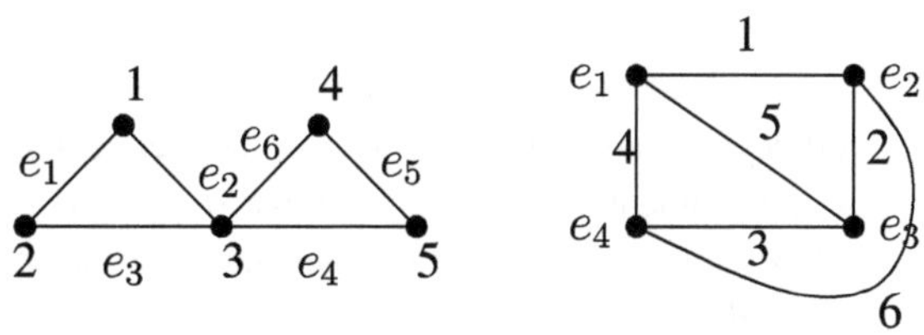

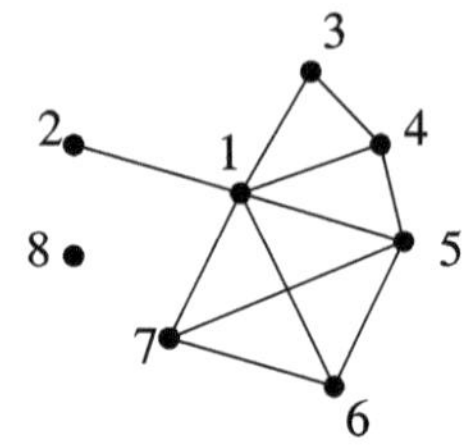

Fig. 13.5 Verification

Theorem 13.10 *Let G be a graph. Then*

$$\sum_{v \in V} d(v) = 2|E|.$$

Definition 13.11 1. Maximum degree $\Delta(G)$ of a graph is the maximum of $d(v)$ for all $v \in V$. i.e.,

$$\Delta(G) = \max_{v \in V} d(v).$$

2. Similarly, minimum degree is denoted as $\delta(G)$ and $\delta(G) = \min_{v \in V} d(v)$.

Definition 13.12 Let $d_1, d_2, \ldots, d_k$ be the degrees of vertices of a graph $G = (V, E)$ where $|V| = k$. Then the sequence $(d_1, d_2, \ldots, d_k)$ is called the degree sequence of G, where $d_1 \leq d_2 \leq \cdots \leq d_k$.

Remark 13.13 A graph $G = (V, E)$ is finite if the graph is with finite set of vertices and a finite set of edges. Otherwise, it will be mentioned as an infinite graph. Example of an infinite graph is a graph on a plane $\mathbb{R}^2$, where $uv \in E$ if and only if $d(u, v) = 1$.

13.3 Some Special Graphs

1. A complete graph is one in which any two vertices are adjacent. A complete graph on n vertices is denoted by K_n (Fig. 13.6).
   Example
2, A graph is called as an empty graph if its edge set is empty (Fig. 13.7).
   Example
3. An independent set in a graph $G = (V, E)$ is a subset $I (\subseteq V)$ of vertices such that no vertices of I are adjacent in G. An independent set I is said to be maximum if there is no independent set I' of G such that $|I'| > |I|$. The number of vertices in a maximum independent set is called the independent number of G and is denoted by α (Fig. 13.8).
   Example $I_1 = \{1, 5, 3\}$ is an independent set.
   $I_2 = \{1, 5, 3, 8\}$ is another independent set. Check whether I_2 is maximum?

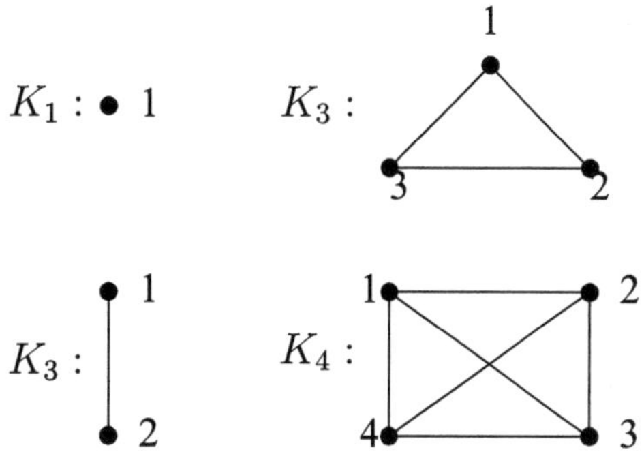

Fig. 13.6 Some complete graphs

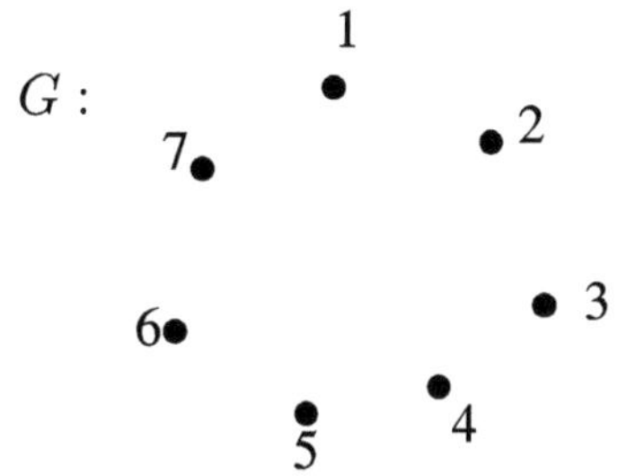

Fig. 13.7 Empty graph

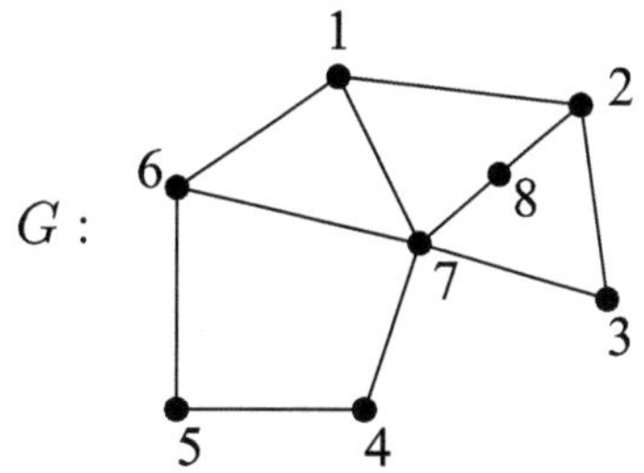

Fig. 13.8 A graph

4. **Bipartite graph:** A graph G is bipartite if the vertex set can be partitioned into two subsets X and Y such that every edge of G has one end vertex in X and the other end vertex in Y. We can denote G as $G = (X \cup Y, E)$. Clearly X and Y are independent sets.
   Example The set $\{1, 2, 3\}$ is an independent set and $\{4, 5, 6, 7\}$ is the maximum independent set (Fig. 13.9).
5. A complete bipartite graph is one in which every vertex in one partition is adjacent with every vertex in the other partition. It is denoted as $K_{m,n}$ where order of $K_{m,n}$ is $m + n$ (Fig. 13.10).
   Example
6. A graph with no edges is called as empty graph or null graph. That is size of the graph is zero.

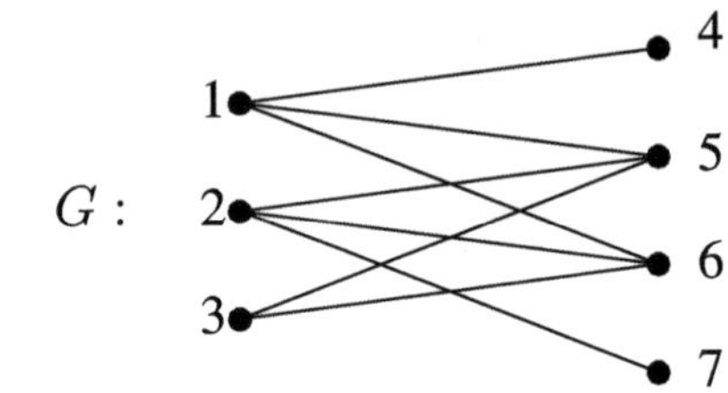

Fig. 13.9 Bipartite graph

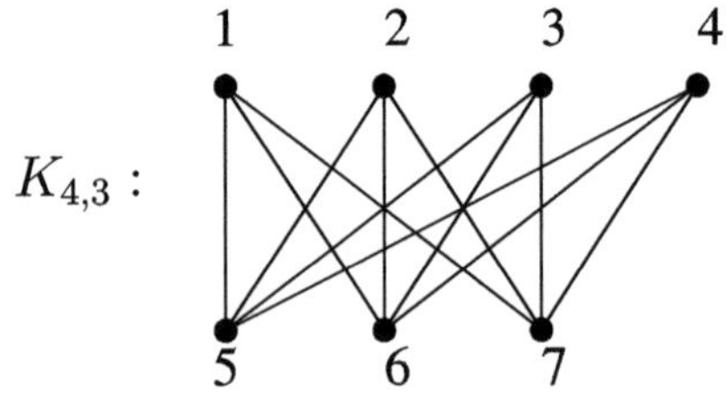

Fig. 13.10 Complete bipartite graph

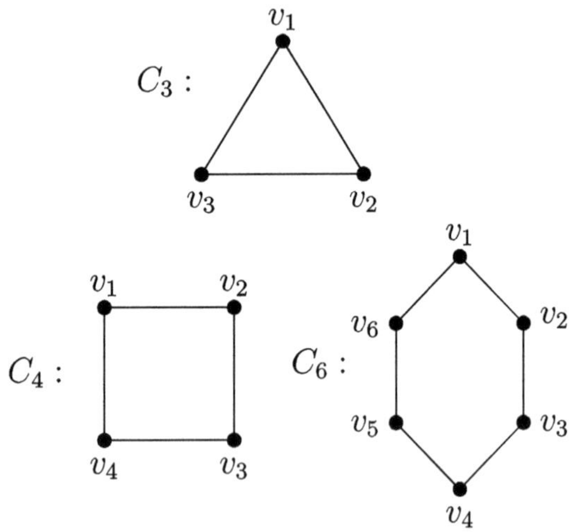

Fig. 13.11 Some cycle graphs

7. In a graph $G = (V, E)$, if the degree of each vertex is a constant r, then G is called as r-regular graph. In general, a regular graph is one in which every vertex is of the same degree.
8. A cycle graph $G = (V, E)$ on n vertices is a simple graph whose vertices can be arranged in a cyclic sequence. For vertices v_i, v_j if $v_i v_j \in E$ then we say they are consecutive. In a cycle graph if any two vertices are adjacent, then they must be consecutive. The definition is valid for all $n \geq 3$ and a cycle graph on n vertices is denoted by C_n (Fig. 13.11).
   Example
   Clearly a cycle C_n has always 'n' edges.
9. Let $G = (V, E)$ be a graph. Then the graph $\overline{G} = (V, \overline{E})$ is said to be the complement of G if the vertices of G and $\overline{G}$ are the same and $uv \in E$ if and only if $uv \notin \overline{E}$, for

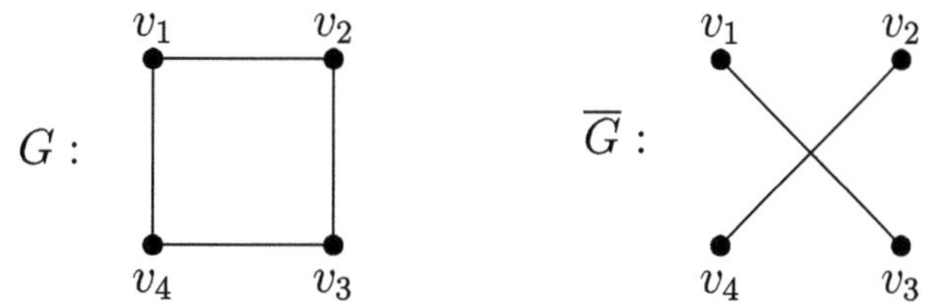

Fig. 13.12 Complementary graphs

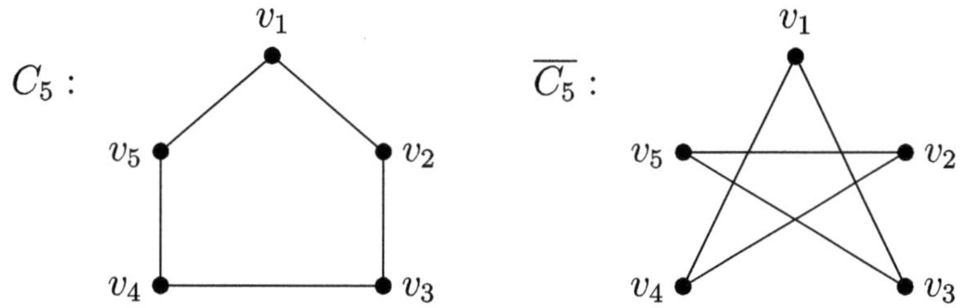

Fig. 13.13 Self complementary graphs

$u, v \in V$. Clearly $\overline{\overline{G}} = G$ (Fig. 13.12).
Example
Consider $G = C_5$ which is (Fig. 13.13)

Exercises

1. Draw all simple graphs on 4 vertices.
2. Find the number edges in $K_{5,2}$, $\overline{K_{5,2}}$. Is $\overline{K_{5,2}}$ bipartite?
3. $G = (V, E)$ is a graph on n vertices. 'k' vertices of G are of degree ℓ and other $|V| - k$ vertices are of degree $\ell + 1$. Show that
$$\ell + 2|E| = (\ell + 1)n$$
4. Draw a graph G on 4 vertices such that $\Delta(G) = 3$, $\delta(G) = 1$.
5. Let $G = (V, E)$ be a graph such that $|V| \geq 2$. Show that G contains two vertices having equal degree.
6. Let $G = (V, E)$ be a graph. A vertex $v \in V$ is said to be an odd degree vertex if $d(v)$ is odd, otherwise $d(v)$ is even and v is called an even degree vertex. Show that G has even number of odd degree vertices.
7. For integers k and m, show that there exists a k-regular graph with n vertices if and only if $0 \leq k \leq n - 1$ and both k and n are not odd.
8. Let $G = (V, E)$ be a bipartite graph on 10 vertices. Find $|E|$ such that $|E|$ is maximum.

9. Let G be graph on 13 vertices and 29 edges. In G there are 4-vertices of degree 4, 4 vertices of degree 3 and the remaining 5 vertices are of same degree. Find the degree of the remaining vertices.
10. Find all 4-regular graphs on 7 vertices.

Paths Connectedness 14

14.1 Paths

Definition 14.1 A path in a graph $G = (V, E)$ is of the form $p : v_0 v_1 v_2 v_3 \dots v_t$ where all v_i's are distinct and for any $i \in \{0, 1, \dots, t\}$ $v_i v_{i+1} \in E$.

Example 14.2 Consider the graph (Fig. 14.1)
$p_1 : v_1 v_2 v_4 v_5 v_6$ is a path.
$p_2 : v_3 v_1 v_2 v_4 v_5 v_6$ is a path.

Remark 14.3 A path graph is a graph whose vertices are arranged linearly to form a single path.

Definition 14.4 A walk in a graph $G = (V, E)$ is of the form

$$w : v_0 v_1 v_2 \dots v_k \text{ where } v_i v_{i+1} \in E$$

and v_0, v_k are distinct.

Example 14.5 In the above graph, $w : v_1 v_2 v_5 v_4 v_2 v_3$ is a walk.

A walk in which the initial and end vertices are same is said to be a closed walk. Note that a walk can have both vertex repetitions and edge repetitions. If in a walk only vertices repeat and edges are distinct then it is named as a trial.

Example 14.6 In the above graph $w : v_1 v_2 v_3 v_4 v_2 v_5 v_6 v_5 v_4 v_3$ is a walk.
$t : v_1 v_3 v_2 v_4 v_5 v_2$ is a trail.

Both w and t are open as the end vertices are distinct.

R. Rama, *Topics in Combinatorics and Graph Theory*,
https://doi.org/10.1007/978-3-031-74252-1_14

Fig. 14.1 A connected graph

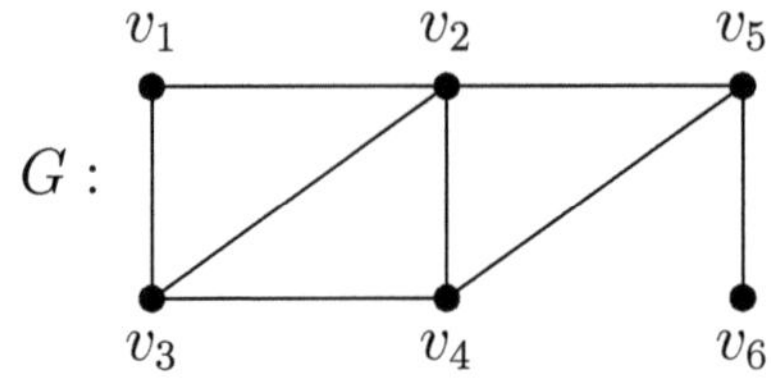

Definition 14.7 A closed trail is one that starts and ends at the same vertex. A trail that visits every edge of a graph is called as Eulerian trail. An Eulerian circuit is a closed trail.

One can define distance metric in terms of number of edges. So we can understand the length of a path to be the number of edges in the path and its notation will be $dis_G(x, y)$, x, y are vertices in G.

Definition 14.8 Let u, v be two vertices in a graph G. Then distance between u, v in G is

$$dis_G(u, v) = \min_{\substack{p:p \text{ is a} \\ x,y \text{ path}}} dis_G(p)$$

if there is a path from u to v in G. Otherwise $dis_G(u, v) = \infty$.

Definition 14.9 Let $G = (V, E)$ be a graph. The eccentricity of a vertex v in G is the distance of a vertex 'u' that is farthest from v. That is

$$e_G(v) = \max\{dis_G(u, v)|u \in V\}.$$

The minimum eccentricity is usually called as radius of G denoted as 'r' and the maximum eccentricity is called as diameter of G, denoted as '$diam$'. That is

$$r = \min\{e_G(v)|v \in V\}$$
$$diam = \max\{e_G(v)|v \in V\}.$$

Remark 14.10 There may be more than one vertex serving for radius and diameter.

Example 14.11 Consider the graph (Fig. 14.2)
$e_G(v_1) = 5, e_G(v_6) = 5, e_G(v_8) = 5, r(G) = 3, diam(G) = 5.$

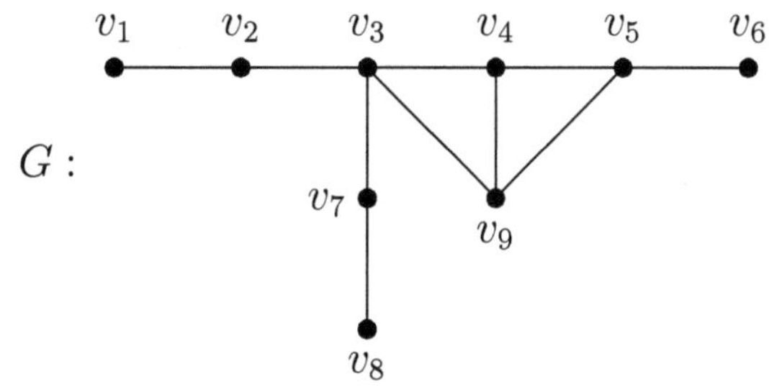

Fig. 14.2 Graph (4)

14.2 Adjacency Matrix and Incidence Matrix

There are two representations of a graph in matrix form. Such representations are very useful from programming point of view.

Definition 14.12 Let $G = (V, E)$ be a graph with $|V| = n$. The matrix $A_G = (a_{ij})$ where

$$a_{ij} = \begin{cases} 1 & \text{if } v_i v_j \in E \\ 0 & \text{otherwise.} \end{cases}$$

is called the adjacency matrix of G. It is denoted by A_G.

Example 14.13 Let G be the following graph (Fig. 14.3).

$$A_G = \begin{array}{c} \\ 1 \\ 2 \\ 3 \\ 4 \\ 5 \end{array} \begin{array}{c} 1\,2\,3\,4\,5 \\ \begin{bmatrix} 0 & 1 & 1 & 1 & 0 \\ 1 & 0 & 1 & 0 & 0 \\ 1 & 1 & 0 & 0 & 1 \\ 1 & 0 & 0 & 0 & 0 \\ 0 & 0 & 1 & 0 & 0 \end{bmatrix} \end{array}$$

We can see the usefullness of such a representation in a later chapter.

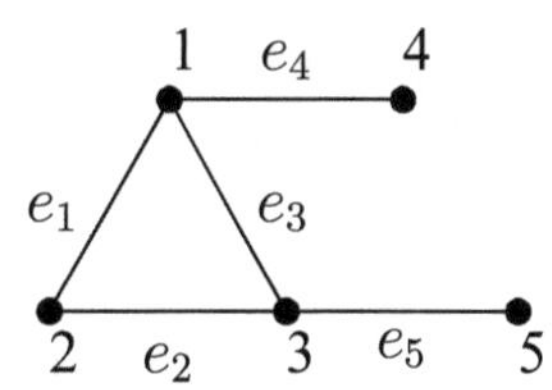

Fig. 14.3 Graph (5)

Definition 14.14 Let $G = (V, E)$ be a graph with $V = \{v_1, v_2, \ldots, v_n\}$, $E = \{e_1, e_2, \ldots, e_m\}$.

The $n \times m$ matrix $I_G = (b_{ij})$ where

$$b_{ij} = \begin{cases} 1 & \text{if } v_i \text{ is incident with } e_j \\ 0 & \text{otherwise} \end{cases}$$

is called the incidence matrix of G.

Example 14.15 For the same example above, the incidence matrix will be

$$I_G = \begin{array}{c} \\ 1 \\ 2 \\ 3 \\ 4 \\ 5 \end{array} \begin{array}{c} \begin{array}{ccccc} e_1 & e_2 & e_3 & e_4 & e_5 \end{array} \\ \begin{bmatrix} 1 & 0 & 1 & 1 & 0 \\ 1 & 1 & 0 & 0 & 0 \\ 0 & 1 & 1 & 0 & 1 \\ 0 & 0 & 0 & 1 & 0 \\ 0 & 0 & 0 & 0 & 1 \end{bmatrix} \end{array}$$

Remark 14.16 Note that the adjacency matrix is always symmetric.

14.3 Subgraphs

Definition 14.17 A subgraph of a graph $G = (V, E)$ is a graph $H = (U, F)$ such that $U \subseteq V$ and $F \subseteq E$. We denote this subgraph of G as $H < G$. Also G is called as super graph of H. If in G and H, $U = V$, then H is called as the spanning subgraph of G and it is denoted by $H \leq G$. Also if F consists of all edges $uv \in E$ then H is called the vertex induced subgraph of G.

Example 14.18 Consider the following graph (Fig. 14.4).

H_1, H_2 are subgraphs of a graph G. H_1 is an induced subgraph of G.

Definition 14.19 A decomposition of a G is a set S of subgraphs of G such that each edge appears in exactly one of the subgraphs of G in the set S (Fig. 14.5).

Example 14.20 Consider K_4.

The two paths

$$S = \begin{Bmatrix} P_1 : 1\ 2\ 3\ 4 \\ P_2 \ \ 2\ 4\ 1\ 3 \end{Bmatrix}$$

form a set of subgraphs and also S is a decomposition set for G.

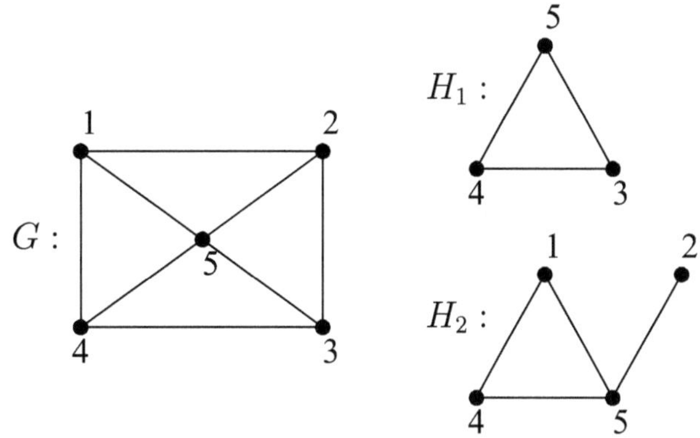

Fig. 14.4 Graphs and subgraphs

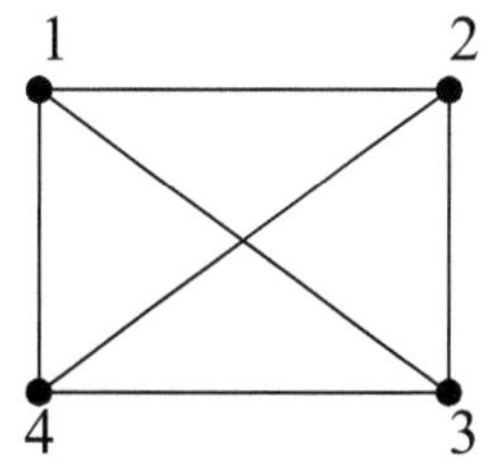

Fig. 14.5 A complete graph

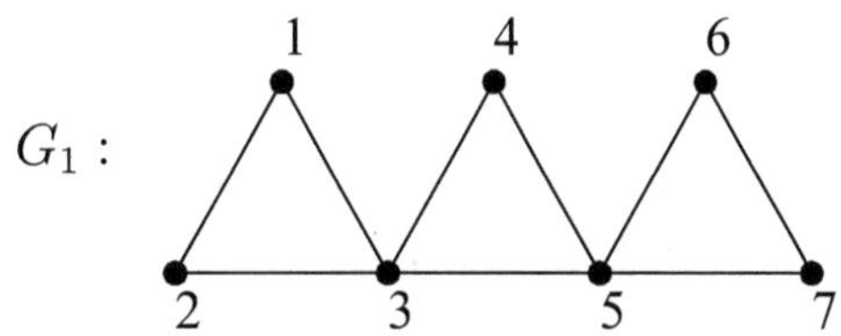

Fig. 14.6 Graph (9)

Decomposing a graph is in general a very difficult problem. One can analyse how to decompose K_n into set of cycles of length 3.

Example 14.21 Consider the graph (Fig. 14.6)

The set

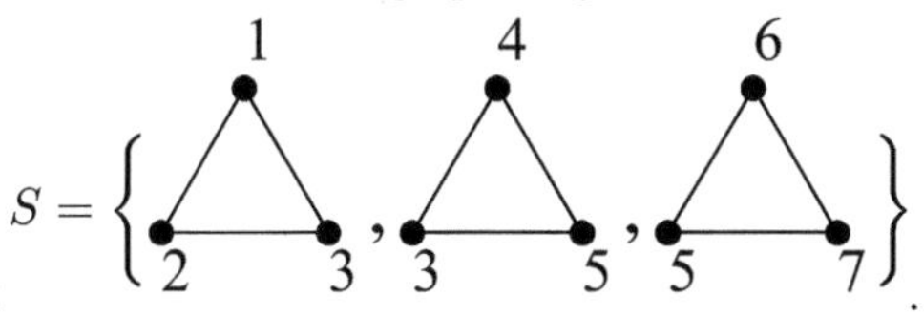

.

Remark 14.22 A graph may or may not have cycle decomposition.

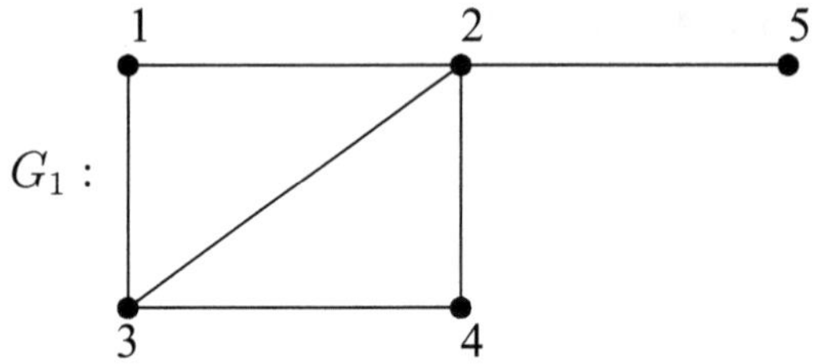

Fig. 14.7 Graph (10-1)

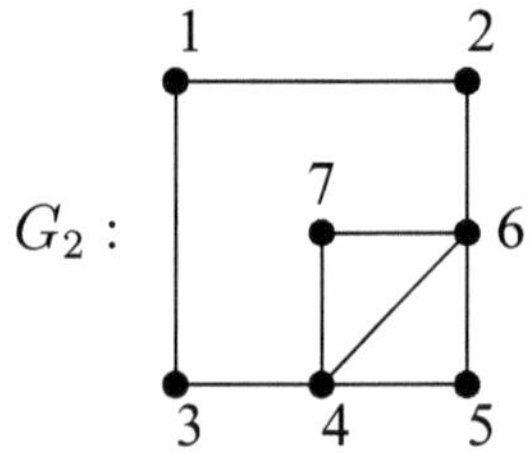

Fig. 14.8 Graph (10-2)

Example 14.23 1. Consider the graph (Fig. 14.7).
G does not have a cycle decomposition (Justify).
2. G_2 has a cycle decomposition (Fig. 14.8).

We state the following theorem.

Theorem 14.24 (Veblen) *A graph G has a cycle decomposition iff all its vertices are of even degree.*

14.4 Isomorphism

Definition 14.25 An isomorphism from a graph $G = (V, E)$ to a graph $G' = (V', E')$ is a bijection f from the vertex set of G to the vertex set of G' such that for $u, v \in V$, $uv \in E$ if and only if $f(u)f(v) \in E'$. Isomorphism between G and G' is denoted by $G \cong G'$.

Remark 14.26 Isomorphism is a binary relation on the graphs. Isomorphism implies the existence of a binary relation.

Example 14.27 Consider the graphs (Fig. 14.9)
$G \cong G'$. Here the binary relation $f : \{1, 2, 3, 4\} \to \{a, b, c, d\}$ is such that $f(1) = a$, $f(2) = b$, $f(3) = c$, $f(4) = d$.

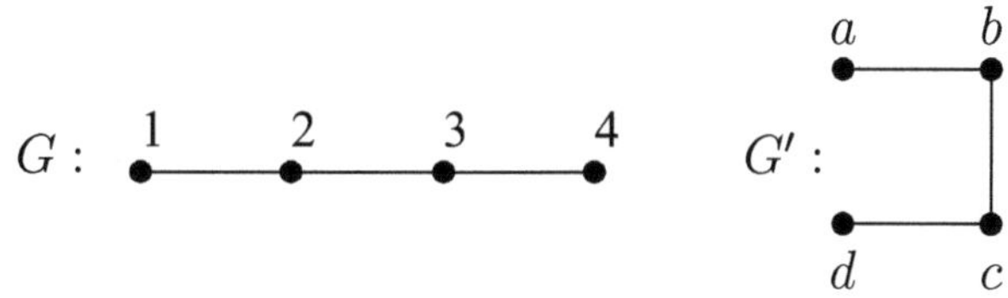

Fig. 14.9 Trees

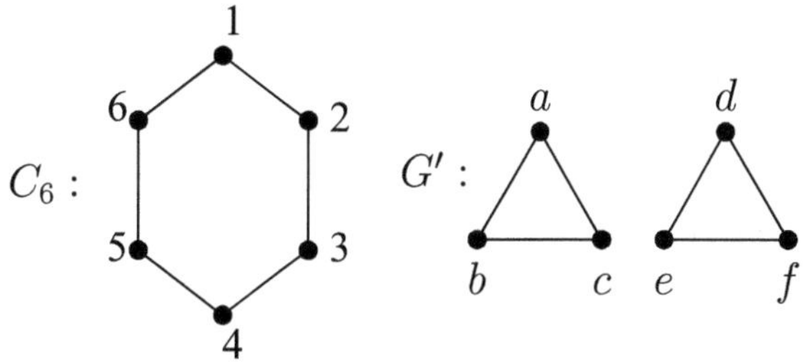

Fig. 14.10 Non isomorphic graphs

Example 14.28 Consider the following two graphs (Fig. 14.10):

$C_6 \ncong G'$. How do we prove this?

Assume C_6 is isomorphic to G'. Then there exists a bijection g between the vertex set of C_6 with that of G'. Note that C_6 and G' have the same number of vertices and edges.

$g(v_i)g(v_j)g(v_k)$ form a triangle in G', then by adjacency preservation, $v_i v_j v_k$ must form a triangle in C_6 which is not true, as C_6 dose not have a triangle. Hence $C_6 \ncong G'$.

Remark 14.29 Two complete graphs K_m and K_n are isomorphic if and only if $m = n$.

Isomorphic graphs in general have the same number of vertices, same number of edges, same length paths, same degree sequences and so on.

Definition 14.30 An isomorphism of a graph G onto itself is called an automorphism.

Remark 14.31 Every graph is isomorphic to itself and guarantees a bijection between the vertex sets preserving adjacency.

Two graphs G and G' being isomorphic does say that their complements are isomorphic.

Definition 14.32 A graph G is self complementary if $G \cong \overline{G}$.

14.5 Connectedness

Let $G = (V, E)$ be a graph. A vertex 'u' is connected to a vertex 'v' if there is a $u - v$ walk in G. Also we say that 'v' is reachable form 'u'.

Remark 14.33 For any graph, reachability relation is an equivalence relation.

Definition 14.34 A graph $G = (V, E)$ is said to be connected if and only if for every pair of vertices $u, v \in V$, 'v' is reachable from 'u'.

Remark 14.35 If in a graph G, for every pair of vertices u, v, G has $u - v$ path, then G is connected and conversely.

Since 'reachability' is an equivalence relation, this partitions the set of vertices into equivalence classes. Each set along with the set of edges on them is called as 'connected' components of the graph.

Example 14.36 Consider the graph G (Fig. 14.11).

Here in G, the 'reachability' of vertices partitions the vertex set as

$$V_1 = \{1, 2, 3, 4, 5, 6\}$$
$$V_2 = \{7, 8, 9, 10, 11, 12\}$$
$$V_3 = \{13\}.$$

Hence the induced subgraphs on V_1, V_2, V_3 are the components of G. Each of this subgraph is also called as 'maximal connected' subgraphs.

Observation 14.37 Adding a new edge between two different components will connect the two components and hence makes a single new component. Hence the new graph $G + e$ will have lesser number of components than G.

Fig. 14.11 Graphs (13)

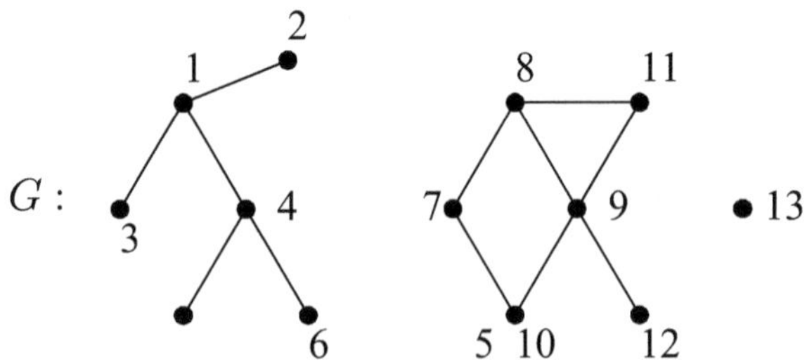

Definition 14.38 A graph G is said to be connected if any vertex 'v' is reachable from any vertex 'u' in G. That is every pair of vertices (u, v) has a walk between them.

Remark 14.39 K_n is connected whereas $\overline{K_n}$ is not connected.

The following theorems are simple and can be easily proved.

Theorem 14.40 *Let G be a graph with 'n' vertices and let $\delta(G) \geq \frac{n-1}{2}$. Then G is connected.*

Theorem 14.41 *A graph $G = (V, E)$ is connected iff for any partition of V as X_1, X_2, there exists an edge between a vertex in X_1 and vertex in X_2.*

We know what a bipartite graph is. A bipartite graph can have more than one component. Note that each component will be a bipartite graph.

Theorem 14.42 *A graph $G = (V, E)$ is bipartite if and only if it contains no odd cycles.*

Proof Let G be a bipartite graph. Then the vertex set V can be partitioned into V_1 and V_2 such that every edge is incident with a vertex in V_1 and a vertex in V_2. Consider a closed walk $v_0 v_1 v_2 \ldots, v_n (= v_0)$ of length n in G. Then such a walk must cross the bipartition an even number of times as there are edges only between the partitions. Then only the walk will become a closed one. Since every cycle is also a closed walk, the result follows that every cycle being of even length.

Conversely let G contain no odd cycles and also assume G is connected. Note that every closed walk of odd length contains a cycle of odd length. This can be proved by the method of induction.

Let $u \in V$. Let w_1 be a walk of odd length from u to v. Then w_1 one can see that every walk from u to v must be of odd length. Suppose this is not true. It must be true that there is a walk w_2 from u to v which is of even length. This means that $w_1 w_2$ is a closed walk of odd length. This means G contains an odd cycle which is not true. So we conclude that every $u - v$ walk in G has same parity for $u, v \in E$. Now partition v based on this parity condition as follows:

Each vertex 'v' with even length will be put in the set X_1 and those with odd length $u - v$ walk will be put in the set X_2. Clearly $X_1 \cap X_2 = \emptyset$ and $X_1 \cup X_2 = V$ is the bipartition. So X_1, X_2 must be independent sets. Otherwise, if $a, b \in X_1, ab \in E$, then by definition of X_1, there exists an even $u - a$ walk and $u - b$ walk, so $a - u - b$ walk has even length. If ab is an edge, then ab together with $a - u - b$ will be a closed walk of odd length. This is a contradiction. Hence X_1 is an independent set. Similarly X_2 can be shown to be independent. Hence the theorem. □

Fig. 14.12 Two regular graphs

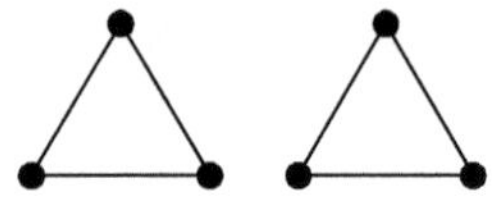

Complete graphs K_n, complete bipartite graphs $K_{m,n}$, path graphs P_n, cycle graphs C_n are all connected for $n, m \geq 0$. In general if in a graph every vertex has sufficiently large degree, then G will be connected.

Example 14.43 Consider the following graph (Fig. 14.12):
This is a 2-regular graph but not connected. If in this graph, there is atleast one vertex whose degree is greater than 2, then the graph becomes connected.

Theorem 14.44 *Let $G = (V, E)$ be a graph on 'n' vertices, $n \geq 1$. For any two non adjacent vertices u, v in V, if $d(u) + d(v) \geq (n-1)$, then G is connected.*

Proof Let $u, v \in V$ and $u \neq v$. If u and v are adjacent, then G is connected. So, consider u, v being not adjacent. Now $d(u) + d(v) \geq (n-1)$ implies that there is a vertex 'w' $\in V$ such that $uw \in E$ and $vw \in E$. Hence $u - w - v$ is a path in G. Hence G is connected. □

Remark 14.45 The lower bound on $d(u) + d(v)$ cannot be $(n-2)$.

Now one can look for the sensitivity of connectedness via their vertices as below.

Definition 14.46 A vertex 'v' of a graph G is said to be a cut vertex, if the removal of 'v' from G, increases the number of components of G.
By removing a vertex 'v' we mean that the vertex 'v' along with all the edges incident with 'v' are to be removed. One can denote this by $G - \{v\}$. By removing an edge e of a graph G means to remove the edge 'e' alone. It is denoted by $G - \{e\}$.

Remark 14.47 If 'v' is a cut vertex of a connected graph G, then $G - \{v\}$ is a disconnected graph.

Theorem 14.48 *Let $G = (V, E)$ be a connected graph. If 'v' is a cut-vertex of G, then there exist vertices $u_1, u_2 \in V - \{v\}$ such that every path between u_1, u_2 passes through 'v' and conversely.*

Proof Let $G = (V, E)$ be a connected graph. Suppose 'v' is a cut vertex then $G - \{v\}$ has more components than G. Let $G_1, G_2, \ldots, G_t$ be such components. Let $G_1 = (V_1, E_1)$ and

Fig. 14.13 No cut vertex

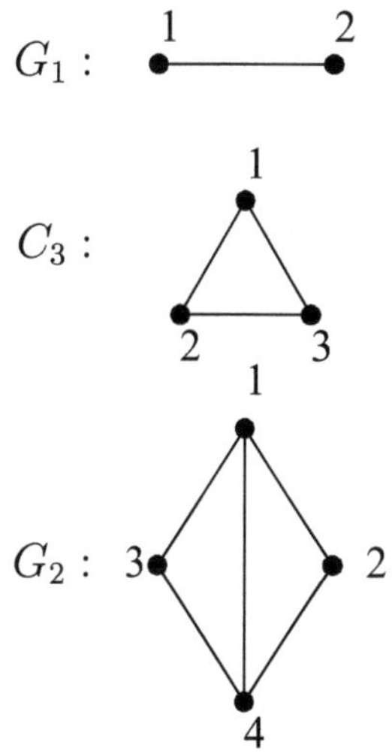

$\overline{G_2} = \bigcup_{i=2}^{t} G_i$. Let $u \in V_1$ and w be a vertex in $G_i, i \neq 1$. Then the $u - w$ path in G definitely passes through 'v' in G. Otherwise $G_1 \cup G_i$ will become a single component.

Conversely if for any two vertices $u_1, u_2 \in G - \{v\}$, u_1 and u_2 are not connected by a path in $G - \{v\}$. This means that they belong to different components in $G - \{v\}$, then $G - \{v\}$ is disconnected. So 'v' must be a cut vertex. □

The following theorem can be proved from the definition of a bridge.

Theorem 14.49 *Let $G = (V, E)$ be a connected graph. The following statements are equivalent.*

(i) *An edge 'e' is a bridge of G.*
(ii) *For an edge $e = u_1u_2$, there is a partition of the edge set $E - \{e\}$ as $E_1 \cup E_2$ such that $G_1 = (V_1, E_1)$, $G_2 = (V_2, E_2)$, where $V_1 \cup V_2 = V$. Also $u_1 \in V_1$, $u_2 \in V_2$. Then every path from a vertex in V_1 to a vertex in V_2 must contain e in G.*
(iii) *There exists vertices $u_1, u_2 \in V$ such that every $u_1 - u_2$ path contains e.*
(iv) *'e' does not lie on a cycle in G.*

In the following example, the graphs do not contain cut vertices and C_3, G_2 have no cut edges.

Example 14.50 See Fig. 14.13.

14.6 Non Separable Graphs

A connected graph may or may not contain cut vertices. We have seen an example for this in the previous section.

Definition 14.51 A connected graph that contains no cut vertices will be called as a non separable graph.

Example 14.52 $K_n, n \geq 2$ are non separable graphs.

Remark 14.53 If G is a non separable graph of order 3 or more, then $\delta(G)$ must be atleast 2.

Next is a theorem that gives a necessary and sufficient condition for a graph to be non separable.

Theorem 14.54 *Let G be a graph of order atleast 3. G is non separable if and only if any two of its vertices lie on a common cycle of G.*

Proof Let $G = (V, E)$ be non separable. We prove the result by induction on the distance between any two vertices of G. Let $u, v \in V$.

For $dis(u, v) = 1$. The edge uv cannot be a bridge. Hence 'uv' must lie on a cycle. Assume the result to be true for any two vertices $u, v \in V$ and $dis(u, v) < k$.

Let $u_1 - v_1$ be a path of length k in G. Let w be a vertex on the u_i-v_1 path such that wv_1 is an edge. Now $u_1 - w$ is a path of length $(k - 1)$ in G. By induction hypothesis u_1 and w must lie on a common cycle C in G and w is not a cut point. Hence $G - \{w\}$ is connected. Hence there exist a $u_1 - v_1$ path P in G not containing 'w.

Let 'v' be the last vertex that lie both on P and C as shown in the following figure (Fig. 14.14).

Such a vertex 'v' exists because 'u_1' is one vertex common to C and P.

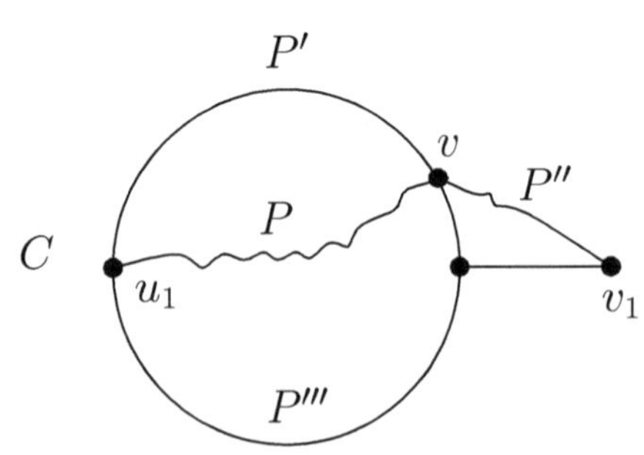

Fig. 14.14 P and C paths

Now the path P' from $u_1 - v$ not containing w on the cycle C, path P'' from v to v_1, edge v_1w and the path P''' from $w - u_1$ on the cycle C form a bigger cycle containing u_1 and v_1.

Conversely let every two vertices of G lie on a common cycle of G. To show G is connected, suppose that G has a cut vertex v. Then every path between any two vertices u_1, u_2 in V must pass through 'v'. As u_1, u_2 also lie on a cycle C in G, there will be path between u_1, u_2 not containing 'v'. This is not possible as 'v' is assumed to be a cut vertex. □

The above theorem has several uses. For example one can define paths between two vertices u_1, u_2 that do not share any internal vertices.

Definition 14.55 Two paths P_1, P_2 between two vertices u_1, u_2 are said to be internally disjoint if u_1, u_2 are the only common vertices between them.

Corollary 14.56 *A connected graph G of order greater than 2 is non separable if and only if for any two vertices u_1, u_2 in G, there are two internally disjoint $u_1 - u_2$ paths.*

Definition 14.57 Let G be a non trivial connected graph. A block of G is a maximal non separable subgraph of G.

Example 14.58 See Fig. 14.15.

Theorem 14.59 *Every connected graph that has cut vertices will contain atleast two blocks.*

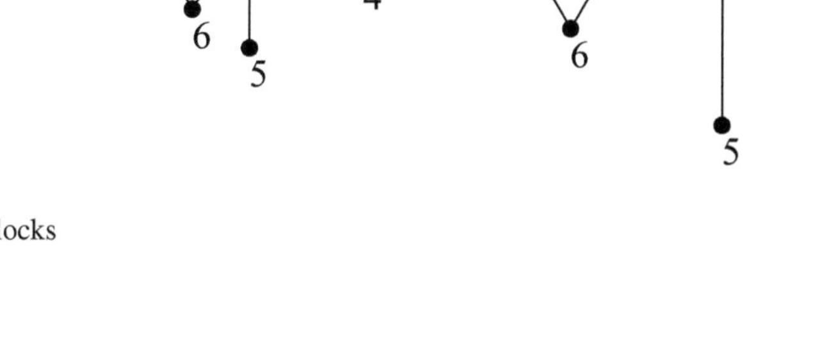

Fig. 14.15 Blocks

Exercises

1. If G is a simple graph on 10 vertices, what is the minimum number of edges that G can have so that G is connected?
2. Say whether the following statements are true or false. Justify your answers.

   (i) If a finite graph G has exactly two vertices of odd degree, then there must be a path from one of these vertices to the other in G.
   (ii) Let G be a graph with $\delta(G) = k$. Then G has a path of length atmost 'k', $k \geq 2$.
   (iii) P_4, path graph on 4 vertices, is an induced subgraph of $K_{4,4}$.
   (iv) If A is an adjacency matrix of a simple graph, then the (i, i)th entry of A^2 is the degree of v_i, v_i is a vertex of G.
   (v) Let A be an adjacency matrix of a simple graph G. Number of 3-cycles in G is $\frac{1}{6}tr(A^3)$ where $tr(A^3)$ is the trace of the matrix A^3.

3. If G_1, G_2 are connected graphs which are 2-regular, then show that $G_1 \cong G_2$.
4. Draw all graphs on 5 vertices that are non isomorphic, self complementary.
5. Show that a graph on 10 vertices and 26 edges cannot be bipartite.
6. Is it true that every graph on n vertices is isomorphic to a subgraph of K_n?
7. Show that a graph G is bipartite if and only if each induced cycle of G is of even length.
8. Show that an induced subgraph of a regular graph is regular.
9. Find two non isomorphic graphs with degree sequence (3, 2, 2, 1, 1, 1).
10. Let 'e' be a bridge of a connected graph G. Show that $G - e$ has exactly two components.
11. Does there exist a connected graph in which every edge is a cut edge?
12. Does there exist a connected graph with one cut vertex and no cut edges?
13. Give an example of a connected graph such that every connected subgraph is an induced subgraph.

Trees 15

15.1 Basic Definitions and Properties

We directly start with the definition of trees. ‘Trees’ are important set of graphs that have several applications.

Definition 15.1 A connected acyclic graph is a tree.
A graph having no cycles is called an acyclic graph.

Example 15.2 G_1 is an acyclic graph but not connected (Fig. 15.1).
G_2 is a tree (Fig. 15.2).

Remark 15.3 Any graph that is acyclic may or may not be connected. Such a graph is in general called as ‘forest’.

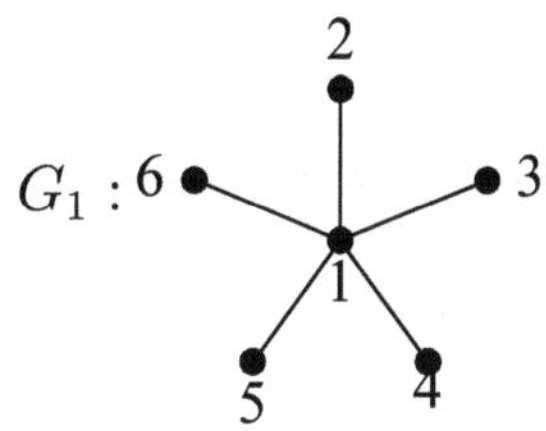

Example 15.4
is a tree called as ‘star’

G_4 : 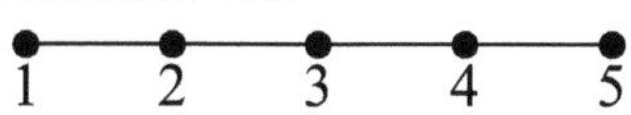

: P_5 is a path graph that is a tree

R. Rama, *Topics in Combinatorics and Graph Theory*,
https://doi.org/10.1007/978-3-031-74252-1_15

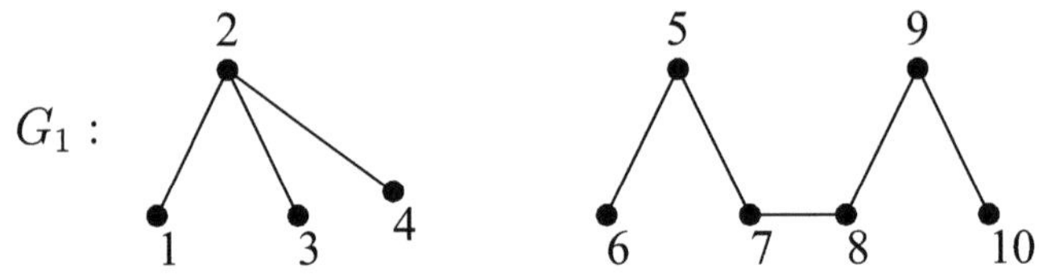

Fig. 15.1 Trees

Fig. 15.2 Another tree

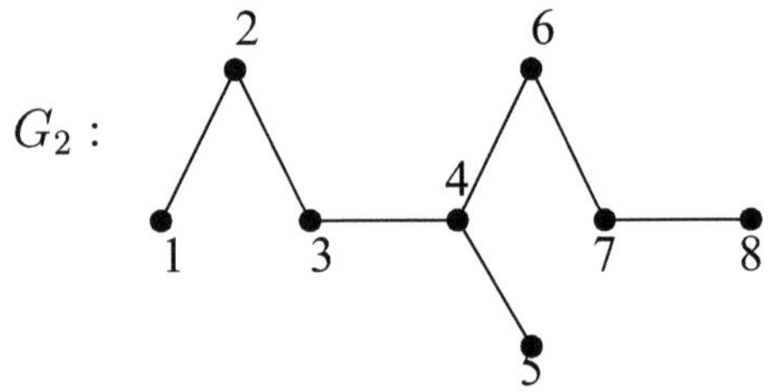

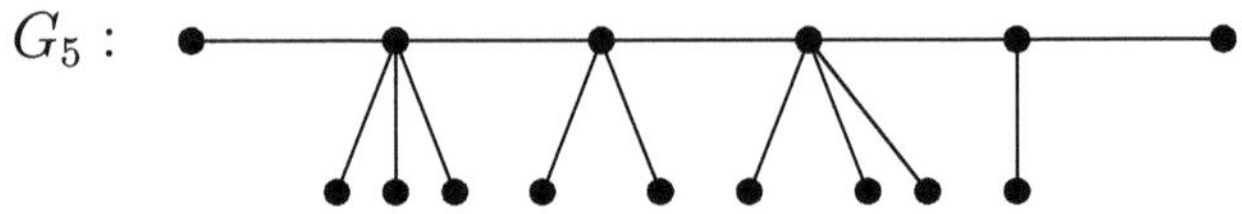

G_5 is a tree called as 'caterpillar'.

Consider the following tree

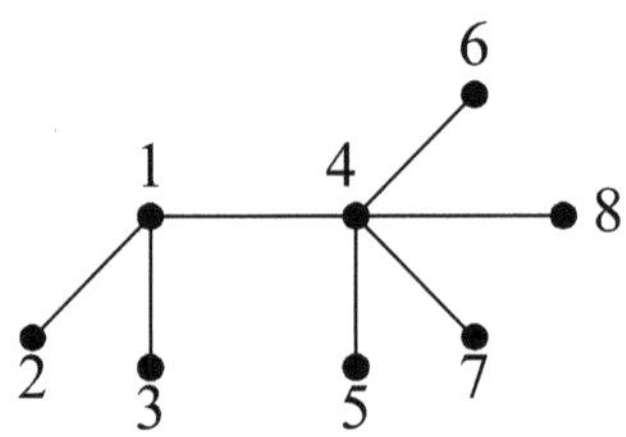

This graph is a tree. The degree sequence of this graph is (5, 3, 1, 1, 1, 1, 1, 1). The vertices with degree 1 are called as 'leaves' in any tree. Also every tree on n vertices, $n \geq 2$ has atleast 2 leaves. This fact is widely used in several properties of trees.

Theorem 15.5 *The following statements are equivalent.*

1. *T is a tree.*
2. *Any two vertices u, v of T are connected by a unique path in T.*
3. *Every edge e of T is a bridge.*
4. *For any two non adjacent vertices u, v in T, $T + uv$ contains a cycle.*

The proof of the above theorem follows directly from the definition and they will serve as a good exercise. Trees are said to be minimally connected and maximal acyclic. The following corollary is simple but a useful one. A tree may be labelled or unlabelled graph.

Corollary 15.6 *Let T be a tree. The vertices of T can always be enumerated. That is if $v_1, v_2, \ldots, v_n$ are the vertices of T, then the list can imply that every v_i, $i \geq 2$ has a unique neighbour in $\{v_1, v_2, \ldots, v_{i-1}\}$.*

Proof The proof is by induction on i. For $i = 1$, choose a vertex as 'v_1'. Its neighbour v with $v_1 v$ will be labelled v_2. Assume the enumeration is complete upto $v_1, v_2, \ldots, v_i$, for $i < n$. Let $T[v_1, v_2, \ldots, v_i]$ be the induced subgraph of T on the vertices $\{v_1, v_2, \ldots, v_i\}$. Now consider a vertex $v \in T - T[v_1, v_2, \ldots, v_i]$. As T is connected, it contains a path P from v_1 to v. Choose the last vertex of P in $T - T[v_1, v_2, \ldots, v_i]$ so that v_{l+1} has a neighbour in $T[v_1, v_2, \ldots, v_i]$. □

Theorem 15.7 *Let T be a tree on 'n' vertices. Then the number of edges will be $(n - 1)$.*

Proof We will give here an intuitive way of proof. Let T be an empty graph on 'n' vertices. So T now has n components. If an edge is added between two different components reduces the number of components by 1. So the number of components of $T + e$ is now $(n - 1)$. Now once again add an edge between any two distinct components of $T + e$. Now the components count is reduced by one and the number of components will be $(n - 2)$. Proceeding like this way of adding edges between any two components $(n - 1)$ times, we get a tree. Hence the number of edges will be $(n - 1)$ in T. □

Corollary 15.8 *If T is a tree and G is any graph with $\delta(G) \geq |T| - 1$, then T is a subgraph of G.*

Proof Using the above vertex enumeration method, trace the tree T inductively through the vertices of G. What one gets is a copy of a tree isomorphic to T. □

15.2 Spanning Trees

We know what a spanning subgraph of a graph is. It is a subgraph that spans over all the vertices of G.

Definition 15.9 A spanning tree of a graph G is a spanning subgraph that is a tree.

Example 15.10 Consider the following graph:

G :

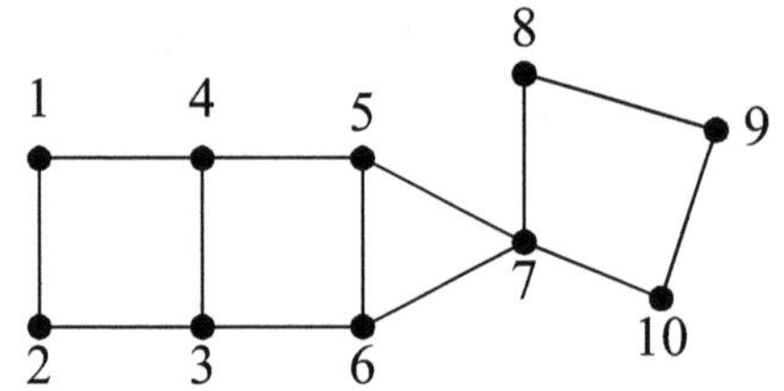

The subgraph

T :

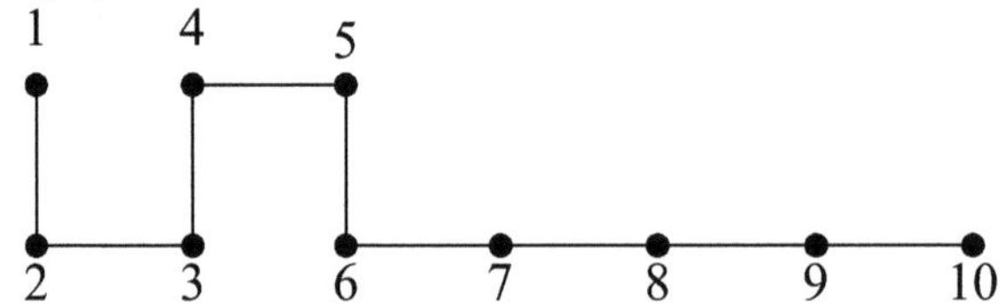

is a spanning tree.

Remark 15.11 Every connected graph has atleast one spanning tree. The tree can be got by deleting an edge on a cycle of the graph for every cycle of the graph.

Definition 15.12 An edge of a tree T can be called as a 'branch' of T and an edge of G that is not on the spanning tree T of G can be called as a 'chord' of G. The definition of branches and chords are only with respect to a given spanning tree.

Example 15.13 Consider the graph G with a spanning tree T (Fig. 15.3).
Branches of T are: 12, 23, 34, 45, 48, 56, 67
Chords of T are: 24, 78, 57.
Clearly the edge set of G is the union of branches and chords of T.
Clearly $\overline{T}$ is the complement of T, whose edges are chords.

G :

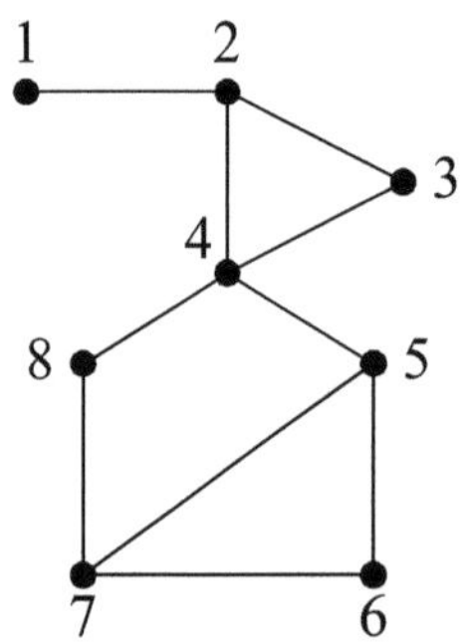

T :

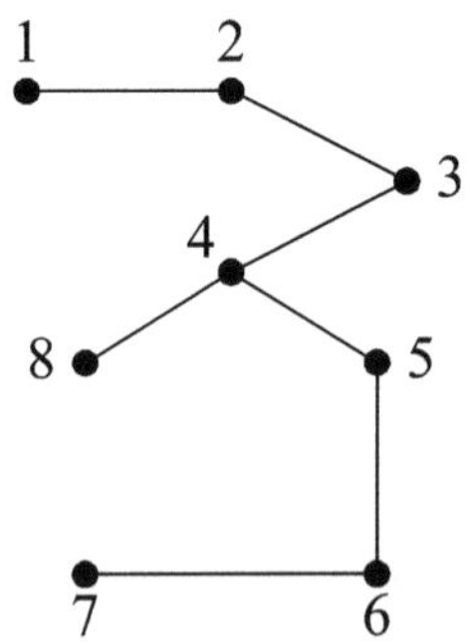

Fig. 15.3 Spanning tree

Theorem 15.14 *Let $G = (V, E)$ be a graph and T be its spanning tree. Then the number of branches of of T will be $(n - 1)$ and number of chords will be $|E| - (n - 1)$.*

Theorem 15.15 *Let G be a connected graph on n vertices. G has exactly one cycle if and only if G has exactly n edges.*

Proof A graph G with n vertices is minimally connected if G has exactly $(n - 1)$ edges.

Suppose G has exactly one cycle. Then if 'e' is an edge on the cycle then $G - \{e\}$ is connected and has $(n - 1)$ edges. So $G - \{e\}$ is a tree. This means G has 'n' edges.

Conversely G is connected and has n edges. Consider a spanning tree T of G. T has $(n - 1)$ branches and one chord say 'e'. This means $T + e$ contains a cycle which is G. □

Remark 15.16 For any spanning tree T of G, $\overline{T}$ will be called as cotree of G.

Proposition 15.17 *Let G be a connected graph and T, T' be two spanning trees of G. Let e be an edge of tree T but not in T'. Then there exists an edge e' in T' not on T such that $T' + \{e\} - \{e'\}$ is a spanning tree of G.*

Proof We know that in T and T' there is unique path between any two vertices. Let $e = uv$, $e' = u'v'$. Clearly $T' - e$ has $(n - 2)$ edges and is disconnected graph on n vertices. So u, v are vertices of $T' - e'$ lying on two different components of $T' - e'$. Hence $T' - e' + e$ will be a tree on $(n - 1)$ edges. □

Definition 15.18 Let u, v be two adjacent vertices of a graph $G = (V, E)$. The edge 'uv' is said to be contracted if it is deleted from G and the vertices u, v are identified. The resulting graph is denoted by G_e.

Example 15.19 Let (Fig. 15.4).

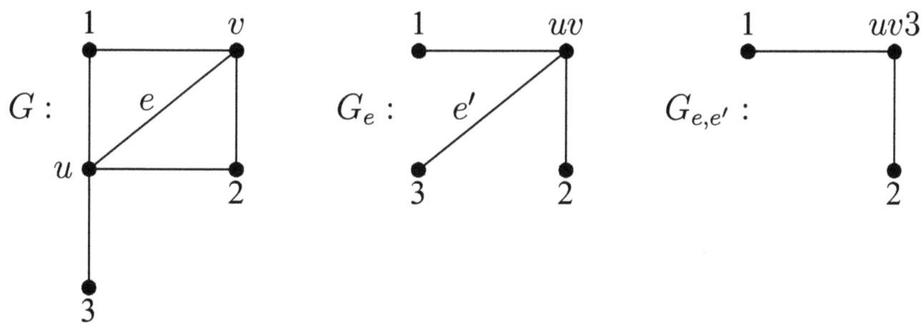

Fig. 15.4 Edge contraction

Remark 15.20 The edge contraction on a graph $G = (V, E)$ can be performed until a graph H is reached. The vertices of H is a set of elements in a partition $\{v_1, v_2, \ldots, v_t\}$ of V where each induced subgraph $G_i = (V_i, E_i)$ is connected and some vertex in V_i is adjacent to some vertex in V_j in G.

Notation: Let $s(G)$ denote the number of spanning trees of G.

Theorem 15.21 *Let e be an edge of graph G. Then* $s(G) = s(G - e) + s(G_e)$.

Proof If T is a spanning tree of G that does not contain the edge e, then it is also a spanning tree of $G - e$. Also $s(G - e)$ denote the number of spanning trees of $G - e$ and hence every such tree does not contain 'e'.

Consider G_e. A spanning tree of G_e will be a tree T_e obtained by contracting e in T. So there is a 1-1 correspondence between such spanning trees and spanning trees of G containing the edge e. Hence the result. □

We know that the number of labelled simple graphs on n vertices and m edges if $C\left(\frac{n(n-1)}{2}, m\right)$. That is the graph on n vertices will have m edges chosen from the set of all possible edges. This m can be in the set $\left\{0, 1, 2, \ldots, \frac{n(n-1)}{2}\right\}$.

Theorem 15.22 *The number of simple graphs on n vertices is* $2^{C(n,2)}$.

Cayley [1889] gave a formula to count the number of distinct labelled trees on vertices. Here he considered the vertices of G being labelled in the beginning to extract the spanning tree from the fixed labelling.

Example 15.23 Consider the labelled graph

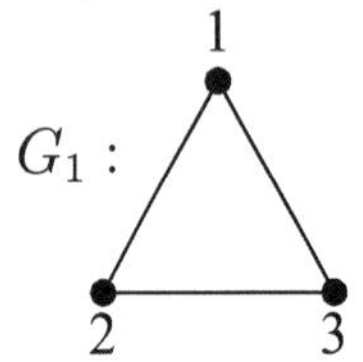

Its spanning trees are

T_1 :
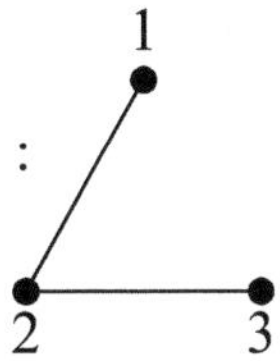

T_2 :
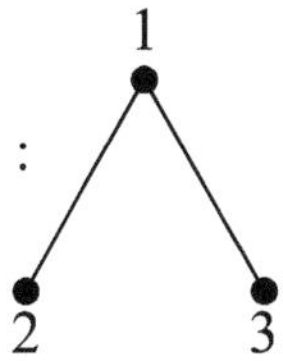

T_3 :
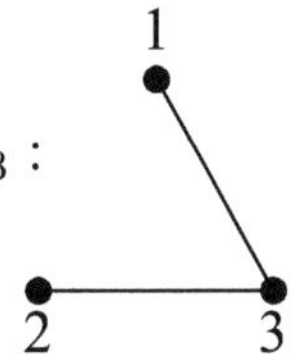

G_2 :
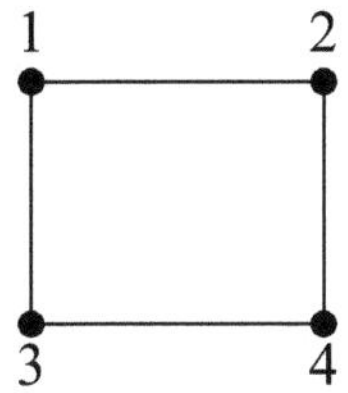

Its spanning trees are

T_1 :
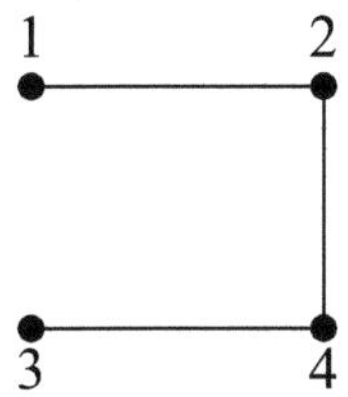

T_2 :
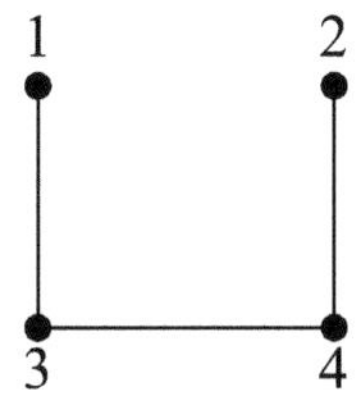

T_3 :
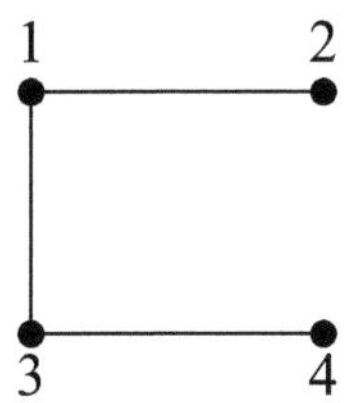

T_4 :
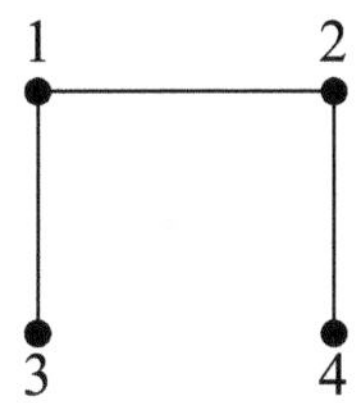

The following theorem gives a formula to find the number of trees.

Theorem 15.24 (Cayley) *The number of distinct labelled trees on n vertices is* n^{n-2}.

Proof We give the proof using Prufer sequences.

Let T be a tree on 'n' vertices. Let the labels of the vertices be from the set $\{1, 2, \ldots, n\}$. We will form $(n-2)$ tuples based on T. Let x_1 be a leaf vertex of T with the smallest label. Let a vertex 'v_1' be adjacent to 'x_1' and remove 'x_1'. In the tree $T - \{x_1\}$, let 'u_2' be the leaf vertex of smallest index, and 'v_2' be the vertex adjacent to u_2. Remove u_2. In the resulting tree again identify the leaf of the smallest index as u_3. Let 'v_3' be adjacent to u_3, remove u_3 and continue this process till a tree with just two vertices result. Let the sequence identified so far be $v_1, v_2, \ldots, v_{n-2}$ and the given tree uniquely identified the $(n-2)$ tuple $(v_1, v_2, \ldots, v_{n-2})$.

Conversely for any $(n-2)$ tuple of $\{1, 2, \ldots, n\}$, a n-vertex tree can be constructed as below:

Let $(v_1, v_2, \ldots, v_{n-2})$ be the $(n-2)$ tuple. Find the first number from $1, 2, \ldots, n$ that do not occur in this $(n-2)$-tuple. Let x_1 be that number. Thus $x_1 v_1$ will be an edge of T. Now remove v_1 from the $(n-2)$-tuple and x_1 from the set $\{1, 2, \ldots, n\}$. Now identify the first number u_2 from $\{1, 2, \ldots, n\} \backslash x_1$ that is not in the new $(n-3)$-tuple resulted by the removal of v_1. Then $u_2 v_2$ will be an edge of T. Continue till all the elements in the tuple $(v_1, v_2, \ldots, v_{n-2})$ are checked out. The last two elements of $\{1, 2, \ldots, n\}$ form an edge in T.

So we can see that for every $(n-2)$ tuple of $\{1, 2, \ldots, n\}$ a unique tree T is identified and conversely for every labelled tree T, there exists a unique $(n-2)$-tuple. Number of $(n-2)$ tuples is n^{n-2}. Hence number of labelled trees is n^{n-2}. □

15.3 Tree Decompositions of a Graph

Decomposition of a graph $G = (V, E)$ is the set of edge-disjoint subgraphs $G_i = (V_i, E_i)$ of G, $1 \leq i \leq k$ such that $V = \bigcup_{i=1}^{k} V_i$, $V_i \cap V_j$ need not be empty. $E = \bigcup_{i=1}^{k} E_i$, $E_i \cap E_j = \emptyset$.

Example 15.25

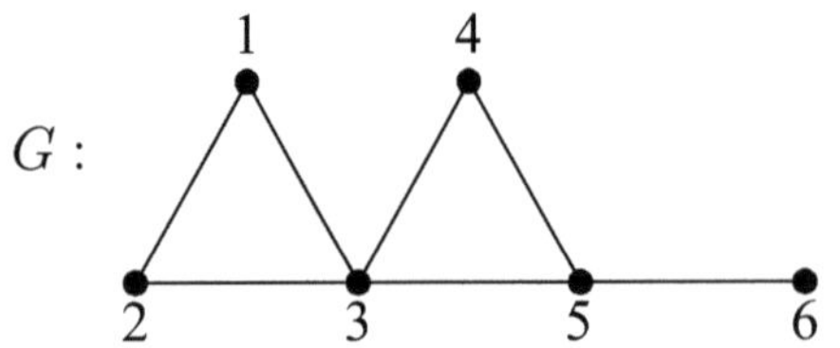

The subgraphs

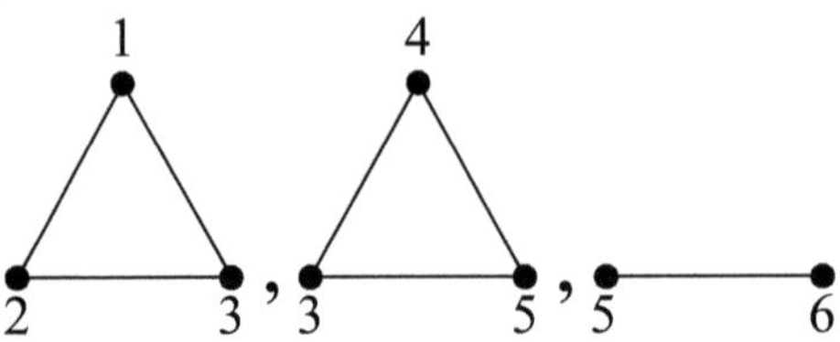

form an edge-disjoint decomposition of G.

A graph can be decomposed into subgraphs. If each of the subgraph in the decomposition is a cycle, then the decomposition will be called as cycle decomposition. If each of the subgraph in the decomposition is a path, then such a decomposition will be called as path decomposition. If each of the subgraph in the decomposition is a tree, then such a decomposition will be called as tree decomposition.

Example 15.26 Consider the graph G

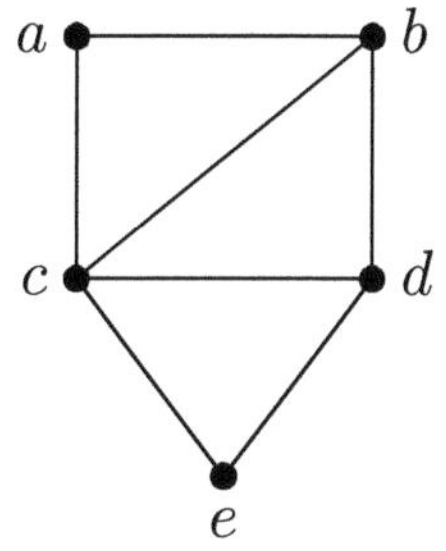

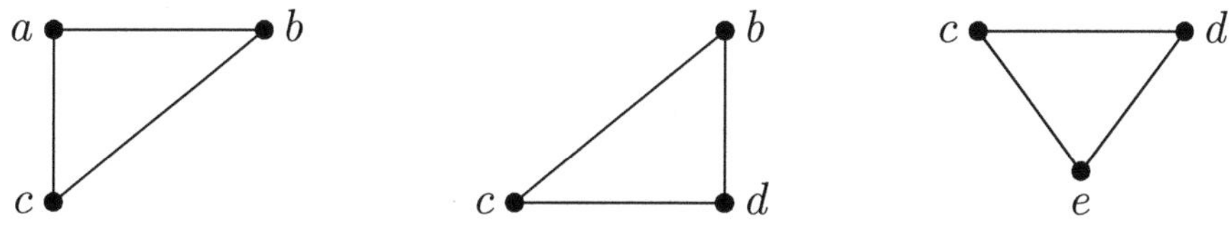

D_1 is not a cycle decomposition. Each cycle is a triangle.

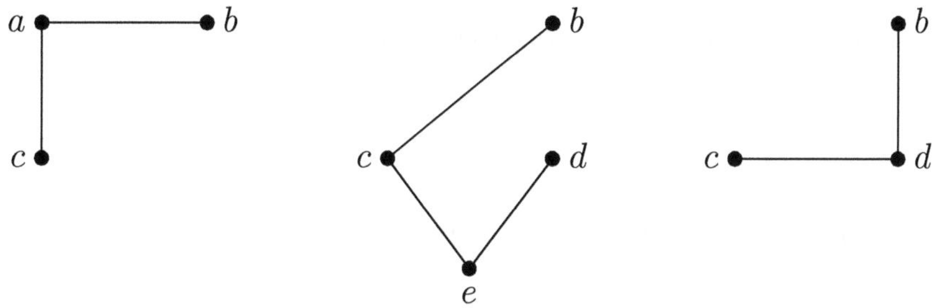

D_2 is a path decomposition. It is also a tree decomposition.

Definition 15.27 A T-decomposition is a decomposition of a graph G if every subgraph in the decomposition is isomorphic to T. If T is a tree, then it is called a tree decomposition.

Some important discussions were done on the decomposition of a complete graph into trees such that each tree in the decomposition is isomorphic to a specified tree. The conjucture of Ringel (1964) gained a lot of interest. The conjecture is:

The complete graph on $(2n + 1)$ vertices can be decomposed into copies of trees such that each tree has n-edges.

Despite several attempts to prove this conjecture, it is still open.

Example 15.28 Consider the following graph:

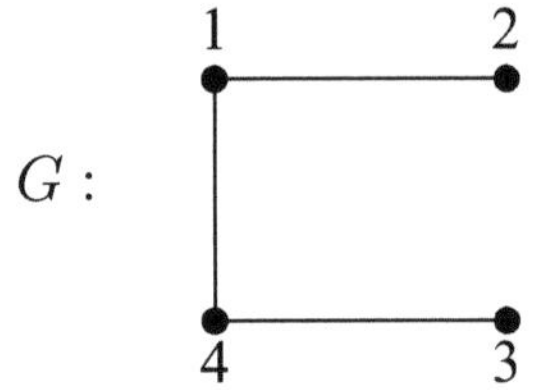

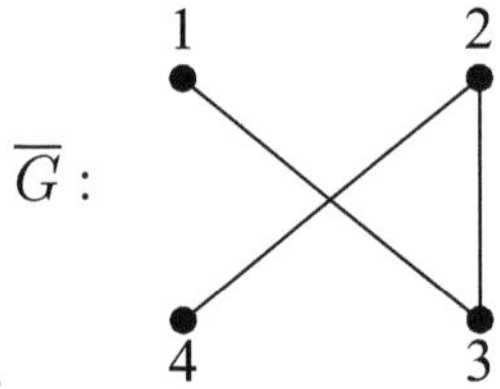

Its complement will be
G is isomorphic to $\overline{G}$.
The two form a tree decomposition of a complete graph.

We state the following proposition without proof.

Proposition 15.29 *If a graph G has a decomposition into copies of a graph H, then*

(i) The number of edges in H divides number of edges in G.
(ii) $\Delta(H) \leq \delta(G)$.

Remark 15.30 The above conditions are not sufficient.

Proposition 15.31 *A k-regular graph G can be decomposed into copies of star graphs with k edges if and only if G is bipartite.*

Proof If G is not bipartite, then G contains an odd length cycle. If G has decomposition into copies of stars with k-edges, the centers of the stars will be some vertices on this odd cycle. If 'v_1' lying on the odd cycle is a center of a star, then all the edges incident at 'v_1' are taken as one copy of the star graph. None of the k vertices appearing as leaves in this star can be the center. Hence, choose the next alternate vertex 'v_2' on the odd cycle as the center of another star with k-edges as the next copy. By this way one can reach alternate vertices on the cycle to identity copies of stars. However, since the cycle is of odd length, there will be some edges left over which cannot appear in a star with k-edges. Hence G cannot be decomposed as star-graphs with k-edges.

Conversely if G is bipartite. Consider one maximal independent set. Every vertex in this set can serve as the center of a star graph with k-edges. Also every such edge appear exactly once in each star. Hence the decomposition is reached. □

Regarding Ringel conjecture on decomposition of a graph, Rosa, 1907 reached a theorem connecting graceful labelling of a tree with the decomposition of complete graph on $(2n + 1)$ vertices. We briefly see the approach here.

Definition 15.32 A graceful labelling of a graph G with m edges is labeling the vertices and edges of G with numbers $\{0, 1, 2, \ldots, m\}$ such that

(i) distinct vertices are labeled distinctly from $\{0, 1, \ldots, m\}$
(ii) edges are distinctly labeled.
(iii) The labeling is graceful if all the numbers $\{1, 2, \ldots, m\}$ appear on the edges of G.

Remark 15.33 Only connected graphs are considered for graceful labeling.

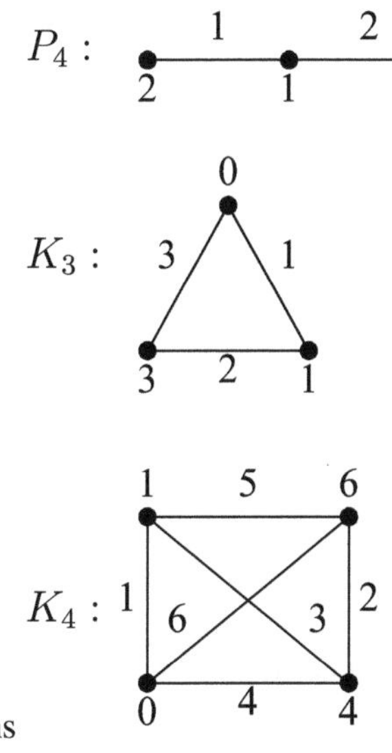

Example 15.34 See graphs .

Remark 15.35 All path graphs and star graphs are graceful.

Graceful Tree Conjecture (Kotzig, Ringel, 1964):
Every tree has a graceful labeling.

Theorem 15.36 *If T is a tree with m edges and admits a graceful labeling, then K_{2m+1} can be decomposed into $(2m + 1)$ copies of T.*

Proof Suppose that T is graceful. Let $T_0, T_1, \ldots, T_{2m}$ be copies of T in K_{2m+1}. Then the vertices of T_k can be labeled as $k, k+1, k+2, \ldots, k+m(\text{mod}(2m+1))$ where $(k+i)$ and $(k+j)$ are adjacent in T_k if and only if vertices i and j are adjacent in the basic tree that is labeled with 0, 1, 2, ..., m. Hence edges of T_k has labeling from $\{1, 2, \ldots, m\}$. The edge between $(k + i)$ and $(k + j)$ in K_{2m+1} will appear in G_k as the corresponding edge. Hence graceful labeling induces a decomposition of K_{2m+1} as copies of trees with m edges. □

Several attempts were made to understand graceful labeling of graphs to make some progress on the Graceful Tree Conjecture. Some classes of graphs that are known to be graceful are as below.

1. Caterpillar graphs.
2. Complete bipartite graphs.
3. Trees with diameter at most five.
4. Trees with atmost 4 leaves.
5. Cycle graphs of length k where $k \equiv 0 \bmod 4$ or $k \equiv 3 \bmod 4$.
6. n-dimensional hypercube Q_n.

Exercises

1. Draw all the unlabelled trees on 5 vertices.
2. Does there exist a tree with 12 edges and the degree sequence (5, 4, 1, 1, 1)? Justify your answer.
3. Show that every tree is bipartite.
4. Does there exist a tree on n vertices with exactly one spanning tree? Justify your answer.
5. For any vertex v of a connected graph G the sequence

$$d'(G, v) = (n_0(v), n_1(v), \ldots, n_e(v))$$

where e is the eccentricity of v and $n_j(v) = |N_j(v)|$ where $N_j(v) = \{u | d(u, v) = j\}$ for $0 \leq j \leq e$ is called the distance degree sequence of v in G (DDS_v).
Find DDS_v of v in the following tree.

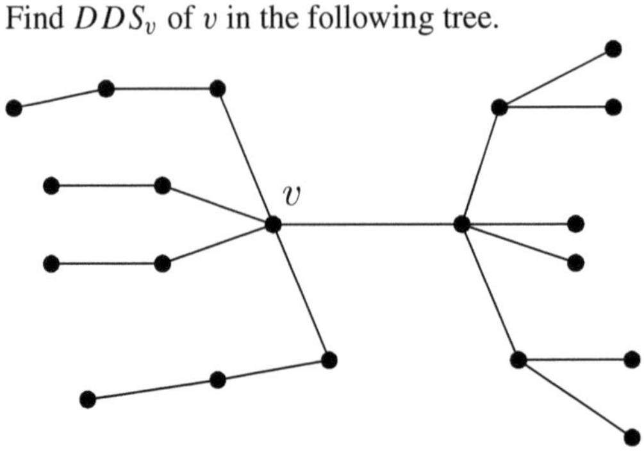

6. Show that the positive sequence $(d_1, d_2, \ldots, d_n)$ is the degree sequence of a tree iff (i) $d_i \geq 1, 1 \leq i \leq n$ (ii) $\sum_{i=1}^{n} d_i = 2(n-1)$.
7. If a sequence (4, 3, 2, 1, 1, 1, 1, 1) is the degree sequence of a tree, then find the number of labelled trees with this degree sequence.
8. Let u be a vertex of a connected graph G. Then show that there exists a spanning tree of G that preserves the distance from u.

Connectivity

16

The two connectivity parameters in graph theory are vertex connectivity and edge connectivity of a given graph. The question here is to find the rigidness of a graph remains connected by removing a single vertex/edge, then we look for minimum number of vertices/edges to be removed resulting in a disconnected graph or trivial graph. The parameters are κ (kappa) and λ (lambda).

16.1 Vertex Connectivity

Definition 16.1 A vertex-cut of a connected graph $G = (V, E)$ is a set $S \subseteq V$ of G such that $G - S$ is disconnected. A minimum vertex cut is one that has minimum number of vertices in S and the cardinality of such a set S is said to be the vertex connectivity or simply connectivity of the graph G. It is denoted by $\kappa(G)$ (κ is the symbol kappa) (Fig. 16.1).

Example 16.2 A complete graph does not contain any cut vertex and $\kappa(K_n) = n - 1$. So we can see that

$$0 \leq \kappa(G) \leq n - 1$$

for any graph G.

Remark 16.3 A vertex-cut is called as separating set.

Proposition 16.4 *Let $K_{m,n}$ be a complete bipartite graph. Then $\kappa(K_{m,n}) = \min\{m, n\}$.* □

Proof In a complete bipartite graph, every induced subgraph of $K_{m,n}$ which has atleast one vertex from each partition of the vertex set must be a connected graph. If A and B are the

R. Rama, *Topics in Combinatorics and Graph Theory*,
https://doi.org/10.1007/978-3-031-74252-1_16

Fig. 16.1 Graph (1)

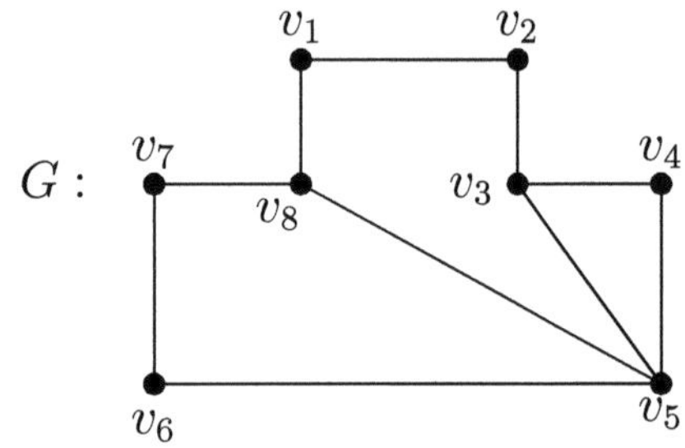

partitions of the vertex set, then A, B are vertex cuts of $K_{m,n}$. Hence $\kappa(K_{m,n}) = \min\{m, n\}$ where $|A| = m$, $|B| = n$. □

Remark 16.5 A connected graph that has vertex connectivity 1 is a separable graph.

Example 16.6 Consider the following graph (Fig. 16.2).

In G removal of any one vertex does not disconnect G. Removal of $S = \{3, 6\}$ disconnects the graph and hence S is a vertex cut.

Remark 16.7 In general it will be more useful to know when and all the graph remains connected rather than knowing its vertex cut.

Definition 16.8 A graph G is said to be k-connected if $\kappa(G) \geq k$, for $k \geq 1$.

So we can infer from the above definition that a k-connected graph is one in which removal of t vertices, $t < k$ never results in a disconnected graph or trivial graph. So a 1-connected graph is simply a non trivial connected graph. A 2-connected graph does not have a cutvertex. A graph with large number of edges will have large connectivity. The following is a sufficient for a graph to be k-connected.

Fig. 16.2 Graph (2)

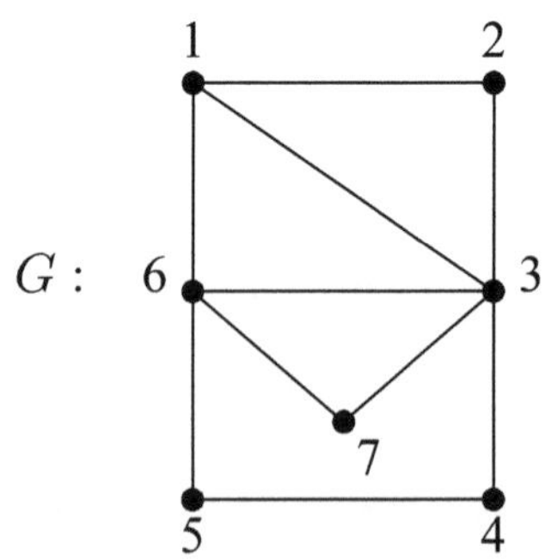

Theorem 16.9 *Let $G = (V, E)$ be a graph with $|V| = n$ and let 'k' be an integer with $1 \leq k < n$. If for every vertex $v \in V$,*

$$d(v) \geq \left\lceil \frac{n+k-2}{2} \right\rceil$$

then G is k-connected.

Proof Suppose G is not k-connected. G is also not a complete graph. So there exists a set $S \subseteq V$ such that S is a vertex-cut of G. Also $|S| = t\ (\leq k-1)$. $G - S$ is disconnected and number of vertices in $G - S$ is $n - \ell$.

Consider a component G_1 in $G - S$ with minimum number of vertices say n_1. Clearly $n_1 \leq \left\lfloor \frac{n-\ell}{2} \right\rfloor$. Let x be a vertex in G_1. This vertex 'x' can be adjacent only to vertices in S and to other vertices in G_1. So its degree will satisfy

$$\begin{aligned} d(v) &\leq \ell + (n_1 - 1) & (16.1)\\ &\leq \ell + \left\lfloor \frac{n-\ell}{2} \right\rfloor - 1 & (16.2)\\ &= \left\lfloor \frac{n+\ell-2}{2} \right\rfloor \leq \left\lfloor \frac{n+k-3}{2} \right\rfloor & (16.3) \end{aligned}$$

which is a contradiction to the degree condition given in the theorem. Hence G is k-connected. □

16.2 Edge Connectivity

The edge connectivity is a parameter that measures the sensitiveness of a connected graph with respect to removal of edges.

Definition 16.10 (i) An edge-cut of a graph $G = (V, E)$ is a subset T of E such that $G - T$ is disconnected.

(ii) Minimum number of edges in T' where $G - T'$ is disconnected is known as edge connectivity of the graph G. It is denoted by $\lambda(G)$.

Example 16.11 (i) If G is a graph with atleast one cut-edge, then $\lambda(G) = 1$.

(ii) For a complete graph K_n, $\lambda(K_n) = n - 1$.

(iii) In general if 't' is the minimum degree of a vertex in a connected graph G, then removal of these 't' edges results in a disconnected graph. So, we can write the following inequality

$$0 \leq \lambda(G) \leq \delta(G) \leq n - 1$$

Fig. 16.3 Graph G_1

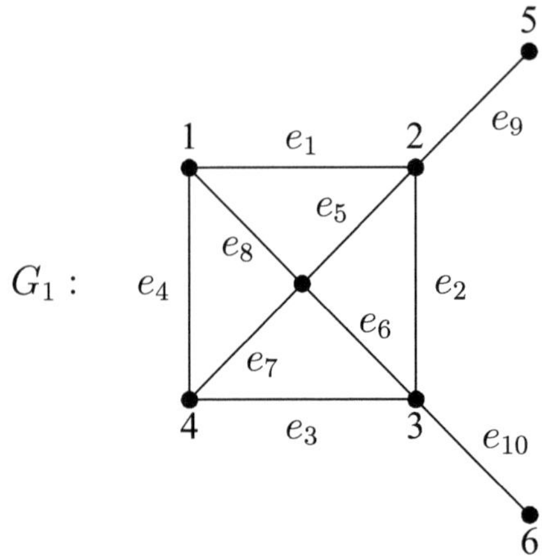

Definition 16.12 A graph G is said to be k-edge-connected for $k \geq 1$ if $\lambda(G) \geq k$. Note that a 1-edge connected graph is a connected graph. A 2-edge connected graph will not contain any cut-edges.

Example 16.13 Consider the graph G_1 (Fig. 16.3):
$\kappa(G_1) = 1, \lambda(G_1) = 1$.
$\kappa(G_2) = 2, \lambda(G_2) = 3$ (Fig. 16.4).
$G_3 : \kappa(C_n) = 2, \lambda(C_n) = 2$.

We can see that edge connectivity of a graph is bounded above by the minimum degree of the graph and bounded below by its vertex connectivity. This is first noticed by Hassler. The proof is established by Whitney.

Theorem 16.14 *For every graph* $G = (V, E)$

$$\kappa(G) \leq \lambda(G) \leq \delta(G).$$

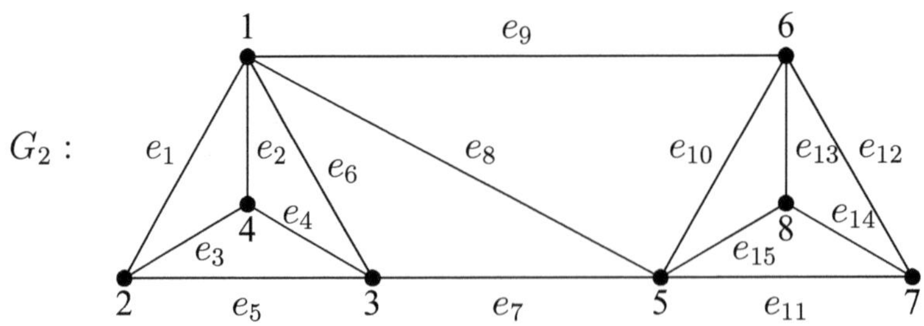

Fig. 16.4 Graph G_2

Proof We have already seen that $\lambda(G) \leq \delta(G)$ earlier.

To show $\kappa(G) \leq \lambda(G)$. We notice that if $\lambda(G) = 0$ or 1, then κ is also 0 or 1. Let $\lambda(G) = k \geq 2$. Let $T = \{e_1, e_2, \ldots, e_k\}$ be the edges such that $G - T$ is disconnected. Let $e_i = u_i v_i$, $1 \leq i \leq k$, some of these u_i, v_i vertices may be identical. Let G_1, G_2 be the components of $G - T$. Let $G_1 = (V_1, E_1)$, $G_2 = (V_2, E_2)$. Here consider the set of vertices $\overline{V} = \{u_1, u_2, \ldots, u_k\}$ and all vertices in $\overline{V}$ lie in the same connected component of $G - T$. If $G - \overline{V}$ is not connected, then clearly $\kappa(G) \leq k$.

Otherwise, a vertex say 'u_1' has atmost 'k' vertices adjacent to it from the set $\overline{V}$ and some vertices from the set $\{v_1, v_2, \ldots, v_k\}$. These vertices form a vertex cut and hence $\kappa(G) \leq k \leq \lambda(G)$. □

Remark 16.15 We observe that vertex-connectivity is bounded above by edge-connectivity. But edge-connectivity cannot be bounded above by vertex connectivity.

We can see that any connectivity parameter is always bounded above by minimum degree of the graph. Now the question is 'Is it possible to have two graphs on $n, n+1$ vertices such that both the graphs are k-connected. We state the following proposition to this effect.

Proposition 16.16 *1. Let G be a k-connected graph on n vertices. If a new vertex 'v' is added to G and made to be adjacent to atleast k vertices of G, then the resultant graph $\overline{G}$ will be k-connected.*

2. Let G be a k-edge-connected graph. If a new graph $\overline{G}$ is constructed by adding a new vertex 'v' adjacent to atleast 'k' vertices of G, then $\overline{G}$ is k-edge connected. □

We know that every 2-edge-connected graph G is one with $\lambda(G) \geq 2$. Every 1-edge-connected graph is also 2-edge-connected.

Proposition 16.17 *Let G be a 2-edge-connected graph. Then every edge of G lies on a cycle.* □

Proof Let $e = xy$ be an edge of G. Since G is 2-edge-connected, $G - e$ is connected. This means there exists a path between x and y in $G - e$. So $e = xy$ must be on a cycle. □

We know every 2-connected graph is also 2-edge-connected. In a 2-edge-connected graph every pair of vertices (x, y) lie on a cycle. x, y need not be adjacent.

Definition 16.18 Let $G = (V, E)$ be graph and $H = (V', E')$ be a subgraph of G. A path $P : uu_1u_2 \ldots u_n v$ with the vertices $u, v \in V'$ and $u_1, u_2, \ldots, u_n \in V \backslash V'$ is said to be H-ear path of G. A H-ear decomposition of the graph G is a sequence of graphs $(C, P_1, P_2, \ldots, P_t)$ such that

(i) $C = G_0$ is a cycle
(ii) for $1 \leq i \leq t$, P_i is a path such that it is G_{i-1}-ear and $G_i = G_{i-1} \cup P_i$
(iii) $G_t = G$.

It can be shown that $H \cup P$ is 2-connected if H is 2-connected for any H-ear of P. So we state the following decomposition theorem for a 2-connected graph without proof.

Theorem 16.19 *Let G be a 2-connected graph, and let C be a cycle. Then G can be decomposed as* $(C, P_1, P_2, \ldots, P_k)$.

16.3 Local Connectivity

A graph is said to be connected if for every pair of vertices u, v there exists atleast one path between u, v. This is generalized as below in the definition of local connectivity.

Definition 16.20 Let $G = (V, E)$ be a graph. Let $u, v \in V$ be any two non-adjacent vertices. A $u - v$ separating set $S \subseteq V - \{u, v\}$ is one that puts u, v in $G - S$ in different components. The minimum cardinality of such a set if said to be the local connectivity of u and v or minimum $u - v$ separating set. It is denoted by $\kappa(u, v)$.

For two adjacent vertices u, v, their local connectivity is defined to be $\kappa_H(u, v) + 1$ where $H = G - uv$. The local edge-connectivity can be defined similarly.

It is true that $\kappa(G) = \min\{\kappa(u, v)/u, v \in V \text{ and } u \neq v\}$.

For two distinct vertices, two paths P_1, P_2 between u and v are said to be uv independent if they have only u, v as the only common vertices. A set of independent uv-paths is a set of paths that are mutually independent. In the next theorem by Menger (1926) we can see the connections between $\kappa(u, v)$ and uv-independent paths.

Theorem 16.21 (Menger) *Let $G = (V, E)$ be a graph. Let $u, v \in V$ be two nonadjacent vertices in G. Then $\kappa(u, v)$ is the maximum number of internally disjoint paths between u and v.*

Proof We give the proof of this theorem given by Dirac (1969).

Let $|V| = n$, $|E| = m$. We use induction on $m + n = g$ (say). That is we prove that if $\kappa(u, v) = t$ is the minimum set that separates the vertices u, v then G has atleast 't' uv-independent path.

For $t = 1$, the result is true for graphs, having n vertices and $(n - 1)$ edges and also some small values of g. This takes care of the basis for induction. Assume the result to be true for $g = k - 1$ that $\kappa(u, v)$ is the maximum number of internally uv disjoint paths in G.

Now let G be a graph with $g = k$ (say).
(1) Suppose $u, v \in V$ have a vertex x such that $ux, vx \in E$. Then x must be in the minimum separating set S of G. In the graph $G - x$, $S - x$ must be the minimum separating set of u, v in $G - x$. Hence by induction hypothesis there will be $(t - 1)$ independent paths between u, v in $G - x$. These $(t - 1)$ paths together with uxv form t independent paths between u and v in G.
(2) Suppose that no vertex of S is adjacent to the vertices u, v in G. S is the minimum separating set for u, v in G. Let G_u, G_v be the connected components containing u, respectively in $G - S$.

Let $S = \{x_1, x_2, \ldots, x_t\}$. Let G_u be the subgraph of G including all $u - x_i$ paths, for $1 \le i \le t$ in G and all these paths have only the vertex x_i from S. Let $\overline{G}_u$ be a new graph obtained from G_u by adding a new vertex $\overline{u}$ and edges $\overline{u}x_i$, $1 \le i \le t$. In a similar manner $\overline{G}_v$ is constructed. The number of edges in $\overline{G}_u$ and $\overline{G}_v$ are definitely smaller than $|E| = m$. Hence by induction hypothesis $\overline{G}_v$ contains t internally disjoint $\overline{v} - v$ paths containing x_i and $\overline{G}_u$ contains t-internally disjoint $u - \overline{u}$ paths containing $\overline{x}_i$. Let these paths be denoted by the sets A^i_u, A^i_v, for $1 \le i \le t$. Then the subpaths from $u - x_i$ of a path in A^i_u and subpaths $x_i - v$ in A^i_v can be combined to make a $u - v$ path in G for $1 \le i \le t$. In fact the combination will be internally disjoint.
(3) For the set S, if u is adjacent to every vertex in S not to v or v is adjacent to every vertex in S not to u, then $d(u, v) \ge 3$. Let P be the shortest path $(u, u_1, u_2, \ldots, v)$. Let $e = u_1u_2$. The minimum separating set S' of $G - e$ must contain atleast $(t - 1)$ vertices by induction hypothesis. We will show now that $|S'| = t$. If $|S'| \ne t$, then let $S' = \{v_1, v_2, \ldots, v_{t-1}\}$ be the minimum separating set of $G - e$. Then $S' \cup \{u_1\}$ is a $u - v$ separating set in G. But $uu_1 \in E$, $u_1v \notin E$. So for each $v_i \in S'$, $uv_i \in E$, $v_iv \notin E$ for $1 \le i \le t - 1$. Similarly for $S' \cup \{u_2\}$ being a minimum $u - v$ separating set in G, each v_i is adjacent to u but not to v means $uu_2 \in E$. This means there a shorter path P' from $u - v$, which is a contradiction to our assumption that P is the shortest $u - v$ path. Hence $|S'| = t$. Hence G has t internally disjoint $u - v$ paths. □

We can see that every block on n vertices, $n \ge 3$ must be 2-connected. So we can write the following theorem.

Theorem 16.22 *Let $G = (V, E)$ be a graph such that $|E| \ge 3$. G is 2-connected if and only if any two vertices u, v of G are connected by atleast two internally disjoint paths.*

Example 16.23 G is a non-separable graph (Fig. 16.5).

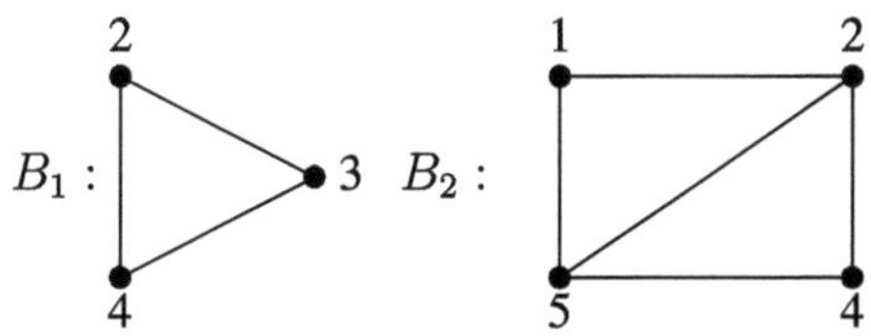

are blocks of G. G is 2-connected.

Fig. 16.5 Non separable graph

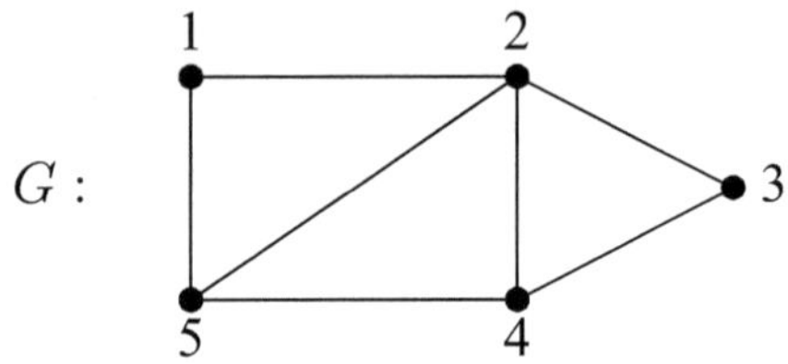

Exercises

1. Let G be a graph such that $\lambda(G) = 2$. Show that any two non-adjacent vertices are connected by atleast 2 edge disjoint paths.
2. Find $\lambda(G), \kappa(G)$ for the following graphs G
   (i) $K_{m,n}$ (iii) C_n
   (ii) $\overline{K}_n$ (iv) P_n
3. Let $G = (V, E)$ be a graph such that $\lambda(G) = t$ and let $|E'| = t$. Then show that $G - E'$ results in a graph that has 1 or 2 components.
4. Let $G = (V, E)$ with $|V| \geq 3$. Let $\delta(G) \geq \frac{|V|}{2}$. Then show that G is δ-edge connected.
5. Let $G = (V, E)$ be a graph such that $G \neq K_n, G \neq C_n$, where $|V| = n$. Show that there exists vertices $u, v \in V$ such that $G - \{u, v\}$ is connected.
6. Let T be a tree and let S be an independent set of T, with $|S| \geq 2$. Is it true that the minimum number of vertices separating the set S is equal to the maximum number of vertex disjoint paths between two vertices, $u, v \notin S$? Justify your answer.

Eulerian and Hamiltonian Graphs 17

In this chapter we are going to see a special class of graphs which gained a lot of interest and importance in graph theory.

17.1 Euler Graphs

We have already seen what a 'Konigsberg Bridge Problem' is. In that problem, the seven land portions can be denoted by vertices, and bridges between them by edges in the graphical representation (Fig. 17.1).

The above graph is a multi-graph.

Euler used graphs and shown the non-existence of a walk only once along the bridges, visiting all the land portions and returning to the same starting position. In fact Euler obtained necessary and sufficient condition for the existence of such a walk. We have defined Euler trail in the earlier chapter.

Remark 17.1 A closed trail is called as a tour. The study of Euler trail or Euler tour is only for connected graphs.

Definition 17.2 A graph that admits an Euler tour is said to be an Eulerian graph.

Theorem 17.3 (Euler 1736) *The necessary and sufficient condition for a connected graph $G = (V, E)$ to be Eulerian is that each of its vertices must be of even degree.*

R. Rama, *Topics in Combinatorics and Graph Theory*,
https://doi.org/10.1007/978-3-031-74252-1_17

Fig. 17.1 Non simple graph

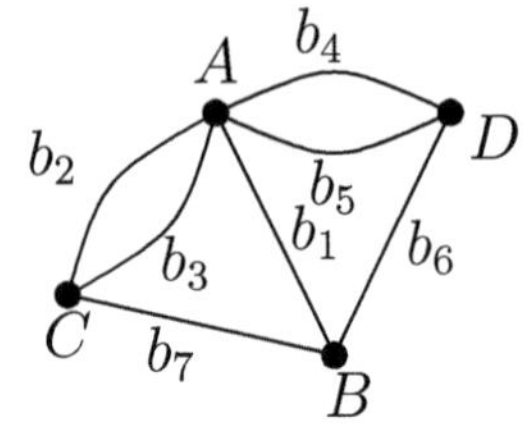

Proof Suppose G is Eulerian. Then G has an Euler tour 'T' starting and ending at the vertex 'v' (say). Every vertex u on T contributes 2 degrees to that vertex as T is a tour and every edge being traversed only once. If 'v' is one of the internal vertices (other than start vertex), $d(v)$ will be even. For the start vertex, the edges that is traversed in the start being different from the edge traversed at the end to complete the tour contribute even degree to it.

Conversely let every vertex of T be an even degree vertex. Suppose there exists no Euler tour in G. Let G be a graph with minimum number of edges. Since $\delta(G) \geq 2$, G has a cycle C in G. Let 'F' be a tour of G of maximum number of edges. The graph with $E - E(F)$ edges where $E(F)$ is the set of edges of F, also has every vertex of even degree. Let F_1 be one of its components. There exists an Euler tour in F_1 say $\overline{F}$ and $\overline{F}$ has a vertex 'v' in common with F. So $F \cup \overline{F}$ becomes a tour in G whose edges are more than F. This is a contradiction. □

Definition 17.4 An open trail that contains every edge of G is an Eulerian trail.

Remark 17.5 All the concepts discussed in this chapter are true for multi-graphs.

Next we give a characterization of connected graphs that contain Euler trial. These results are also given by Euler.

Theorem 17.6 *Let $G = (V, E)$ be a connected graph. G contains Euler trail if and only if G has exactly two vertices of odd degree.*

Proof Let G contain an Euler-trail say $T : x - y$. Now construct $G' = G + w$ where w is a new vertex added to G such that w is adjacent only to the vertices x and y. Now T together with wx, wy form an Euler tour in $G + w$. Hence every vertex of $G + w$ is even. This means x, y are the only two vertices of odd degree in G.

Conversely let G be connected graph with exactly two vertices x, y of odd degree. Now to G, add a new vertex w such that xw, yw are the new edges added to make a new graph H. Now H has an Euler tour Z as every vertex of H being even. This tour must contain the edges xw, wy as these are the only edges connecting to x and y. Now removal of w from Z, results in an Euler trail. □

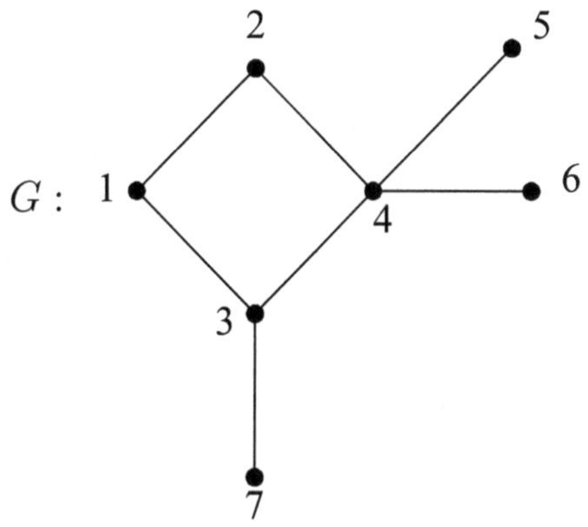

Fig. 17.2 Graph (1)

Now the graph G may contain more than two odd degree vertices. If G contains more than two odd degree vertices then G will not have an Euler tour or Euler trail .

Example 17.7 G does not have Euler Trail. This can be seen by listing all the trails of G.

The following result talks about trails in such graphs (Fig. 17.2).

Theorem 17.8 *If G be a connected graph having $2n$ vertices of odd degree ($n \geq 1$), then G contains k pair wise edge disjoint open trails connecting two odd degree vertices such that every edge lies on one of these trails.*

Proof Let $x_1, x_2, \ldots, x_n$ and $y_1, y_2, \ldots, y_n$ be the $2n$ odd degree vertices of G. Now add 'n' new vertices $u_1, u_2, \ldots, u_n$ to G such that $u_i x_i, u_i y_i$, $1 \leq i \leq n$ are new set of edges added to make a new graph G' for $1 \leq i \leq n$. G' is connected and every vertex of G' is now even. So G' has an Euler tour Z. Now removal of the n vertices $u_1, u_2, \ldots, u_n$ from Z splits Z in to n open trails from x_i to y_i for $1 \leq i \leq n$. Since every edge appeared only once in Z, these are edge disjoint open trails. □

Next is a characterization of an Eulerian graph whose edge set can be partitioned in to disjoint cycles.

Theorem 17.9 *Let G be a non-trivial connected graph. G is Eulerian if and only if G contains pairwise edge disjoint cycles such that every edge of G lies on one of its cycles.*

Proof Suppose $G = (V, E)$ is Eulerian. We prove the cycle partitioning the edges by method of induction on the number of edges of G. If $|E| = 3$, the result is true. Assume for $|E|$ such that $|E| < m$ (say) $m \geq 4$. G is Eulerian imply that every vertex of G is of even degree.

Then G contains a cycle say C_1. If $G = C_1$, the result is complete. Otherwise consider G_1 obtained from G by removing edges of C_1. Now in G_1, every vertex is again of even degree. Thus the non trivial component G_1 has fewer edges than m, and hence has a cycle partition of its edges by our induction hypotheses. Now all the cycles of the components of G_1 and C_1 form a cycle partition of the edges of G.

Conversely suppose the edges of G can be partitioned by cycles, means every vertex of G is even. Hence G is Eulerian. □

Definition 17.10 Let $G = (V, E)$ be a graph. G is said to be arbitrarily traversable from a vertex $v_1 \in V$ if the following procedure gives a trail which is Eulerian.

Start from the vertex v_1 and move to the next vertex v_2 adjacent to v_1. From v_2 move to its adjacent vertex v_3 via an edge not used so far and continue to travel till all the edges in E are used exactly once.

Note that if the procedure ends in the vertex v_1, then what is traced is an Eulerian tour. The following is the algorithm by Fleury to construct an Eulerian trail.

Fleury's Algorithm

1. Choose an arbitrary vertex 'v_0' and let $T_0 = v_0$.
2. For $T_1 = v_0 e_1 v_1 e_2 v_2 e_3 \ldots v_i$ such that $e_1, e_2, \ldots, e_i$ are distinct choose edge $e_{i+1} = v_i v_{i+1}$, not a bridge unless no other option prevails.
3. Continue till step-2 becomes impossible.

Remark 17.11 Note that Fleury's algorithm just traces one trail of the graph G. It will trace an Eulerian trail if it exists in G.

17.2 Non Eulerian Graphs

We saw that an Eulerian graph will always have all its vertices being even degree. A graph that is not Eulerian will have even number of odd degree vertices. We also saw that if a graph has exactly two odd degree vertices then graph will contain an Eulerian trail.

In a non-Eulerian graph G one can attach paths between pair of vertices to obtain a graph H which is Eulerian. H will be called as super Eulerian graph of G. The graph G then is a spanning subgraph of H.

Definition 17.12 A graph G is said to be sub-Eulerian graph if it is a spanning subgraph of some Eulerian graph.

Remark 17.13 Every graph being a spanning subgraph of an Eulerian graph is not true in general.

The following theorem characterizes sub Eulerian graphs.

Theorem 17.14 (Boesch, Suffel, Tindell) *Let $G = (V, E)$ be a connected graph. G is sub-Eulerian if and only if G is not spanned by a complete bipartite graph.*

Definition 17.15 A non-Eulerian graph G is said to be super Eulerian if it contains a subgraph that is Eulerian.

The following is a result for super-Eulerian graphs which just gives a sufficient condition.

Proposition 17.16 *Let $G = (V, E)$ be graph on atleast 6-vertices. Also any two non adjacent vertices $u, v \in V$, $d(u) + d(v) \geq (n - 1)$; then G is super-Eulerian.*

17.3 Hamiltonian Graphs

Eulerian graphs are one that contain a closed trail including every edge exactly once. Now we will see graphs that contain cycles which contain every vertex of the graph.

Definition 17.17 Let $G = (V, E)$ be a graph.

(i) If there exists a path P in G that contain every vertex of G, then P is said to be Hamiltonian path of G.
(ii) If there exists a cycle in G including all the vertices of G, then such a cycle is said to be a Hamilton cycle of G.
(iii) A graph that has a Hamilton cycle is said to be Hamiltonian graph. Otherwise it is said to non-Hamiltonian.

Example 17.18 G_1 is non-Hamiltonian. But it has a Hamilton path (Fig. 17.3).

The only way to check whether a graph G_1 is Hamiltonian or not is by trial and error technique. There are some characterizations proved for Hamiltonian graphs. To show a graph to be non-Hamiltonian is more difficult if the number of edges are large. So in general showing existence or non existence of a Hamilton cycle in a graph demands only exhaustive search in all possible ways.

Following are some famous graphs that are non-Hamiltonian (Figs. 17.4 and 17.5).

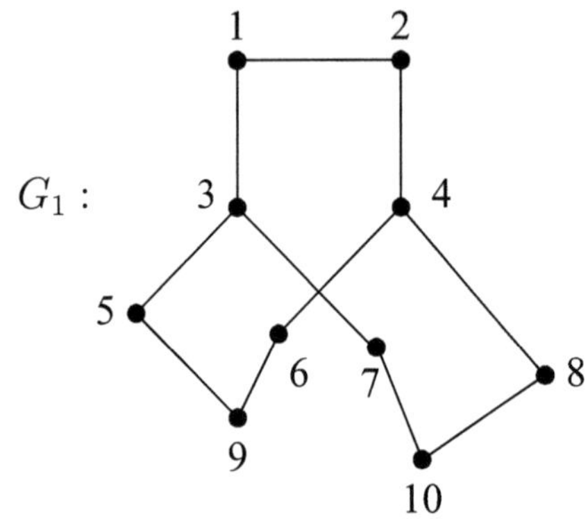

Fig. 17.3 Graph (2)

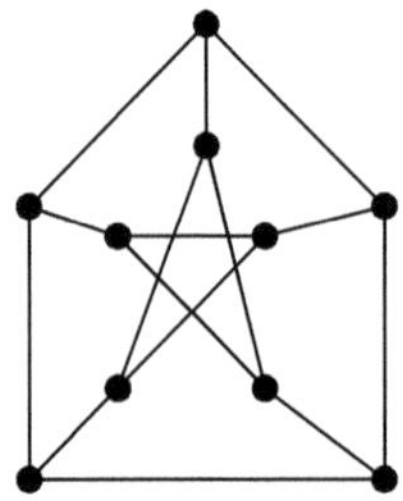

Fig. 17.4 Non Hamiltonian graphs

Finding whether a graph is Eulerian or not is simple when compared to finding Hamilton cycle in a graph. There are several characterizations of Eulerian graphs. But there is no efficient way to analyze the graph being Hamiltonian or not. These are some necessary conditions and some sufficient conditions to this analysis. It is more likely for a graph to posses a Hamilton cycle if it has more number of edges.

Theorem 17.19 (Ore's theorem) *Let* $G = (V, E)$ *be a graph with* $|E| = n$, $n \geq 3$. *For any two non-adjacent vertices* $u, v \in V$, *if* $\min_{u,v \in V}\{d(u) + d(v)\} \geq n$, *then G is Hamiltonian.*

Proof Suppose the condition given for G' is satisfied but G' is not Hamiltonian. Then add as many edges as possible between non adjacent vertices of H resulting in graph G. But G is non-Hamiltonian. We can say that G is maximal non-Hamiltonian. It means $G + uv$, for $u, v \in V$ being non adjacent becomes Hamiltonian. G cannot be a complete graph here (Fig. 17.6).

Let uv be added to G enabling to get a Hamilton cycle C (say). C certainly contains the edge 'uv' and G certainly has a path $P : u = x_1, x_2, \ldots, x_n = v$. P is a Hamilton path. Here if u is adjacent to a vertex x_i for $2 \leq i \leq n - 1$, then v cannot be adjacent to x_{i-1}. Otherwise will be a Hamilton cycle in G. This is not possible. So for every vertex of G adjacent to 'u', there will be vertex of $V - \{v\}$ not adjacent to 'v'. This means

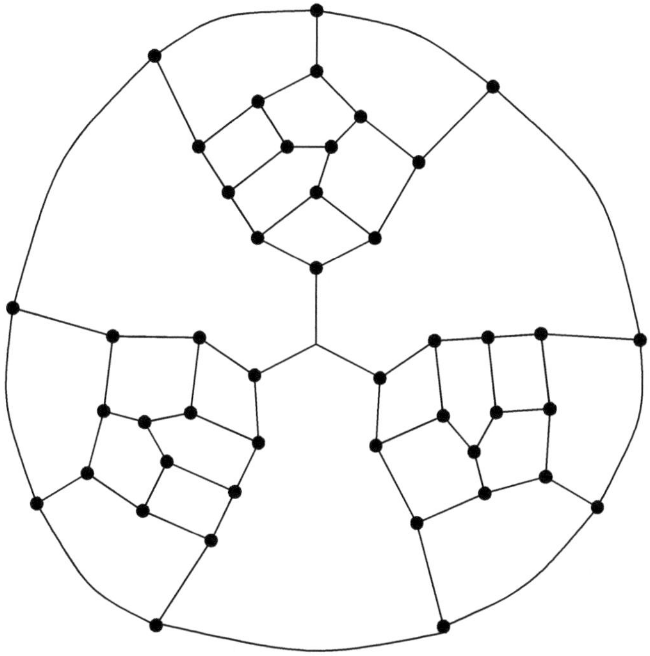

Fig. 17.5 Non Hamiltonian graphs

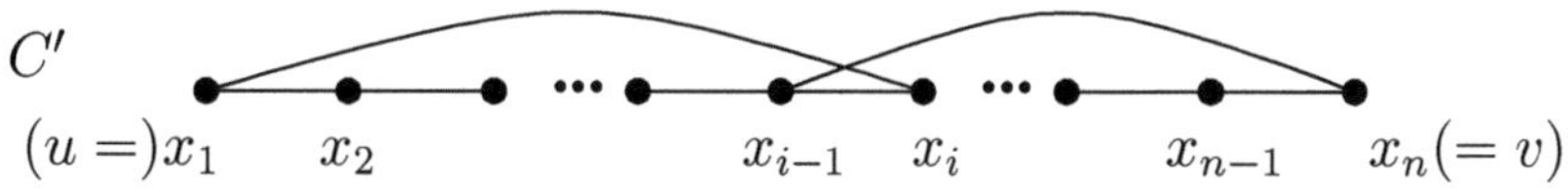

Fig. 17.6 A situation

$$deg_G(v) \leq (n-1) - deg_G(u)$$
$$\Rightarrow deg(u) + deg(v) \leq n-1,$$

which is a contradiction. Hence G has a Hamilton cycle. □

Definition 17.20 Let $G = (V, E)$ be graph on n-vertices. If H is a graph obtained from G by repeatedly adding edges between two non-adjacent vertices whose degree sum is atleast n. That H is the smallest super graph of G having n-vertices such that $d_H(u) + d_H(v) \geq n$ for all $u, v \in V$, $uv \notin E(H)$.

Note that we can denote by $cl(G) = H$ to indicate the closure of G. $cl(G)$ is unique.

Example 17.21 Consider the following graph (Fig. 17.7).

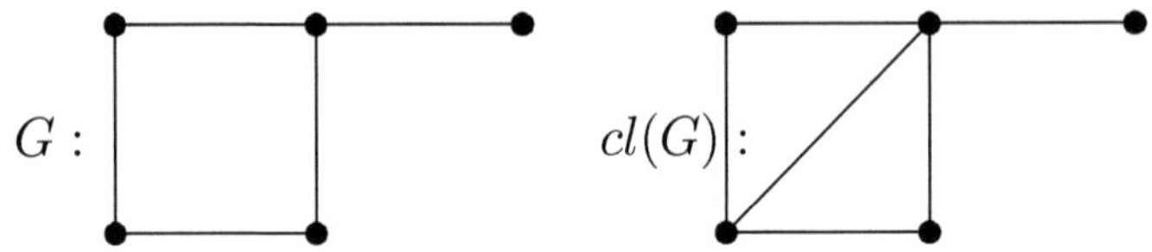

Fig. 17.7 Closure

Theorem 17.22 *G is Hamiltonian iff* $cl(G)$ *is Hamiltonian.*

Proof Let $e_1, e_2, e_3, \ldots, e_k$ be the set of edges to construct $cl(G)$. That is $G_1 = G + e_1$, $G_2 = G_1 + e_2$, ..., $G_k = G_{k-1} + e_k = cl(G)$. Now if G is Hamiltonian, then all $G_1, G_2, \ldots, G_k = cl(G)$ is Hamiltonian and conversely. Hence the theorem. □

Remark 17.23 For a graph G, $cl(G)$ may or may not be a complete graph.

Example 17.24 Consider the graph (Fig. 17.8).

Many sufficient conditions for a graph to be Hamiltonian are based on the degrees of vertices.

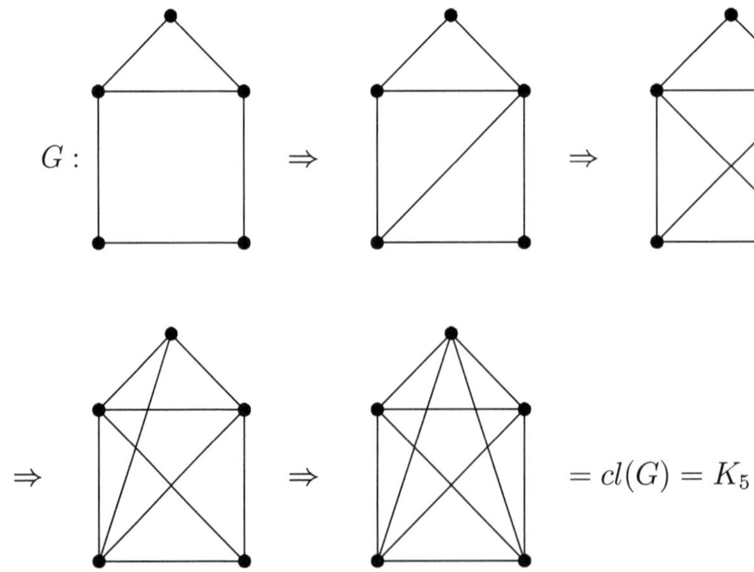

Fig. 17.8 Closure

Theorem 17.25 (Chvatal, 1972) *Let $G = (V, E)$ be a graph with atleast 3 vertices. Let $(d_1, d_2, \ldots, d_n)$ be the degree sequence of G such that $d_1 \leq d_2 \leq d_3 \leq \cdots \leq d_n$. If for every value $t < \frac{n}{2}$ either $d_t \geq t$ and $d_{n-t} \geq (n - t - 1)$. Then G is Hamiltonian.*

Proof Let $cl(G) = H$. We show that H is a complete graph which implies that G is Hamiltonian by the previous theorem. Assume that H is not complete. Then H contains two non-adjacent vertices u, v with the condition that $d_H(u) + d_H(v)$ is as large as possible. Also $d_H(u) + d_H(v) \leq n - 1$. Let $d_H(u) \leq d_H(v)$ and $d_H(u) = t$.

So $d(v) < n - t$. Also $t < \frac{n-1}{2} < \frac{n}{2}$ and $d_h(v) \leq n - 1 - t$.

Let U be the set of vertices in $V - \{u\}$ which are not adjacent to v in H. Let X be the set of vertices in $V - \{v\}$ that are not adjacent to u in H. Then

$$|U| = n - 1 - d_H(u) = n - 1 - t \quad \text{and}$$
$$|X| = n - 1 - d_H(v).$$

For every vertex $x \in U$,

$$d_G(x) \leq d_H(x) \leq d_H(v) \leq n - 1 - t$$
$$\Rightarrow d_{n-t-1} \leq n - t - 1.$$

But $d_G(u) \leq d_H(u) \leq d_H(v) \leq n - t - 1$ which means $d_{n-t} \leq n - t - 1$.

This is raising a contradiction to the hypotheses of the theorem.

Hence $cl(G)$ is complete. □

Note that the above theorem is stronger than many theorems on Hamiltonian graphs. There are degree sequences of a graph that does not satisfy the Chvatal's Theorem, but they specify a Hamilton cycle in the graphs. This section gave only a few theorems that specify sufficient conditions for a graph to be Hamiltonian.

17.4 Toughness

It is not difficult to see that every Hamiltonian graph is 2-connected. There are several sufficient conditions for a graph G to be Hamiltonian. Regarding necessary conditions a few of them are widely used. Regarding the number of components of a Hamiltonian graph, it is always connected and removal of a single vertex from it results in a connected graph. The following necessary condition of a Hamiltonian graph is a very popular one.

Theorem 17.26 *If $G = (V, E)$ is a Hamiltonian graph, then*

$$c(G - S) \leq |S|$$

for every subset S of V where $c(G - S)$ is the number of components of $G - S$.

Proof Let C_n be the Hamilton cycle of G where $|V| = n$. Let $C_n = v_1 v_2 v_3 \ldots v_n v_1$.

Suppose $S \subset V$ and $G - S$ may be connected or not connected. If $G - S$ is connected then $C(G - S) \leq |S|$. Suppose $G - S$ is not connected. Let $c(G - S) = t$ and let $H_1, H_2, \ldots, H_t$ be that t components of $G - S$. We can assume that H_1 contains $v_1 \in V$. Let v_l be the vertex in G_l. If 'l' is the highest index of C_n such that v_l being a vertex of G_l then $v_{l+1} \in S$ for $1 \leq l \leq t$. This means $|S| \geq t$. □

Example 17.27 G is Hamiltonian. If $S = \{3\}$ then $G - S$ is connected. So $c(G - S) \leq |S|$. If $S = \{3, 4\}$, then $G - S$ is not connected as seen below (Figs. 17.9 and 17.10).

$c(G - S) = 2$ and $c(G - S) \leq |S|$

Example 17.28 Consider the following graph which is non-Hamiltonian (Fig. 17.11).

Let $S = \{u, v\}$. Then $c(G - S) = 3 \geq |S| = 2$ (Fig. 17.12).

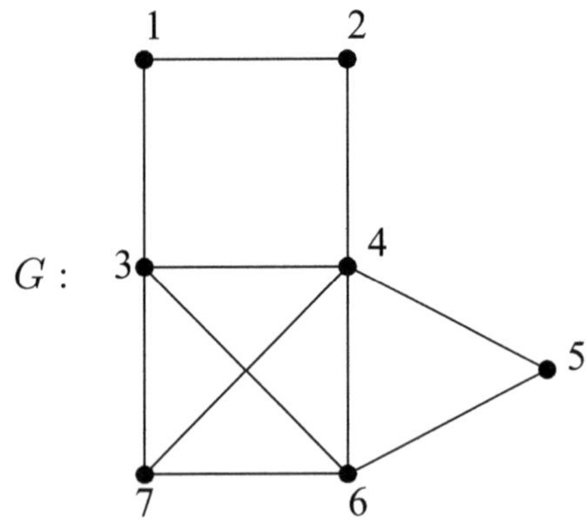

Fig. 17.9 Hamiltonian graph

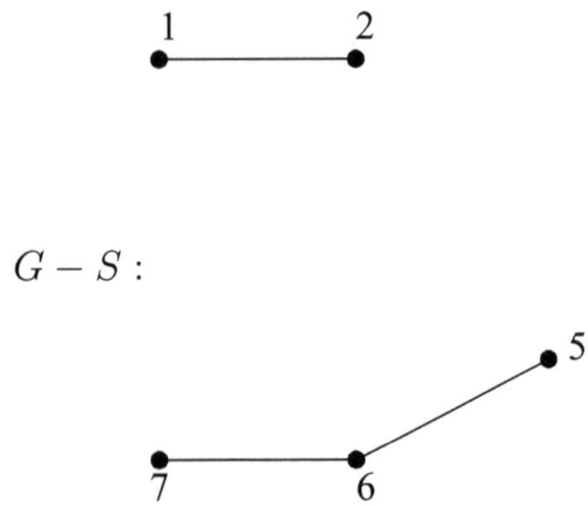

Fig. 17.10 Components

Fig. 17.11 Graph (6)

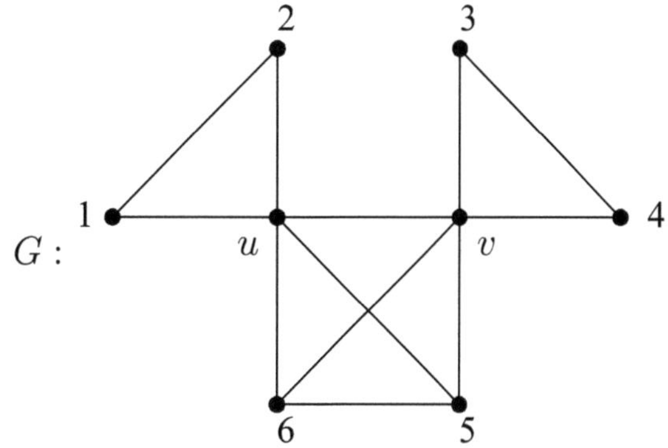

Fig. 17.12 Components

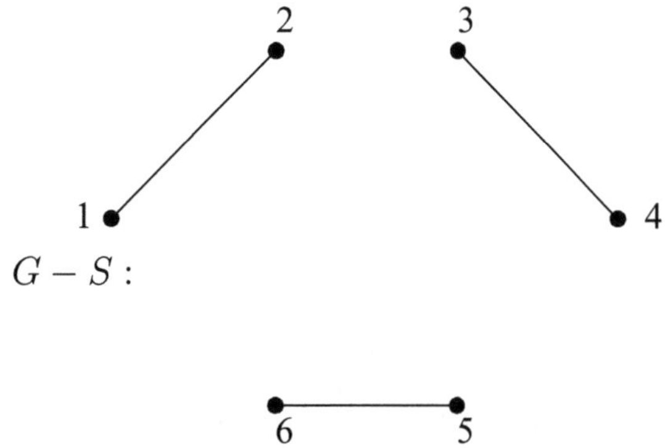

Now observe that if the graph $G = (V, E)$ is Hamiltonian, then by the above theorem we can write

$$\frac{|S|}{c(G - S)} \geq 1$$

for any nonempty subset S of V.

Definition 17.29 A graph $G = (V, E)$ which is not complete is said to be k-tough if the ratio

$$\frac{|S|}{c(G - S)} \geq k$$

for any vertex cut S of V.

Remark 17.30 A graph G that is k-tough is also 't' tough if $t < k$. Any Hamiltonian graph is 1-tough.

There exists a 3-connected 3-regular graph which is 1-tough as shown below.

Fig. 17.13 Peterson graph

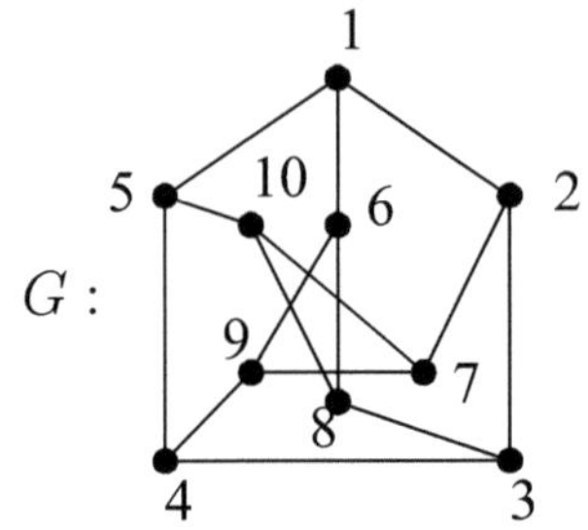

Example 17.31 Consider the Peterson graph which is 3-regular, 3-connected graph (Fig. 17.13).

For every set S, consider

$$\frac{|S|}{c(G-S)}.$$

Let $S \subset V$ such that $|S| \geq 3$. Here in Peterson graph, $|S| = 3, 4, 5, 6, 7, 8, 9, 10$. For $|S| \geq 5$, it is obvious that

$$\frac{|S|}{c(G-S)} \geq 1$$

So it remains to show the above inequality only for $|S| = 3, |S| = 4$. Let $A = \{1, 2, 3, 4, 5\}$ and $B = \{6, 7, 8, 9, 10\}$. Choosing the set S such that $|S| = 3$ or 4 can be as follows.

Case i: All vertices of S are completely in A or completely in B. Then clearly $G - S$ remains connected and hence

$$\frac{|S|}{c(G-S)} > 1.$$

Case ii: If one vertex of S is from B, remaining vertices from A, then $G - S$ contains a component with minimum 5 vertices. Hence

$$\frac{|S|}{c(G-S)} \geq 1.$$

Case iii: If two vertices of S are from B and remaining vertices from A, then consider $G - S$. The maximum number of components can be only 4 and this can be checked directly. Hence

$$\frac{|S|}{c(G-S)} \geq 1.$$

So we can see that Peterson graph is 1-tough.

The ratio $\frac{|S|}{c(G-S)}$ play some role in determining the property of a graph G containing a Hamilton cycle. The following are some definitions that talk about the ratio and its relation to the neighbouring vertices of the vertices in S.

Definition 17.32 1. Let $G = (V, E)$ be a graph. For a vertex $v \in V$, neigbourhood of 'v' is defined to be the set

$$N(v) = \{u \in V \mid d(u, v) = 1\}.$$

The following theorem gives a bounding condition on $N(S)$ for a graph to be Hamiltonian. We omit the proof as it is too much involved.

Theorem 17.33 *If a graph $G = (V, E)$ is 2-connected with $\delta(G) \geq \frac{(n+2)}{3}$ and $|N(S)| \geq \frac{1}{3}(n + |S| - 1)$ for every subset S of V, then G is hamiltonian.*

Definition 17.34 (i) The toughness of a graph $G = (V, E)$ is

$$t(G) = \min\left\{\frac{|S|}{c(G - S)} \mid S \text{ is a vertex cut of } G\right\}$$

(ii) The binding number of a graph $G = (V, E)$ is denoted by $b(G)$ and

$$b(G) = \min\left\{\frac{|N(S)|}{|S|} \mid S \subseteq V, S \neq \emptyset, S \neq V\right\}$$

The two measures defined above have some important role in characterising Hamiltonian graphs. We state a few results on this topic without proof.

Theorem 17.35 *1. Every Hamiltonian graph is 1-tough.*
2. The binding number $b(G)$ is the largest number 'k' such that

$$|N(S)| \geq \min\{k|S|, |V| \mid S \subseteq V(G)\}.$$

The t_0-tough conjecture[Chvatal]
There exists a constant t_0 such that every t_0-tough graph is Hamiltonian.

Remark 17.36 If G is a graph such that $t(G) \geq t_0$, then G is Hamiltonian.

17.5 Other Hamiltonicity

The problem of determining whether a graph contains a Hamilton cycle is a difficult problem. Hence finding some sub-class of graphs containing Hamilton cycle became important. Such class of graphs are generally called as 'highly Hamiltonian graphs'.

Hamiltonian-connected graphs are a special type of graphs having a property on Hamilton paths that they contain.

Definition 17.37 Let $G = (V, E)$ be a graph. If for any two vertices $u, v \in V$, there is a path which is Hamiltonian, then G is said to be Hamiltonian-connected. A Hamilton path is one that visits all the vertices of the graph.

Example 17.38 G_1 is not Hamiltonian-connected.
G_2 is Hamilton connected.

The following theorem gives a sufficient condition for a graph to be Hamiltonian-connected (Fig. 17.14).

Theorem 17.39 *Let $G = (V, E)$ be a graph on n vertices. For any two non-adjacent vertices $u, v \in V$, if $d(u) + d(v) \geq n + 1$, then G is Hamiltonian-connected.*

Proof Let u, v be two non-adjacent vertices in G. Let $\overline{G}$ be a graph obtained by adding a new vertex 'x' to V and adding edges xu, xv to E. Let $\overline{G} = (\overline{V}, \overline{E})$ where $\overline{V} = V \cup \{x\}$ and $\overline{E} = E \cup \{xu, xv\}$. Construct $G' = cl(\overline{G})$. In the graph $\overline{G}$, for any two non adjacent vertices $s, t \in V$, $d_{\overline{G}}(s) + d_{\overline{G}}(t)$ must be atleast $(n + 1)$. This means in $cl(H)$, the induced subgraph on the vertices V of G will be a complete graph K_n. Also the new vertex x of $\overline{V}$ is adjacent to every vertex $l \in V - \{u, v\}$ in the closure graph $cl(\overline{G})$. Also $xu, xv \in \overline{G}$ and $xv, xv \in cl(\overline{G})$. So $cl(\overline{G})$ must be a complete graph on $|\overline{V}|$ vertices. Hence $\overline{G}$ is Hamiltonian. The Hamilton cycle C of $\overline{G}$ must contain the edges xu, xv. Removing x from C results in a Hamiltonian path between the two vertices u, v. Hence G is Hamiltonian connected. □

We next consider a stronger property of a graph than that of hamiltonian-connected.

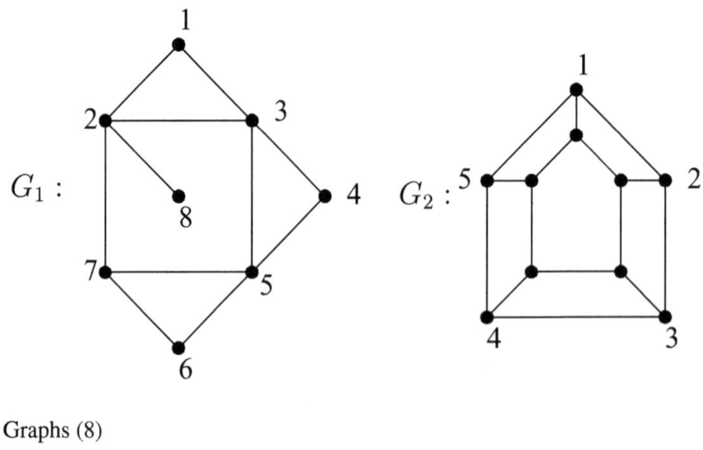

Fig. 17.14 Graphs (8)

Definition 17.40 Let $G = (V, E)$ be a connected graph. If for every pair of distinct vertices $u, v \in V$, there exists a $u - v$ path of length k for each k, $dis(u, v) \leq k \leq n - 1$, then G is said to be panconnected.

Remark 17.41 Every panconnected graph is Hamiltonian connected. There are graphs which are Hamiltonian-connected but are not panconnected.

Example 17.42 (i) K_n, $n \geq 1$, is panconnected
(ii) $K(m, n)$, $1 \leq m \leq n$ is not Hamiltonian connected (for $m \neq n, m, n \neq 1$). $K(m, n)$ is also not panconnected.

The following result gives a sufficient condition for a graph to be panconnected. We state the theorem without proof.

Theorem 17.43 *Let $G = (V, E)$ be a graph with $|V| \geq 4$. If for every vertex $v \in V$, $d(v) \geq \frac{|V|+2}{2}$, then G is panconnected.*

Remark 17.44 The above theorem gives the best condition for panconnectivity.

Definition 17.45 Let $G = (V, E)$ be graph with $|V| \geq 3$. If G contains a cycle for every possible length k, $3 \leq k \leq |V|$, then G is said to be pancyclic.

Example 17.46 Consider the graph, G is pancyclic (Fig. 17.15).

Definition 17.47 A graph $G = (V, E)$ is said to be vertex pancyclic if every vertex $v \in V$ belongs to every cycle of length k, $3 \leq k \leq |V|$

Remark 17.48 The above graph G in the example is not vertex pancyclic.

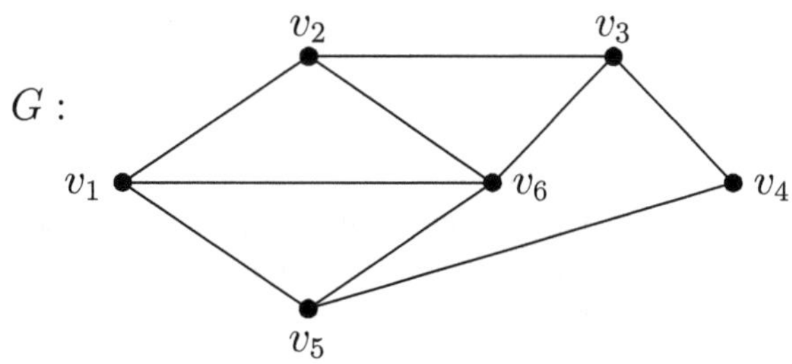

Fig. 17.15 Pancyclic graph

The results and study on pancyclic graphs was initiated by Bondy. Bondy studied the importance of Ore's sufficient condition for a graph G to be Hamiltonian. The Ore's condition that $\delta(G) \geq \frac{n}{2}$ imply much more information on the structure of the graph G. Following are some sufficient conditions on G being pancyclic.

Theorem 17.49 *1. Let $G = (V, E)$ be a graph with $|V| \geq 3$. Also $d(u) + d(v) \geq n$ for every two non adjacent vertices $u, v \in V$. Then G is either pancyclic or else complete bipartite graph $K_{\frac{n}{2},\frac{n}{2}}$. In particular, if G is Hamiltonian and $|E| > \frac{n^2}{4}$, then G is pancylic.*

2. If $G = (V, E)$ is not a complete bipartite graph with $|V| \geq 3$, and for every pair of non adjacent vertices $u, v \in V$, $d(u) + d(v) \geq |V|$, then G is pancyclic.

Exercises

1. Give examples of graph G such that

   (i) G is Eulerian and Hamiltonian
   (ii) G is Eulerian but not Hamiltonian.
   (iii) G is non Eulerian but Hamiltonian.
   (iv) G is both non Eulerian and non Hamiltonian.

2. Show that for an Eulerian graph edge connectivity is always even.
3. Give an example of an Eulerian graph with $\kappa = 1, \lambda = 2, \delta = 4$.
4. Show that a connected graph G on n vertices, $n > 1$ is Eulerian if and only if every edge of the graph lies on an odd number of cycles of G.
5. Show that if G is odd bipartite graph then G is not Hamiltonian.
6. Does there exist a graph G such that $cl(G)$ is not a complete graph? Justify your answer.
7. Construct if possible Eulerian graphs with n number of vertices and m number of edges where

n	4	5	6
m	5	8	10

   Justify your answers.
8. Construct a non-Hamiltonian graph on S-vertices.
9. Show that $K_{m,n}$ is non-Hamiltonian for (i) $m + n$ being odd (ii) $m < n$.
10. Does there exist a graph G which is not pancyclic whose $cl(G)$ is complete? Justify your answer.

Planar Graphs 18

Planar graphs is an important class of graphs that has very nice applications. A graph can be drawn in many ways. Some representations bringout the properties of the graphs immediately, whereas other representations may not. A planar graph is one that can be represented on a plane in such a manner that no two edges cross over each other.

18.1 Embedding of Graphs

A graph is said to be embedded on a surface if its vertices are mapped to distinct points on the surface and the edges to simple curves between the corresponding pairs. The drawing is done such that no two curves intersect each other at any interior position. It means the only intersecting points are the end points of the curves.

Definition 18.1 Let G be a graph and $F(G)$ be an embedding of $G = (V, E)$ on a surface say S.

(i) For any two points u, v on the surface S such that $u, v \notin V$ and $u, v \in (S - F(G))$, the points u, v can be related as below:
$u \sim v$ if and only if there is a Jordan curve joining u and v on S but does not contact any position of $F(G)$. Note that '$\sim$' is an equivalence relation.
(ii) The equivalence classes of $\sim$ are the regions of $F(G)$.
(iii) The points, curves and regions of $F(G)$ is one mapping of G on the surface S.
(iv) The surface 'S' can be a plane surface or a sphere.
(v) When G is embedded on a plane, it will be called as planar embedding. In the planar embedding $F(G)$ of G all the regions will be bounded except only one unbounded region.
(vi) A graph that admits an embedding on a plane is called a 'planar graph'.

R. Rama, *Topics in Combinatorics and Graph Theory*,
https://doi.org/10.1007/978-3-031-74252-1_18

Remark 18.2 From the above definition it is immediate that not all graphs admit planar embedding.

We will consider only embedding of graphs on a plane. We will not consider any embedding of a graph on a spherical surface.

For any graph embedded on a plane, the points, curves and regions will be referred as vertices, edges and faces of the graph. When we say planar graph, it means a graph represented on a plane.

We now give the first result on a planar graph known as Euler's formula [?].

Theorem 18.3 *If a graph $G = (V, E, F)$ is a planar graph where $|V| = n$, $|E| = m$, $|F| = f$, then*

$$n - m + f = k + 1 \tag{18.1}$$

where G has k components. Here F denotes the faces of G.

Proof Since the graph has n vertices, it can be constructed by adding edges to $\overline{K}_n$.

For $\overline{K}_n$, $k = n, m = 0, f = 1$. Hence (18.1) is true.

Let G_{i-1} be the planar graph with m edges, with k components, f faces and (18.1) is true.

Let G_i be obtained from G_{i-1} by adding the ith edge say e.

Case i: If the ith edge connects two different components of G_{i-1}, then in G_i number of edges increase by 1, number of components decrease by 1. So,

$$\begin{aligned} n - (m+1) + f &= k \\ \Rightarrow n - m + f &= k + 1. \end{aligned}$$

Case ii: If the new edge 'e' joins any two non-adjacent vertices of the same component without cross over, then number of edges increase by 1, number of components remain the same, number of faces increase by 1. So,

$$\begin{aligned} n - (m+1) + f + 1 &= k + 1 \\ \Rightarrow n - m + f &= k + 1. \end{aligned}$$

Hence the theorem. □

We now state two corollaries which are immediate consequences of the above theorem.

Corollary 18.4 *If G is a connected planar graph, then*

$$n - m + f = 2$$

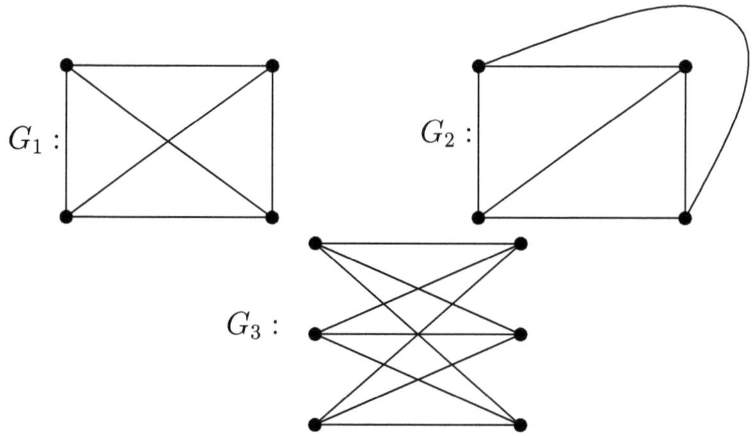

Fig. 18.1 Graphs (1)

Definition 18.5 A planar graph embedded on a plane is a plane graph.
A graph that is not planar is called as non planar graph (Fig. 18.1).

Example 18.6

G_1 is a planar graph.
G_2 is a plane graph. G_3 is a non-planar graph.

The concept of Jordan curve theorem helps to sketch a planar graph on a plane to get a plane graph.

Definition 18.7 Let x, y be any two points on a plane. A Jordan curve from x to y is a continuous non-self-intersecting curve in the plane. It is a closed Jordan curve if $x = y$.

We now state the Jordan curve theorem.

Theorem 18.8 *A closed Jordan curve J on a plane S divides the plane into three parts. The 3 parts are interior of J, exterior of J, boundary of J.*

Remark 18.9 Every edge of a plane is a Jordan curve. The boundary of every region/face is a closed Jordan curve.

Definition 18.10 The degree of a face 'f' of a plane graph G is the number of edges bounding the face where cut edges if any are counted twice. We use the notation $deg_G(f)$ to denote the degree of a face f in G the plane graph G (Fig. 18.2).

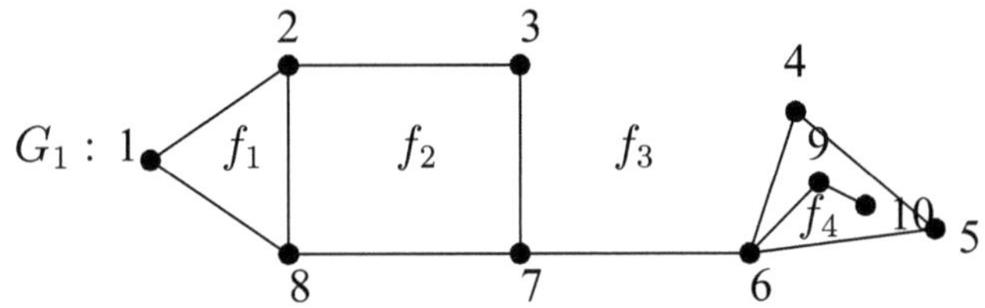

Fig. 18.2 Faces

Example 18.11

In G_1 boundary of $f_1 : b(f_1) = \{12, 28, 18\}$
boundary of $f_2 : b(f_2) = \{23, 28, 37, 87\}$
boundary of $f_3 : b(f_3) = \{12, 18, 23, 37, 87, 67, 46, 45, 65\}$
boundary of $f_4 : b(f_4) = \{45, 46, 65, 69, 910\}$
$deg(f_1) = 3, deg(f_2) = 4, deg(f_3) = 11, deg(f_4) = 7.$

Definition 18.12 Let G be a plane graph. If G has t_i faces of degree i, $2 \leq i \leq k$, then the sequence $(2^{t_1}, 3^{t_2}, \ldots, k^{t_k})$ is the face-degree sequence of G.

Remark 18.13 If a graph G has two different embeddings say G_1, G_2 then their face-degree sequence will be same.

Definition 18.14 A planar graph with each face of it being a triangle is called as planar triangulation.

For a graph G, the embedding of G be $F(G)$. We have a few observations as below:

Let $F(G)$ be a plane embedding of a planar graph G.

(i) For two faces f_1 and f_2, they are disjoint if their boundaries share no edge. If f_1, f_2 share an edge 'e' then f_1, f_2 intersect only on the edge e.
(ii) Exterior face of $F(G)$ is unique.
(iii) Every edge e can be on atmost two different faces.
(iv) A cut-edge can belong to only one face of $F(G)$.
(v) A non cut-edge can belong to exactly two faces of $F(G)$.

18.2 Some Simple Properties

We discussed the basic idea of plane drawing of a graph. We had Euler's formula that relates basic parameters of a planar graph. This formula has some nice applications as well.

Theorem 18.15 *Let G be a simple connected plane graph. Let $deg_G(f) \geq k$ for every face f of G. Then*

$$m \leq \frac{k(n-2)}{k-2}.$$

Proof G is a connected plane graph. So every face of G is a cycle. Each edge belongs to exactly two faces. Hence the face-degree sum of G will be $2m$ where m is the number of edges of G. So,

$$2m = \sum_{f \in \overline{F}(G)} deg(f) \geq k.\ell$$

where $\overline{F}(G)$ is the faces of G and $|\overline{F}(G)| = \ell$. By Eulers formula

$$n - m + \ell = 2$$
$$2m \geq k\ell$$

Solving we get the result. □

If G is a connected plane graph with atleast 3 vertices, the following result gives an upper bound on the number of edges.

Theorem 18.16 *If $G = (V, E)$ is a planar graph with $|V| \geq 3$, then*

$$|E| \leq 3\,V| - 6.$$

Proof For $|V| = 3$, the bound is true that $|E| \leq 3$. Let $G = (V, E)$ be a graph with $|V| \geq 4$ that is a planar graph. This graph must be connected as well. Otherwise edges can be added between the components to make it as a connected one. Suppose there are F faces for this graph, then by Euler's identity,

$$|V| - |E| + |F| = 2. \tag{18.2}$$

Also we know that every edge must be shared by exactly 2 faces and so

$$\sum_{i=1}^{k} |F_i| \leq 2|E| \tag{18.3}$$

where F_i is a face of G and there are k-faces for this graph. Now each face F_i must be bounded by 3 or more edges. So

$$3k \leq \sum_{i=1}^{k} |F_i|. \tag{18.4}$$

From (18.2), (18.3) and (18.4) we have

$$\begin{aligned} 3(2 - |V| + |E|) &\leq 2|E| \\ \Rightarrow |E| &\leq 3\,V| - 6. \end{aligned}$$

□

From the above theorem, we can formulate the following result for a graph with more than 5 vertices and that does not satisfy the bound $|E| \leq 3|V| - 6$.

Theorem 18.17 *If $G = (V, E)$ is a graph with atleast 5 vertices and $|E| > 3|v| - 6$, then G is not a planar graph.*

Corollary 18.18 *Let $G = (V, E)$ be a planar graph. Then G contains a vertex of degree 5 or less.*

Proof Since G is planar, we can write $|E| \leq 3|V| - 6$. If $|V| \leq 6$, the result is obvious. Suppose $|V| > 6$ and suppose degree of every vertex of such a graph is atleast 6. Then

$$\begin{aligned} 2|E| &= \sum_{v \in V} d(v) \geq 6\,V| \\ \Rightarrow |E| &\geq 3\,V| \\ \Rightarrow |E| &> 3\,V| - 6. \end{aligned}$$

Then by the above theorem, G is not a planar graph. Hence G contains atleast one vertex of degree 5 or less. □

Corollary 18.19 *K_5 is non planar.*

Proof Suppose K_5 is planar

$$10 \leq 15 - 6 = 9$$

which is not true. □

Corollary 18.20 *$K_{3,3}$ is non planar.*

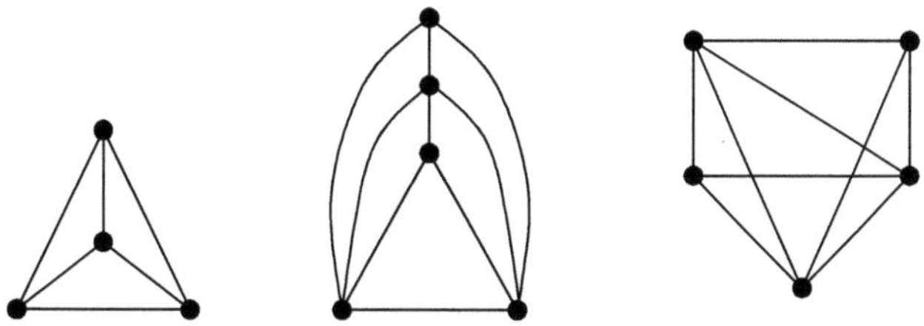

Fig. 18.3 Maximal planar graphs

Proof Suppose $K_{3,3}$ is a planar graph. We know it is a bipartite graph on 6 vertices and so for every face f of $K_{3,3}$, $deg_{K_{3,3}}(f) \geq 4$. Using this in the inequality,

$$m \leq \frac{k(n-2)}{(k-2)}$$

of Theorem 18.17, we get

$$9 \leq \frac{4(4)}{2} = 8$$

This is not true. Hence $K_{3,3}$ is non planar. □

Definition 18.21 A planar graph G is said to be maximal planar if adding a new edge between two non adjacent vertices of G results in a non planar graph (Fig. 18.3).

Example 18.22

G_1, G_2, G_3 are maximal planar graphs.

Remark 18.23 Every face of a maximal planar graph will be a triangle or C_3. So every maximal planar graph is a called triangulated graph.

Theorem 18.24 *The number of edges in any maximal planar graph* $G = (V, E)$ *is* $3|V| - 6$.

Remark 18.25 Any planar graph is a spanning subgraph of a maximal planar graph. Every maximal planar graph is 2-connected.

Definition 18.26 A graph $G = (V, E)$ is said to be outer planar if it has an embedding in a plane such that all its vertices lie on the boundary of the exterior face (Fig. 18.4).

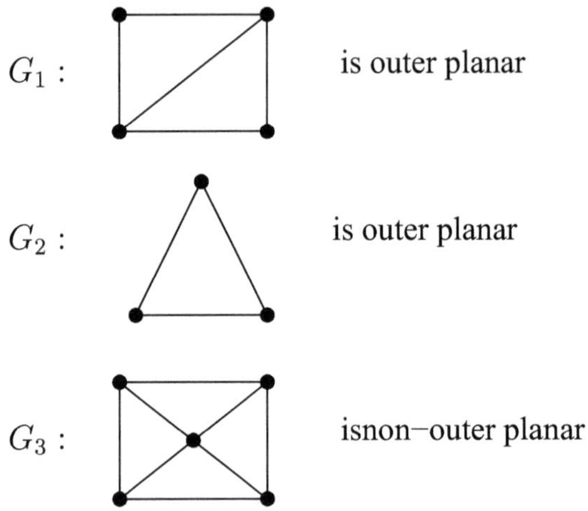

Fig. 18.4 Graphs (4)

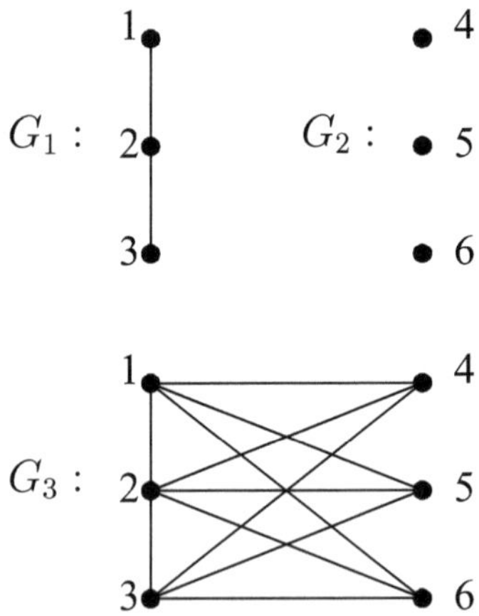

Fig. 18.5 Graphs (5)

Definition 18.28 Let $G_1 = (V_1, E_1)$, $G_2 = (V_2, E_2)$ be two simple graphs. Then join of G_1, G_2 is defined to be a graph $G = (V, E)$ such that $V = V_1 \cup V_2$, $E = E_1 \cup E_2 \cup \{xy \mid x \in V_1, y \in V_2\}$. Join of G_1, G_2 is denoted by $G_1 \vee G_2$ (Fig. 18.5).

Example 18.29 Let

The following is a necessary and sufficient condition for a graph to be outerplanar.

Theorem 18.30 *A graph $G = (V, E)$ is outer planar if and only if $G \vee K_1$ is planar.*

Proof Let G be outerplanar. Then it can be embedded in a plane such that all its vertices lie on the boundary of the outer region. Then a vertex 't' can be placed in the exterior region and $G \vee K_1$ can be seen as a planar graph.

Suppose $G \vee K_1$ is planar. Then in the construction of $G \vee K_1$ a vertex 'x' that is joined to every vertex of G can be deleted to see the embedding of G such that all its vertices are in the exterior region. Hence the theorem. □

Example 18.31

K_4 :

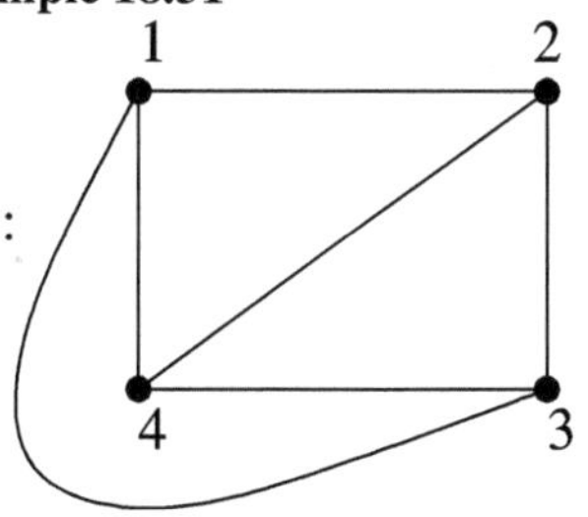

K_4 is not outer planar.

K_4 is not outer planar.

$K_{2,3}$:

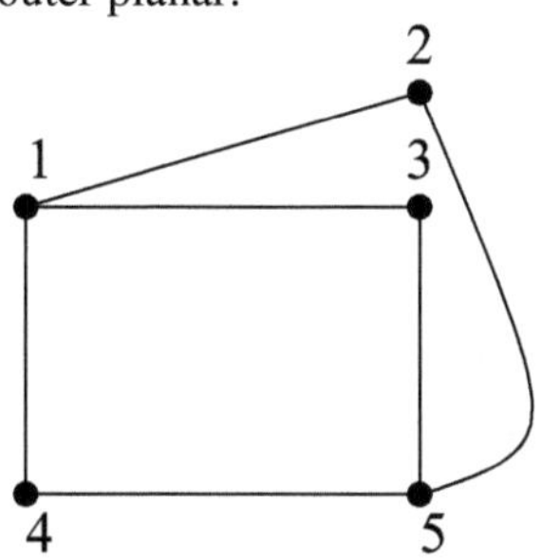

$K_{2,3}$ is not outer planar.

We state the following necessary and sufficient condition for a graph to be outer planar without proof.

Theorem 18.32 *A graph G is outerplanar if and only if it has no subgraph homeomorphic to either K_4 or $K_{2,3}$.*

18.3 Characterization of Planar Graphs

In this section we will see a characterization of planar graphs by Kuratowski.

We have seen that K_5 and $L_{3,3}$ are non planar. So if any graph that contains either K_5 or $K_{3,3}$ as its subgraph will naturally be nonplanar. Here we make a statement that every subgraph of a planar graph is planar.

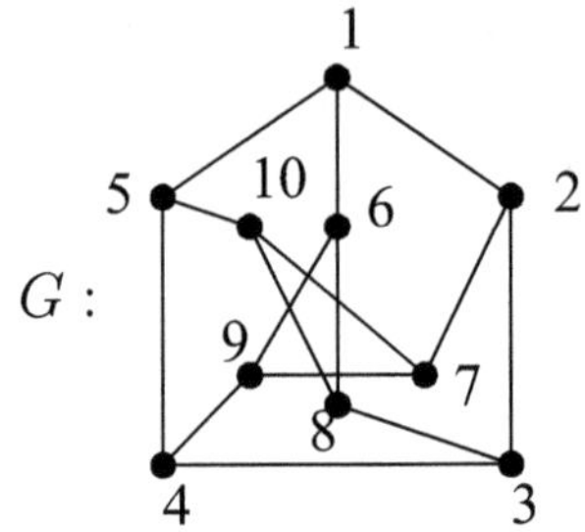

Fig. 18.6 Graphs (7)

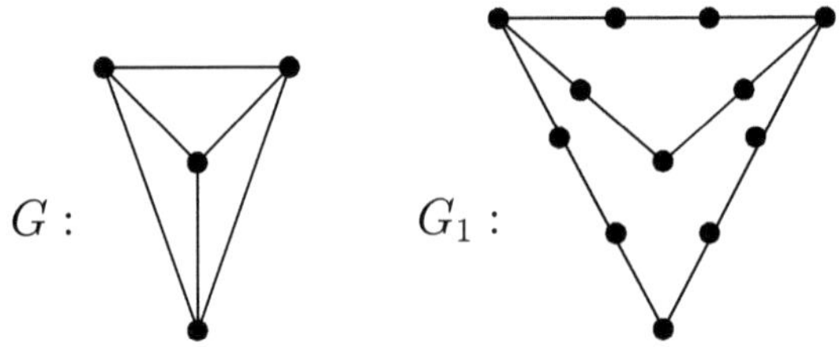

Fig. 18.7 Graphs (8)

Example 18.33 Consider the following graph (Fig. 18.6).

This is the Peterson graph and it is non-planar. But it does not contain K_5 or $K_{3,3}$ directly as a subgraph.

Definition 18.34 An edge $uv \in E$ of a graph $G = (V, E)$ can be replaced by a path $uu_1u_2u_3 \ldots u_nv$ by introducing vertices $u_1, u_2, \ldots, u_n$ into the edge $uv \in E$. This subdivision may be done in some or all the edges of G. The resultant graph is called a subdivision graph of G denoted by $s(G)$.

G_1 is a subdivision of G and $G_1 = s(G)$. The following are some observations on planar graphs (Fig. 18.7).

1. If G is a planar graph, then $s(G)$ is also a planar graph.
2. If G is a nonplanar graph, then $s(G)$ is also a non planar graph.
3. A graph G is planar means that every component, subgraphs, blocks are planar.
4. $s(K_5)$, $s(K_{3,3})$ are non planar.
5. If G is planar, then G is not a subdivision of K_5 or $K_{3,3}$ or G does not contain a subgraph H which is a subdivision of K_5 or $K_{3,3}$.

Example 18.36 Consider the peterson graph (Fig. 18.8).

By direct argument one can show that G is non planar (Fig. 18.9).

$s(K_{3,3})$ is a subgraph of G.

We now state and prove Kuratowski's characterization of planar graphs.

Fig. 18.8 Graphs (9)

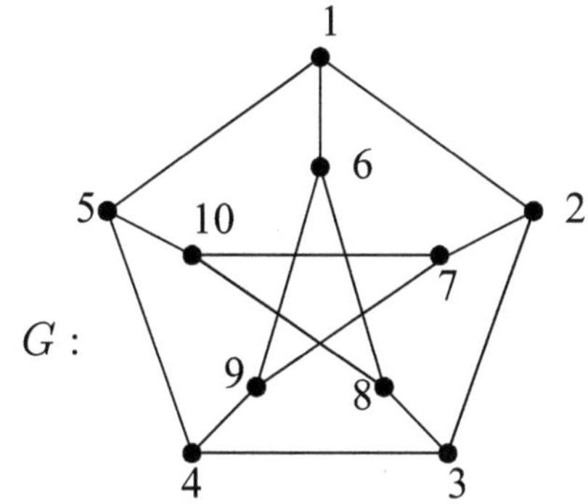

Fig. 18.9 Division graph

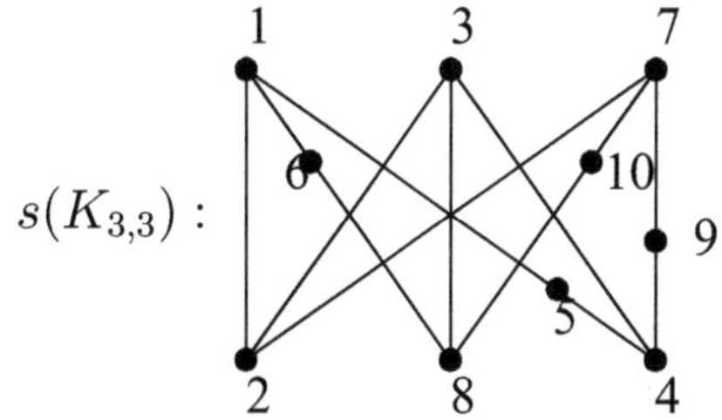

Definition 18.37 Let $G = (V, E)$ be a graph with $xy \in E$. The graph obtained by contracting the edge xy means, the vertices x and y are removed and a new vertex 'w' is added in such a manner that every vertex that is adjacent to x or y is now adjacent to w in the modified graph. This contraction of an edge is known as elementary contraction of an edge of G. If H is a graph obtained from a graph G through a sequence of elementary contractions then we say G is contractable to H. In notation we write $H = \zeta(G)$.

Example 18.38 Consider the peterson graph (Figs. 18.10 and 18.11).

Fig. 18.10 Graph (10)

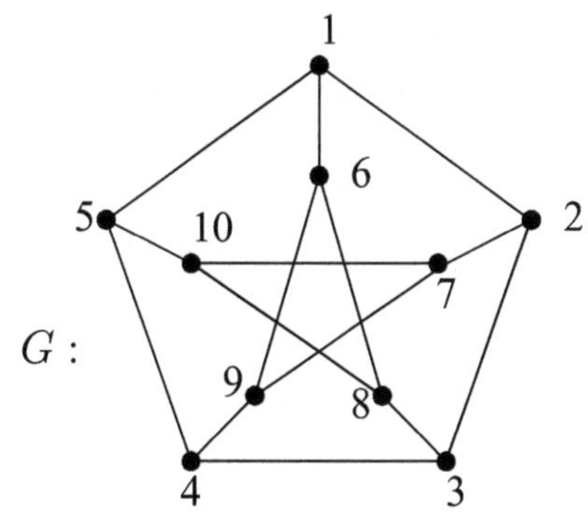

Fig. 18.11 Contraction

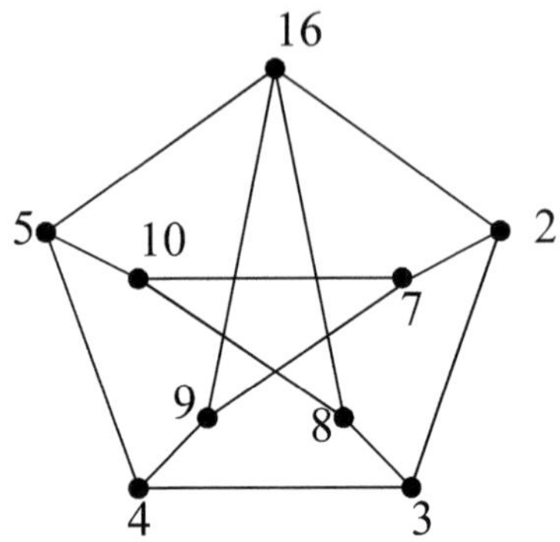

Contracting 27, 510, 49, 38 in sequence we reach K_5 which is non-planar.

Theorem 18.39 *A graph G is planar if and only if it does not have a subgraph contractable to either K_5 or $K_{3,3}$.*

Definition 18.40 Let $G = (V, E)$ be a plane graph. Let $V = \{v_1, v_2, \ldots, v_n\}$ and $E = \{e_1, e_2, \ldots, e_m\}$ and $F = \{f_1, f_2, \ldots, f_r\}$ be the sets of vertices, edges and faces of G respectively. The dual $G^* = (V^*, E^*)$ of G is a graph with

$$V^* = \{f_1^*, f_2^*, \ldots, f_r^*\}$$
$$E^* = \{e_1^*, e_2^*, \ldots, e_m^*\}$$

where $e_j^* = f_s^* f_t^*$ if and only if faces f_s, f_t are adjacent or e_j is the common edge for f_s and f_t of F.

The following diagram illustrates the understanding of G^* where the dotted lines are edges of G^* (Fig. 18.12).

Remark 18.41 (i) G^* is a plane graph.

(ii) G^* is always connected.

(iii) If e is a cut edge in G, then e^* is a loop in G^* and conversely.

(iv) G^* is always connected.

(v) Number of edges of G and G^* are equal. Degree of each face f will be the degree of the vertex f^* in G^*.

The following lemma can be seen geometrically using geometric inversion.

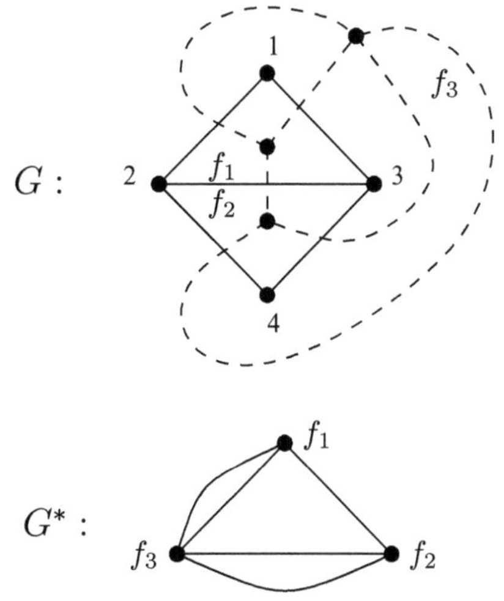

Fig. 18.12 Graphs of faces

Lemma 18.42 *Let $G = (V, E)$ be a plane graph. Let $\overline{E}$ be the edges bounding a face 'f' of G. Then there exists a plane embedding of the graph G where the edges $\overline{E}$ are lying on the exterior face or outer region.*

Lemma 18.43 *Let $G = (V, E)$ be a non planar graph without the subdivision of K_5 or $K_{3,3}$ as a subgraph. Let G be a minimal such graph on $|V| + |E|$. Then G is 3-connected.*

Proof G is connected by the assumption that $|V| + |E|$ being minimal.

G is also 2-connected. If not, then G must have a cut vertex 'v' (say) and let $G - v$ have its connected components as $G_1, G_2, \ldots, G_t$. Let $G_\ell = (V_\ell, E_\ell)$, $! \le \ell \le t$. Each of the induced subgraph on $V_\ell \cup \{v\}$ will have a plane embedding due to minimality on E and G being non-planar. Now joining the plane embeddings of each of $G_\ell \cup \{v\}$ at 'v', we arrive at a plane embedding of G which is not true. Hence G is 2-connected.

We continue the argument to show the 3-connectedness as below.

Let $S = \{u, v\}$ be the vertex separating set of G. Let $G_1 = (V_1, E_1)$ and $G_2 = (V_2, E_2)$ be two subgraphs of G such that $E = E_1 \cup E_2$ and $V_1 \cap V_2 = \{u, v\} = S$. Clearly G_1, G_2 will contain a connected component of $G - S$. As G is 2-connected there exists a path P from u to v in both G_1 and G_2. Also u and v are adjacent to a vertex in each of the connected components of $G - S$. Now let $\overline{G_i} = G_i + uv$. Here uv may be an edge in E. If $\overline{G_1}$ and $\overline{G_2}$ are planar, then by Lemma 18.42 above, it is possible to to have an embedding of $\overline{G_i}$ such that the edge 'uv' lies on the exterior face or outer region. Now join $\overline{G_1}$ and $\overline{G_2}$ together at the edge uv. We get a planar embedding of G as seen below (Fig. 18.13):

If one of $\overline{G_1}$ or $\overline{G_2}$ is non-planar, say $\overline{G_1}$ is non planar, then $|E_1 \cup \{uv\}| < |E|$ and by minimality condition on $|V| + |E|$. $\overline{G_1}$ must contain either subdivision of K_5 or subdivision of $K_{3,3}$. Since $\overline{G_2}$ contains a path from u to v, this path can be taken as a subdivision of

Fig. 18.13 A situation

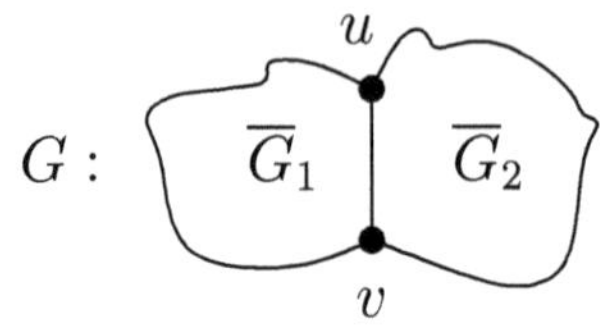

Fig. 18.14 Graph (11)

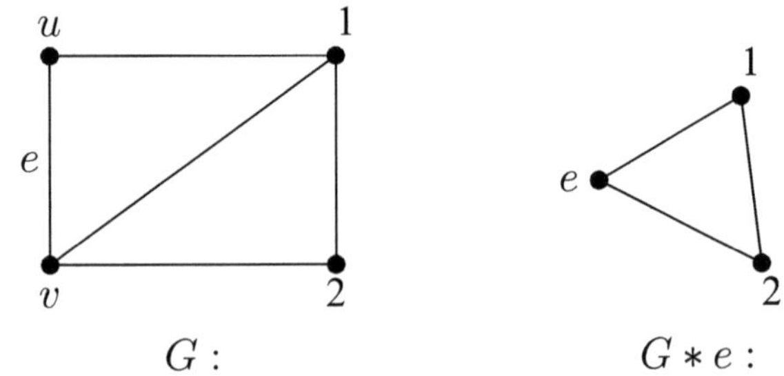

the edge uv and hence G must include subdivision of K_5 or $K_{3,3}$. This is a contradiction. Hence G must be 3-connected. □

Definition 18.44 Let $G = (V, E)$ be a graph and let $e = uv \in E$. By contracting the edge we mean that edge uv contracts as a vertex 'x' (say) such that $G * e$ is a new graph obtained from G such that (Fig. 18.14)

$$V_{G*e} = (V - \{u, v\}) \cup \{x\}$$
$$E_{G*e} = \{\overline{e}/\overline{e} \in E, \overline{e} \text{ is not incident with } u \text{ or } v\} \cup \{wx/wu \in E \text{ or } wv \in E\}.$$

The following lemma can be proved by method of contradiction and we leave its proof as an exercise to the readers.

Lemma 18.45 *Let $G = (V, E)$ be a 3-connected graph such that $|V| \geq 5$. Then $G * e$ is 3-connected for some edge $e \in E$.*

Lemma 18.46 *Let $G = (V, E)$ be a graph. If $G * e$, for some edge $e \in E$ contains either a subdivision of K_5 or subdivision of $K_{3,3}$, then G also contain such subdivisions.*

Proof An elaborate proof can be given using case by case method such subdivisions of K_5 or $K_{3,3}$. We will not get into these details. However, we can see that if $\overline{G}$ is a subdivision of K_5 or $K_{3,3}$ in $G * e$, then if $d(w) = 2$ in $\overline{G}$ for $w = uv$, $uv = e$, then the result follows.

If $d(w) = 3$ or 4 in $\overline{G}$. If there is atmost one edge wx in $\overline{G}$ such that ux or $vx \in E$, then also G will contain a subdivision of K_5 or $K_{3,3}$, if $G * e$ does.

If $\overline{G}$ is a subdivision of K_5, and both the vertices u, v are adjacent to 3 vertices in the subgraph of G that corresponds to $\overline{G}$, then G must contain a subdivision of $K_{3,3}$. □

Lemma 18.47 *Let $G = (V, E)$ be a 3-connected graph that contains no subdivision of K_5 or $K_{3,3}$. Then G is planar.*

Proof We will prove this lemma by the method of induction on the number of vertices of the given graph.

Let $G = (V, E)$ be a graph with $|V| \geq 5$. For $|V| \leq 4$, the result is true as K_4 is the only 3-connected graph which is planar.

Since G is 3-connected, by Lemma 18.45, there exists an edge $e - uv$ such that $G * e$ is also 3-connected. By Lemma 18.46, $G * e$ contains no subdivision of K_5 or $K_{3,3}$. So by induction hypothesis $G * e$ is planar and $\overline{G * e}$ be the plane embedding of it where $e = \{u, v\}$. In the embedding, consider the embedding portion of $(G * e) - \{x\}$ where x is the contracted vertex. Let C be the face of that part in the embedding $G * e$ excluding the vertex x which included the vertex 'x' on it. Clearly C is a cycle as G is 3-connected.

Infact embedding of $G * e$ excluding the vertex 'x' is nothing but an embedding of $G - \{u, v\}$. Every neighbouring vertex of u will be a vertex on 'C' as well as vertex 'v'. Similarly every vertex adjacent to v will be a vertex on C as well as u.

Suppose $d(v) \leq d(u)$ in G and $v_1, v_2, \ldots, v_t$ are the vertices in order on C such that $v_i \neq u, 1 \leq i \leq t$. Suppose P_{ij} is the path on C from vertex i to vertex j, then an edge vv_i can be added so that $G - \{u\}$ has a plane embedding for $1 \leq i \leq t$.

Case i: If the adjacent vertices of u other than v occur on a path $P_{k(k+1)}$ for some k, then $k + 1$ modulo t will be taken for the path. Here a plane embedding of G excluding u is taken first. Then the vertex u will be planted in the interior region of the face bounded by edges $vv_i, v_iv_{i+1}, v_{i+1}v$ and then the edges to u are supplied.

Case ii: Suppose there are two vertices x, y that are adjacent to u, $x, y \neq v$. If x is a vertex on the path P_{ij} and y is not on P_{ij} for some i, j, and $x, y \notin \{v_i, v_j\}$. Then the vertices $\{u, v_i, v_{i+1}\} \cup \{v, x, y\}$ form a subdivision of $K_{3,3}$.

Case (i) and (ii) imply that the set of vertices adjacent to u in G other than 'v' are also adjacent to the vertex 'v'. Hence all the vertices adjacent to 'u' are also adjacent to 'v' in G as $d(v) \leq d(u)$. By Case (i) above $d(v) = d(u)$ in G. So u, v, v_1, v_2, v_3 will form a subdivision of K_5. Hence the result. □

Theorem 18.48 (Kuratowski's Theorem) *A graph is planar if and only if it does not contain $s(K_5)$ or $s(K_{3,3})$.*

Proof If G is planar, then G cannot contain $s(K_5)$ or $s(K_{3,3})$ is discussed earlier.

Conversely we show that every non planar graph must contain either $s(K_5)$ or $s(K_{3,3})$. Suppose not for a graph G that has minimum number of edges. Then by Lemma 18.43, G is 3-connected. By Lemma 18.47, G is planar, which is a contradiction. □

Exercises

1. Show that every subgraph of a planar graph is planar.
2. Show that Peterson graph is non planar.
3. Show that a graph is planar if and only if every block of it is planar.
4. Let G be a connected planar graph. Will G always have an unique embedding? Justify your answer.
5. Show that a simple planar graph cannot be such that all its vertices are of degree atleast 6.
6. Let G be a simple planar graph on six vertices. What is the maximum number of edges that it can have? Represent the graph on a plane.
7. Let G be a planar graph on 100 faces where each face is a triangle. Find the number of vertices and edges of such a graph.
8. Let G be a planar graph with all its faces are pentagons and all the faces of its dual G^* being triangles. Find the number of vertices, edges and faces of G.
9. Show that a planar graph is 2-connected iff the boundary of every face is a cycle in its plane embedding.
10. Show that a planar graph with triangulation having atleast 4 vertices is 3-connected.

Independent Sets, Coverings and Matchings 19

In this chapter we see an important graph parameter that give rise to what is known as coverings and matchings. Vertex independents sets, edge independent sets often can be seen to occur in many applications.

19.1 Independent Sets

Definition 19.1 Let $G = (V, E)$ be a graph. A subset S of V is called a vertex independent set of G if for any two vertices $u, v \in S$, $uv \notin E$. That is no two vertices in S are adjacent in G. The set S is said to be maximum, if there exist no vertex independent set S' of V such that $|S'| > |S|$. The cardinality of a maximum vertex independent set of a graph G will be called as independent number of G and is denoted by $\alpha(G)$.

Example 19.2 Consider the following wheel graph, W_7.
Here $\{v_7\}, \{v_1, v_3\}, \{v_1, v_3, v_5\}$ are vertex independent sets. $\{v_1, v_3, v_5\}$ is a maximum vertex independent set and $\alpha(W_7) = 3$ (Fig. 19.1).

Definition 19.3 Let $G = (V, E)$ be a graph. A subset K of V such that every edge $e \in E$ is incident with a vertex in K is called a vertex covering of G.

A vertex covering K is called a minimum vertex covering of G if there is no covering K' of G such that $|K'| < |K|$. The number of vertices in a minimum vertex covering of G is called as vertex covering number of G and is denoted by $\beta(G)$.

Example 19.4 Consider the wheel graph W_7 of the above example.

The set $K = \{v_1, v_2, v_3, v_4, v_5, v_6\}$ is a vertex covering of W_7. The subset $K' = \{v_1, v_3, v_5, v_7\}$ is the minimum vertex cover and $\beta(W_7) = 4$.

R. Rama, *Topics in Combinatorics and Graph Theory*,
https://doi.org/10.1007/978-3-031-74252-1_19

Fig. 19.1 Wheel graph

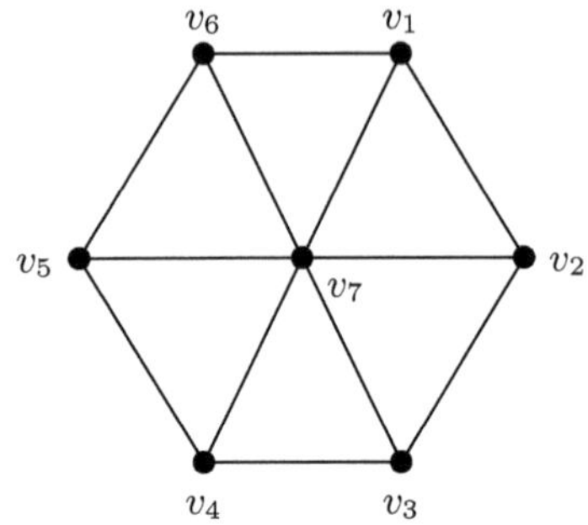

Theorem 19.5 *Let $G = (V, E)$ be a graph. A subset S of V is a vertex independent set of G if and only if $V - S$ is a vertex covering of G.*

Proof. S is a vertex independent set if and only if no two vertices in S are adjacent in G. This means every edge $e \in E$ must be incident with a vertex in $V - S$. By definition, then $V - S$ is a vertex covering. □

Corollary 19.6 *For any graph $G = (V, E)$, $\alpha(G) + \beta(G) = |V|$.*

Proof. If S is a maximum vertex independent set of G, then $\alpha(G) = |S|$.
If K is a minimum vertex covering of G, then $\beta(G) = |K|$.
By definition and above theorem,

$$\begin{aligned} |K| \leq |V - S| &\Rightarrow \beta(G) \leq |V| - \alpha(G) \\ &\Rightarrow \alpha(G) + \beta(G) \leq |V| \end{aligned} \tag{19.1}$$

If $V - K$ is a vertex independent set and S being maximum,

$$\begin{aligned} |S| \geq |V - K| &\Rightarrow \alpha(G) \geq |V| - \beta(G) \\ &\Rightarrow \alpha(G) + \beta(G) \geq |V| \end{aligned} \tag{19.2}$$

From (19.1) and (19.2), we get $\alpha(G) + \beta(G) = |V|$. □

Definition 19.7 1. Let $G = (V, E)$ be a graph. A subset M of E is called an edge independent set if no two edges are adjacent in G. The number of edges in a maximum edge independent set is called as edge independent number and is denoted by $\alpha'(G)$.
2. A subset L of E is said to form an edge cover if every vertex of G is incident to some edge in L. The minimum number of edges in an edge cover set is called as edge covering number and is denoted by $\beta'(G)$.

3. Also a matching is a set of independent edges. $\alpha'(G)$ is also called as maximum matching number.

Theorem 19.8 *For any graph $G = (V, E)$, $\alpha'(G) + \beta'(G) = |V|$.*

Proof. Let M be the set of independent edges where M is maximum and $|M| = \alpha'(G)$. Suppose L is the set edges, satisfying the property that it contains one edge incident for each of the $(|V| - 2\alpha')$ vertices in of G that are not covered by any edge in M. So, $M \cup L$ will be an edge covering of G. That is

$$
\begin{aligned}
|M \cup L| &\geq \beta'(G) \quad \text{by definition of } \beta' \\
\Rightarrow \alpha' + |V| - 2\alpha' &\geq \beta' \\
\Rightarrow |V| &\geq \beta' + \alpha'
\end{aligned}
\tag{19.3}
$$

Suppose $\overline{M}$ is a minimum edge cover of G and then $|\overline{M}| = \beta'(G)$. Since $\overline{M}$ is minimum, it cannot have an edge $\rho = uv$ such that u and v are also incident with an edge in $\overline{M}$ other than ρ. This means $G[\overline{M}]$ is a edge spanning subgraph of G induced by $\overline{M}$. So each edge of $\overline{M}$ will be incident with atleast one end vertex of an edge in $G[\overline{M}]$. Let Q be such vertices where the incidency is exactly with one end vertex of $\overline{M}$. This means $|Q| = |\overline{M}| = \beta'(G)$. Each of the induced subgraph $G[\overline{M}]$ will have exactly one vertex that is not in Q. So we can write,

$$|V| = |Q| + (\text{number of induced subgraphs } G[\overline{M}]) \tag{19.4}$$

Now one can form a set of independent edges of G by picking one edge from each of the induced subgraphs $G[\overline{M}]$. This means

$$\alpha'(G) \geq (\text{number of induced subgraphs } G[\overline{M}])$$

Hence

$$|V| \leq \beta'(G) + \alpha'(G) \tag{19.5}$$

From (19.3) and (19.5), we have

$$\alpha'(G) + \beta'(G) = |V|.$$

□

19.2 Matchings

We have seen that any set M of independent edges is also called as matching of that graph.

Definition 19.9 1. A matching M is said to be a maximum matching of a graph G if G has no matching M' such that $|M'| > |M|$.
2. A matching M is said to be maximal matching of G if there is no matching M' of G such that $M \subset M'$.
3. If $e = uv \in M$, a matching of a graph G, we can say that the vertices u and v are matched under M. We say the vertices u, v are M-saturated.
4. A set S of vertices of a graph G is said to be M-saturated for a matching M of G if every vertex in S is incident with some edges in M.
5. A vertex u of a graph G is said to be M-unsaturated for a matching M if it is not incident with any edge in M.

Example 19.10 Consider the following graph (Fig. 19.2).

The matching $M = \{v_1v_2, v_4v_5\}$ is a maximal matching.

The matching $M' = \{v_1v_6, v_2v_7, v_3v_4\}$ is a maximum matching. The vertices v_1, v_2, v_3, v_4, v_6, v_7 are M'-saturated where as v_5 is M'-unsaturated.

Definition 19.11 1. Let M be a matching of a graph G. A path P in $G = (V, E)$ called an M-alternating path if the edges in P are alternating between M and $E - M$.
2. Let M_1, M_2 be two matchings of a graph $G = (V, E)$. Let $M_1 \Delta M_2$ denote the symmetric difference of M_1 and M_2. That is

$$M_1 \Delta M_2 = \{e | e \in M_1, e \notin M_2\} \cup \{e' | e' \notin M_1, e' \in M_2\}.$$

The subgraph induced by $M_1 \Delta M_2$ of G will be a graph whose components are either paths or cycles in G of even length whose edges alternate between M_1 and M_2.

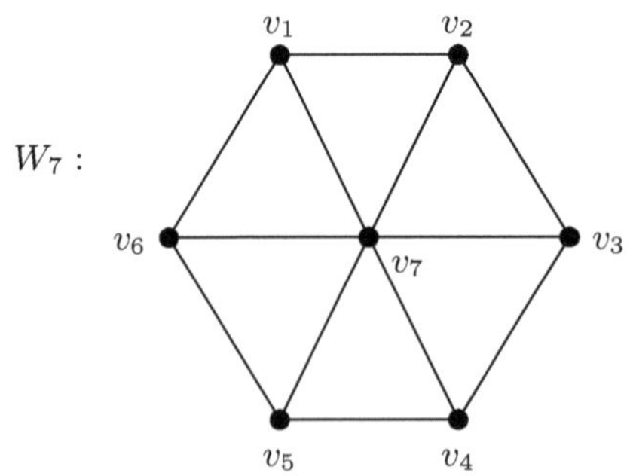

Fig. 19.2 Maximal Matching

3. A path whose edges alternate between M and $E - M$, for a matching M, having start and end vertices being M-unsaturated is called as M-augmenting path.

Example 19.12 In W_7, $M' = \{v_1v_6, v_2v_7, v_3v_4\}$ $P : v_1v_6v_7v_2v_3v_4$ is an alternating path, but not an augmenting path.

For $M'' = \{v_1v_2, v_3v_4\}$, $\overline{P} : v_6v_1v_2v_3v_4v_5$ is an augmenting path.

Following is a characterization by Berge for maximum matchings in graphs.

Theorem 19.13 *Let $G = (V, E)$ be a graph. A matching M of G is maximum if and only if G has no M-augmenting path.*

Proof. (i) Suppose for a matching M there is a M-augmenting path P. Then delete from the matching M, the edges of M that are on P and adding to M those edges of P not in M, we get a new matching $\overline{M}$. Here $\overline{M}$ has one extra edge than M. So M cannot be maximum.

(ii) Conversely, suppose M is a matching of G without M augmenting path in G. If $\overline{M}$ is a maximum matching of G, then by (i), there is no $\overline{M}$-augmenting path in G. Now for the two matchings, $M, \overline{M}$, consider $M \Delta \overline{M}$. The components of $G[M \Delta \overline{M}]$ can be only single vertices, alternating paths of even length, or even cycles that has alternate edges between M and $\overline{M}$. This means $|M| = |\overline{M}|$ which means M is maximum.

□

Definition 19.14 Let $G = (V, E)$ be a graph. A spanning subgraph of G is a factor G. A factor of a graph G that is k-regular will be called as k-factor of G. So a 1-factor of G is a matching which saturates every vertex of G. A 1-factor of G is called a perfect matching of G. A 2-factor of G is a 2-regular spanning subgraph and hence it is the disjoint union of cycles.

Example 19.15 Consider the graph (Fig. 19.3)

G_1 does not have 1-factor. G_2 has 1-factor.

19.3 Matching in Bipartite Graphs

An institute has $k_1, k_2, \ldots, k_n$ laboratries and there are t-courses $\ell_1, \ell_2, \ldots, \ell_t$ to be conducted in a particular slot in the labs. Suppose the course ℓ_m can be conducted in say 'f' labs. The problem is to assign the t-courses in the n labs so that no two labs are assigned to the same course.

The above assignment activity can be seen as a graph theory problem. Consider the n labs to be the vertices of one partition and t-courses as vertices in the other partition. If the lab k_s

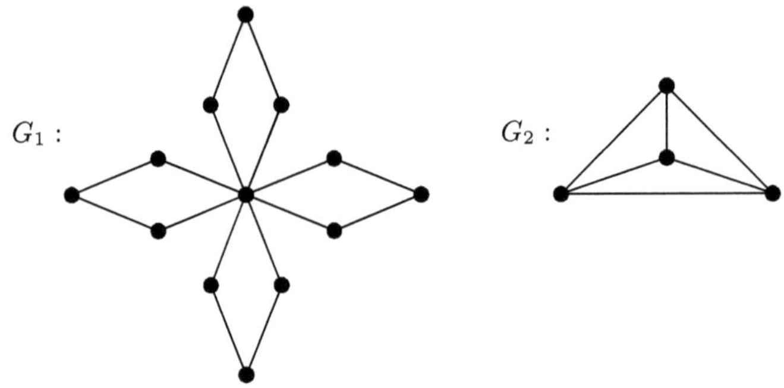

Fig. 19.3 Graphs (5)

is suitable for a course ℓ_m, then make an edge between them. Once all the assignments are complete, we can see whether there is a matching that saturates every vertex of the n-labs set. A solution to this problem was given by Hall. We have the following theorem to this effect.

Theorem 19.16 *Let $G = (V, E)$ be a bipartite graph with $V = X \cup Y$, $X \cap Y = \emptyset$. Then G has a matching M saturating all the vertices of X if and only if*

$$|N(S)| \geq |S|$$

for every subset S of X. Here $N(S)$ denotes all the vertices each of which is adjacent to atleast one vertex in S.

Proof. (i) Suppose G has a matching saturating all vertices of X. Let S be any subset of X. Every vertex of S is matched to different vertices in the neighbourhood of S which lies in Y. So $|S| \leq |N(S)|$.

(ii) Conversely suppose for every subset S of X, $N(S)| \geq |S|$. Suppose G has no matching saturating all the vertices in X. Let M be a maximum matching and that M saturate maximum number of vertices in X. Let x be a M-unsaturated vertex in X. Let Z be the set of vertices of G that are reachable from x by M-alternating paths. There will be no M-augmenting path in these as M being maximum. Let $S = Z \cap X$ and $T = Z \cap Y$. Clearly every vertex in T is matched by M to a vertex in $S - \{x\}$ and conversely. This means $N(S) = T$ and $|T| = |S| - 1$. Also $|S| = |T| + 1$ as $S \subseteq X$. This means $|S| > |T| = |N(S)|$ which is not true. Hence G has maximum matching.

□

We have already seen system of distinct representatives in chap ??. We just recall the definition here once again.

Definition 19.17 Let $V = \{a_1, a_2, \ldots, a_n\}$ be a given set. Let $\zeta = (A_1, A_2, \ldots, A_m)$ be a class of subsets of V. i.e., $A_i \subseteq V$ for $1 \leq i \leq m$. Then a subset $R = \{x_1, x_2, \ldots, x_m\}$ of V is a system of distinct representations (SDR) of the class ζ if $x_i \in A_i$, $1 \leq i \leq m$ and $x_i \neq x_j$ for $i \neq j$.

We now have Hall's SDR theorem as a corollary to the above theorem.

Corollary 19.18 *A necessary and sufficient condition for the existence of a SDR for a class* $\zeta = (A_1, A_2, \ldots, A_m)$, *each* A_i *being subset of a set* V *is that* $|R| \leq \left|\bigcup_{i=1}^{m} A_i\right|$ *for every set* $R = \{x_1, x_2, \ldots, x_m\}$, $x_i \in A_i$, $1 \leq i \leq m$, $x_i \neq x_j$ *for* $i \neq j$.

Proof. Let $G = (\overline{V}, E)$ be a bipartite graph of the given situation where
$\overline{V} = X \cup Y$ and
$X = \{A_1, A_2, \ldots, A_m\}$
$Y = V = \{a_1, a_2, \ldots, a_n\}$
A_i is incident with a_j if $a_j \in A_i$. Now for $R = (x_1, x_2, \ldots, x_m)$,

$N(X) = \bigcup_{i=1}^{m} A_i \subseteq V$ and $|X| = m$. By Hall's theorem the system has SDR if and only if G has a matching that saturates evey vertex of X. □

Remark 19.19 Hall's theorem is also known as the marriage theorem. This theorem has several equivalent formulations by different authors.

19.4 The Perfect Matching Theorem

Now, let G be any graph. We will see under what condition G has a 1-factor. 1-factor of a graph G is also known as perfect matching. We have the following result by Tutte.

Theorem 19.20 *Let* $G = (V, E)$ *be a graph.* G *has a 1-factor if and only if*

$$k_o(G - S) \leq |S| \; \forall \; S \subseteq V \tag{19.6}$$

where $k_o(G - S)$ *denotes the number of odd components of* $G - S$.

Proof. Suppose G has perfect matching M. Let $S \subseteq V$ be any subset. For each of the odd components of $G - S$, a vertex in such a component must be incident with an edge in M whose other end vertex being in S. All such vertices must be distinct. So 'S' is as large as the number of odd components and may be larger.

Conversely suppose $k_o(G - S) \leq |S|$ for every $S \subseteq V$. Suppose $S = \phi$, then G has no odd components. This means $|V|$ is even, say $|V| = 2n$. Here we take G to be a connected graph.

The result is proved by induction on n. For $n = 1$, $G = K_2$ and the result is trivial. Let the result be true for all the graphs with even number of vertices whose count is less than $2n$. Consider a graph $G = (V, E)$ with $|V| = 2n$. For any S, $k_o(G - S) \equiv |S| \bmod 2$. That is $k_o(G - S)$ and $|S|$ are both either odd or even.

Now we have two cases.

Case i: Suppose $k_o(G - S) < |S|$ for every subset S of V and $2 \leq |S| \leq 2n$. By the parity matching of $k_o(G - S)$ and $|S|$, we have $k_o(G - S) \leq |S| - 2 \ \forall \ S \subseteq V$. Consider an edge $e = uv \in E$. Let $V' = V - \{u, v\}$ and $G' = G - \{u, v\}$. By induction assumption G' satisfies (19.6). If not, $k_o(G' - T) > |T|$ for some $T \subseteq V'$. This means $k_o(G - S) = k(G' - T) > |T|$ and $|T| = |S| - 2$. This means $k_o(G - S) \geq |S| - 1$, which is a contradiction. So G' must have a 1-factor M'. $M' + uv = M''$ is a 1-factor for G.

Case ii: Suppose $k_o(G - S) = |S| \geq 2$. Let $S' = \{v_1, v_2, \ldots, v_t\}$ be a subset of V with maximum number of vertices such that $k_o(G - S') = |S'|$. Let $H_1, H_2, \ldots, H_s$ be the odd components of $G - S'$.

By Hall's theorem for bipartite graphs we can see that $|S|$ odd components of $G - S$ can be attached with vertices in S. Tutte's condition implies that any k such odd components must be joined in G with minimum 'k' vertices in S. If this is not true, then we get that k' number of vertices in S are joined to k odd components with $k' < k$. This means $k_(G - T) \geq k > |T|$ for $T \subseteq S$, which is a contradiction.

Hence a vertex of an odd component can be matched to a vertex of S in a distinct manner. Suppose $\overline{G}$ is an odd component with a vertex 'x' being matched with $\overline{x}$ of S. Then $G' = \overline{G} - x$ has even number of vertices. Tutte's condition must be true for G'. i.e., $k_o(G' - T) \leq |T|$, for T, a subset of vertex set of G'. If this is not true, then $k_o(G' - T) > |T| \Rightarrow k_o(G' - T) > |T| + 2$. Consider the set $\overline{S} = S \cup T \cup \{x\}$. Then $k_o(G - S') =$

$$\begin{aligned} k_o(G' - T) + k_o(G - S) - 1 &= k_o(G' - T) + |S| - 1 \\ &> |T| + |S| + 1 \\ &= |\overline{S}| \end{aligned}$$

This is a contradiction. Hence Tutte's condition must be true for G. G' has a 1-factor.

For an even component E of $G - S$ and x, a vertex of E, we have $G - (S \cup \{x\})$.

$k_o(G - (S \cup \{x\})) \geq k_o(G - S) + 1$ violating Tutte's condition. So $G - S$ has no even components.

Hence a vertex in the odd component of $G - S$ is distinctly matched with a vertex of S and the 1-factor of $\overline{G} - \{x\}$ of the odd components $\overline{G}$ together make a 1-factor for G. This completes the proof. □

Corollary 19.21 *Let $G = (V, E)$ be a connected 3-regular graph and having no cut edges. Then G has a 1-factor.*

Proof. Let $S \subseteq V$ and let the odd components of $G - S$ be $H_1, H_2, \ldots, H_t$. Let $H_i = (V_i, E_i)$, $1 \leq i \leq t$. Let k_i be the number of edges of G having end vertices u_i, v_i such that $u_i \in V_i$ and $v_i \in S$. Then we have

$$k_i \geq 2 \text{ as } G \text{ is 3-regular.}$$

Also

$$\sum_{v \in V_i} d(v) = 3.|V_i| = 2.|E_i| + k_i \tag{19.7}$$

$$\sum_{v \in S} d(v) = k|S| \geq \sum_{i=1}^{t} k_i + 2m \tag{19.8}$$

where m is the number of edges in the induced subgraph $G[S]$.

From (19.7) $k_i = 3|V_i| - 2|E_i| \Rightarrow k_i$ is odd and $k_i > 2$ implies $k_i \geq 3$.

Now summing over all the t odd components, we get

$$3t \leq \sum_{i=1}^{t} k_i \leq \sum_{i=1}^{t} k_i + 2m \leq 3\,S|.$$

So $k_o(G - S) = t \leq |S|$. By the above theorem G has a perfect matching. □

Remark 19.22 A 3-regular graph with cut edges will not have a perfect matching.

Example 19.23 This graph G is 3-regular, with 16 number of vertices. But G does not have a 1-factor. (Verify) (Fig. 19.4).

19.5 Domination

Domination in a graph studies about the nature of a subset of the vertex set in relevance to its neighbours. A vertex v of a graph is said to dominate itself as well as its adjacent vertices.

Definition 19.24 (1) Let $G = (V, E)$ be a graph. A subset S of V is called a dominating set of G if every vertex in $V - S$ has an adjacent vertex in S. A vertex 'v' dominates itself.

(2) Every vertex of G is either in S or in the neighbourhood $N(S) = \bigcup_{v \in S} N(v)$ in G.

Fig. 19.4 Graphs (6)

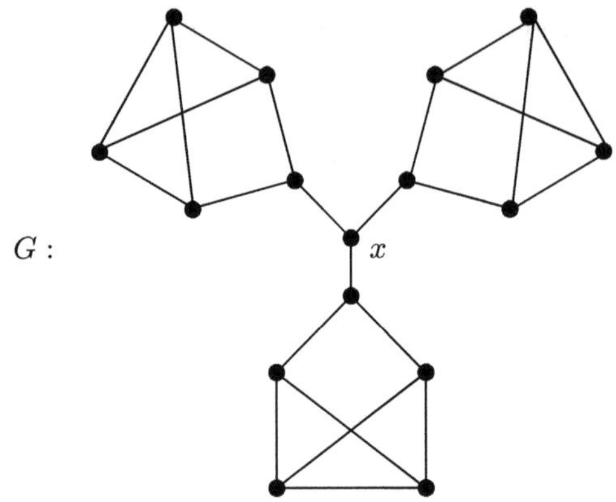

(3) The minimum dominating set of G is said to be the γ-set of G. A γ-set has minimum cardinality such that it is a dominating set of G.
(4) A minimal dominating set of S of G is one such that there exist no dominating set S' with $S' \subset S$.
(5) The domination number of a graph G is denoted by $\gamma(G)$.

Example 19.25 Consider the following Peterson graph (Fig. 19.5).
$\gamma(G) = 3$. $S = \{v_1, v_8, v_9\}$ is a γ-set.
The set $S_1 = \{v_6, v_7, v_8, v_9, v_{10}\}$ is a minimal dominating set.

The 'domination in graphs' was introduced by Ore. The research on this topic has become important because of its applications. In the following theorem we can see a necessary and sufficient condition for a set S of a graph G to be a minimal dominating set.

Fig. 19.5 Graphs (7)

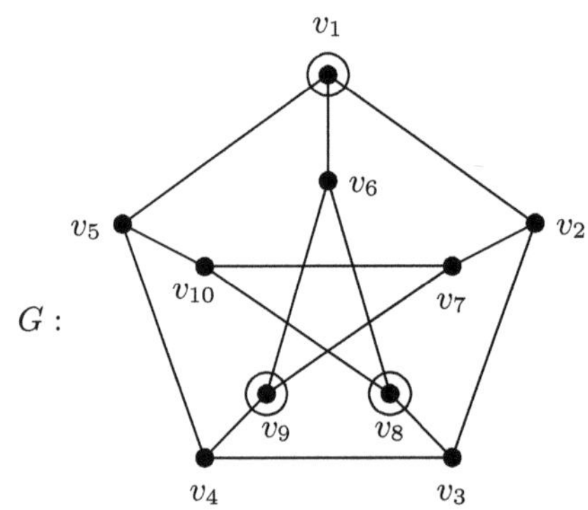

Theorem 19.26 *Let $G = (V, E)$ be a graph. Let $S \subseteq V$ be a dominating set of G. S is minimal dominating set if and only if one of the following holds for each vertex $d \in S$:*

(i) d is not adjacent to any vertex in S.
(ii) There exists a vertex $s \in V - S$ such that d is the only vertex of S adjacent to s.

Proof. If S is minimal, then $S - \{d\}$ not a dominating set. Hence either d or a vertex s in the set $V - S$ is not dominated by $S - \{d\}$. This covers conditions (i) and (ii) above.

Converse is true and left as an exercise. □

Corollary 19.27 *Let $G = (V, E)$ be a graph without isolated vertices. Suppose S is a minimal dominating set of G. Then $V - S$ is a dominating set of G.*

Proof. If $V - S$ is not a dominating set, then a vertex $d \in S$ is not adjacent to any vertex in $V - S$. This means $S - \{d\}$ is a dominating set which is a contradiction. □

Exercises

1. Determine the parameters $\alpha, \beta, \alpha', \beta'$ for the following graphs.

   (i) $K_{m,n}$

   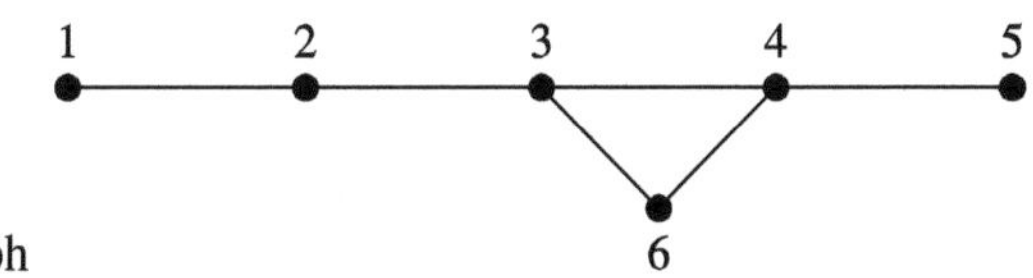

   (ii) Petersen graph

2. Say whether the following are true or fasle with justifications.

   (i) Every vertex covering of a graph contains a minimum covering.
   (ii) For a bipartite graph G, $\alpha\beta > |E|$ where $G = (V, E)$ if and only if G is complete bipartite.

3. Give an example of a graph that has 2-factor.
4. Find the number of 1-factors of K_{2n}.
5. Let G be a k-regular, $(k - 1)$-edge connected graph with even number of vertices. A graph G' is obtained from G by removing $(k - 1)$-edges. Then show that G' has a 1-factor.

Graph Coloring 20

In this chapter we introduce the concepts of vertex coloring, edge coloring of graphs. We can see some bounds on the number of colors used on the vertices or edges. One can see that main research topic on graph theory was on graph coloring in the last two decades.

20.1 Vertex Coloring

Definition 20.1 (1) Let $G = (V, E)$ be a graph. By k-vertex coloring, we mean that the vertex set V is partitioned as $V_1 \cup V_2 \cup \cdots \cup V_k$ such that the vertices in V_i are assigned the same color. Each V_i is non empty independent set and V_i's are called the color classes. That is a function $f : V \rightarrow \{1, 2, \ldots, k\}$ such that $f(v) = i$, $1 \leq i \leq k$ will be the defined color function on V. If the graph G has a k-coloring of its vertices, we say G is k-vertex-colorable.

(2) The chromatic number of a graph G is the minimum number of independent subsets $V_1, V_2, \ldots, V_k$ such that $V = V_1 \cup V_2 \cdots \cup V_k$, $V_i \cap V_j = \emptyset, i \neq j, 1 \leq i, j \leq k$. Here k is called the chromatic number of G and is denoted by $\chi(G)$. Any such partition will be called as chromatic partition of V.

(3) A k-coloring of a graph means that the graph G used almost k-colors for coloring its vertices.

(4) If $G = (V, E)$ is a k-chromatic graph i.e., $\chi(G) = k$ but $\chi(G - v) = (k - 1)$ for every vertex $v \in V$, then G is said to be a k-critical graph.

Remark 20.2 (1) If every k-coloring of a graph G with no color class is empty, gives rise to the same partition of the vertex set, then we can say G is uniquely k-colorable.

(2) In the computation of chromatic number, only adjacent vertices are taken into account. So multiple edges and loops are discarded.

R. Rama, *Topics in Combinatorics and Graph Theory*,
https://doi.org/10.1007/978-3-031-74252-1_20

Example 20.3 1. $\chi(K_n) = n$
2. $\chi(K_{m,n}) = 2$
3. $\chi(C_n) = \begin{cases} 2 & \text{if } n \text{ is even} \\ 3 & \text{if } n \text{ is odd} \end{cases}$.

Theorem 20.4 *Let $G = (V, E)$ be a graph such that $|V| = n$ and its independent number be $\alpha(G)$. Then $\frac{n}{\alpha(G)} \leq \chi(G) \leq n - \alpha(G) + 1$*

Proof. Let $V = V_1 \cup V_2 \cdots \cup V_{\chi(G)}$ be the chromatic partition of V. As V_i's are independent, $|V_i| \leq \alpha(G)$, $1 \leq i \leq \chi(G)$. Hence

$$n = \sum_{i=1}^{\chi(G)} |V_i| \leq \alpha(G).\chi(G) \tag{20.1}$$

Let S be a maximum independent set of G, so $|S| = \alpha(G)$. Let $V - S = \overline{V}_1 \cup \overline{V}_2 \cdots \cup \overline{V}_\ell$ be a partition of $V - S$ with $|\overline{V}_i| = 1\ \forall\ i, 1 \leq i \leq \ell, \ell = n - \alpha(G)$. This partition together with S form a partition of V and there are $n - \alpha(G) + 1$ independent subsets of V. So we have

$$\chi(G) \leq n - \alpha(G) + 1 \tag{20.2}$$

Combining (20.1) and (20.2) we get the result. □

A proper k-coloring of a graph G is also a proper k-coloring of any of its subgraph H. So $\chi(H) \leq \chi(G)$. A complete graph K_n is also called as a clique of size n. So a clique in a graph G is a complete subgraph of G. We know that an independent set of G is an empty induced subgraph in G.
Notation: Let G be graph. A subgraph of G that is a clique of maximum order will be denoted by $\omega(G)$.

Proposition 20.5 *Let $G = (V, E)$ be a graph. Then $\chi(G) \geq \omega(G)$.*

Proof. If $\omega(G) = m$, then G contains K_m as a subgraph. So $\chi(K_m) = m$ and $\chi(G) \geq \chi(K_m) = m$. □

Example 20.6 Consider W_7 (Fig. 20.1).
$\omega(W_7) = 3$ $\chi(W_7) = 3$.
For W_6, $\omega(W_6) = 3$ $\chi(W_6) = 4$.

Theorem 20.7 *Let $G = (V, E)$ be a graph and let G be k-critical. Then $\delta(G) \geq k - 1$.*

Fig. 20.1 W_7

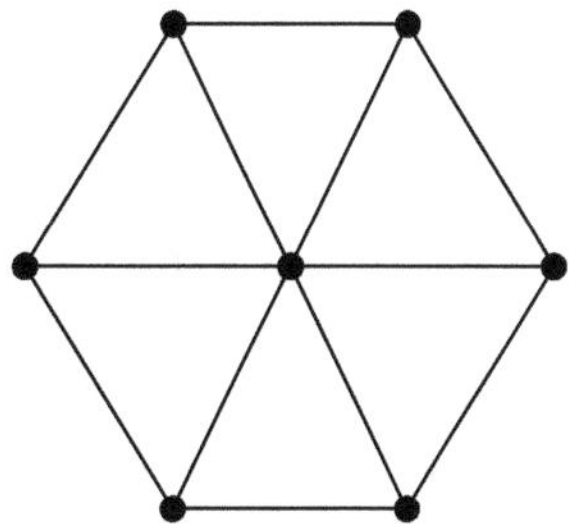

Proof. Since $G = (V, E)$ is k-critical, we have $\chi(G - v) = k - 1$ for any vertex $v \in V$. Suppose there exists a vertex $v \in V$ such that $d(v) < k - 1$. Then the $(k - 1)$-coloring of $G - v$ can be extended such that G is $(k - 1)$-colorable. As 'v' has less than $(k - 1)$-neighbours, one can assign a color to v that is assigned to any of its neighbours of v. So G is $(k - 1)$ colorable which is not true. □

Remark 20.8 Suppose G is a graph with $\chi(G) = k$ and H being a subgraph of G with minimum number of edges having no isolated vertices. H is a k-critical subgraph of G. So every k-chromatic graph will contain a k-critical subgraph.

Corollary 20.9 *1. Let $G = (V, E)$ be a graph with $\chi(G) = k$. Then G has atleast k vertices of degree minimum $(k - 1)$.*
2. For any graph G, $\chi(G) \leq \Delta + 1$.

Theorem 20.10 *Let $G = (V, E)$ be a k-critical graph. Suppose $S \subseteq V$ is a vertex cut of G, then $G[S]$ cannot be a clique.*

Proof. Let $G = (V, E)$ be a k-critical graph. Let $S \subseteq V$ be any vertex cut of G. We want to show $G[S]$ is not a clique. Suppose $G[S]$ is a clique. Let $G_1, G_2, \ldots, G_t$ be the components of $G - S$. Let H_i be the induced subgraph of G on $V(G_i) \cup S$, $1 \leq i \leq t$. This H_i will be a proper subgraph of G and so every H_i is $(k - 1)$ colorable. In particular the vertices of S will be distinctly colored because S being vertices of K_k. So fixing the colors for the vertices of 'S' and coloring $V(H_i)$ gives a $(k - 1)$-coloring for H_i, $1 \leq i \leq t$. This turns out to be a $(k - 1)$-coloring of G. This is a contradiction. Hence $G[S]$ cannot be a clique. □

20.2 Brooks' Theorem

It is not difficult to see for any graph G, $\chi(G) \leq \Delta + 1$. If G is an odd cycle, then $\chi(G) = 3$ and if G is K_n, $\chi(K_n) = k = \Delta(G) + 1$. These two are the only graphs where the bound $\Delta(G) + 1$ for its chromatic number is achieved. The following theorem by Brooks gives a better upper bound for the chromatic number.

Theorem 20.11 *If G is a connected graph where $G \neq C_{2n+1}, K_m$, then $\chi(G) \leq \Delta(G)$.*

Proof. Let $G = (V, E)$ be a graph such that $G \neq C_{2n+1}, K_m$. Let the maximum degree in G be k. Without loss of generality we can take $k \geq 3$. [When $k = 1$, G is a clique, and when $k = 2$, G is either an odd cycle or a bipartite graph].

If G is not a regular graph then there is a vertex v in G such that $d(v) < k$. As G is connected, one can look for a spanning tree of G from the vertex 'v' in an order of decreasing indices based on the distance from v. Let every such vertex be labelled as $v_1, v_2, \ldots, v_{n-1}, v_n, (v_n = v)$.

Also each v_i has atmost $(k - 1)$ lower indexed neighbours. Therefore if we color vertices v_i, $1 \leq i \leq n - 1$, in that order with colors from the set $\zeta = \{1, 2, \ldots, \Delta\}$ using a least integer from ζ that has not been used to color any of its adjacent vertices that are assigned colors so far. Here there is definitely one color available to color v_i. As $d(v_n) = d(v) < \Delta$, again one color is available for $v_n = v$. So this gives a Δ-coloring for G.

(i) Suppose G is k-regular, clearly $k \geq 3$ as G is not an odd cycle.
(ii) Suppose G has a cut vertex 'v'. Consider the component $\overline{G}$ in $G - v$. Let G' be the graph obtained by adding the vertex 'v' to $\overline{G}$ along with the edges connected to 'v'. Clearly $d_{G'}(v) < k$ and so G' can be properly colored with k-colors. By permuting colors in such a subgraph, one can reach the k-coloring of G such that 'v' has the same color in both G and G'.
(iii) Suppose G has a vertex v with $vu, vw \in E(G), uw \notin E(G)$. Also suppose $G - \{u, w\}$ is connected. Then $G - \{u, w\}$ has a spanning tree and we can name the vertices of this tree as $v_3, v_4, \ldots, v_n$ in the non-decreasing order of their distances from v. As in the above description, each vertex before v will have maximum $(k - 1)$ lower indexed vertices as their neighbors. Now the vertices u and w are non-adjacent and can be given same color and $v_i, 3 \leq i \leq n - 1$ with the least color from the set $\{1, 2, \ldots, k\}$ that has not been assigned so far to any of its neighbours. This sequential assignment always leaves a color for the vertex 'v'. So G is δ-colorable. (iv) below shows the existence of two such vertices u, w in G.
(iv) Let G be a 2-connected, k-regular graph. Let $v \in V(G)$. If $G - v$ is also 2-connected, then let $v_1 = v$ and v_2 be a vertex at a distance two from v_1. This is possible as G is connected and not a complete graph. If $G - v$ has a cut-vertex, then 'v' is adjacent to a vertex in every subgraph $G_1, G_2, \ldots, G_\ell$ each of them being a block. Also the

neighbours of v_1, v_2 of v lie in two such blocks and they are not adjacent. Algo $G - \{v_1, v_2, v\}$ is a connected graph as each block has no cut vertices. So $d(v) \geq 3$ implying that $G - \{v_1, v_2\}$ is connected.

□

Remark 20.12 The maximum degree bound on the chromatic number is really poor. For example, for any bipartite graph it is 2.

20.3 Planar Graphs Coloring

In this section the concept of coloring vertices of a planar graph will be considered. We know that any planar graph will have a plane embedding. For any plane graph the coloring of faces can be defined.

Definition 20.13 Let $G = (V, E)$ be a plane graph with $F = \{f_1, f_2, \ldots, f_t\}$ its faces. Let $\overline{F} = F_1 \cup F_2 \cup \cdots \cup F_\ell$ be the partition of F. Each set F_i of faces is such that no two faces in F_i share a common edge on the boundary. Some F_i's may be empty. That is faces in F_i are non adjacent with each other. Now each face 'f_i' can be assigned same color and each F_i get different coloring. That is a color function $f : F \to \{1, 2, \ldots, k\}$ such that $f(f_s) = t$ for every $f_s \in F_t$, F_i's are called as color classes. If the plane graph has k-face coloring, then it will be called as k-face colorable and the smallest such k will be called as face chromatic number of G. It is denoted by $\chi''(G)$.

Example 20.14 Consider the plane graph (Fig. 20.2)
Each face is adjacent to all other faces. So $\chi''(K_4) = 4$.

The proof of the following propositions are easy to prove and left as an exercise.

Proposition 20.15 *1. Let $G = (V, E)$ be a plane graph. G is k-vertex colorable or k-face colorable if and only if all its components have this property.*
2. G is k-vertex colorable or k-face colorable if and only if all its blocks have this property.

Fig. 20.2 Graph (3)

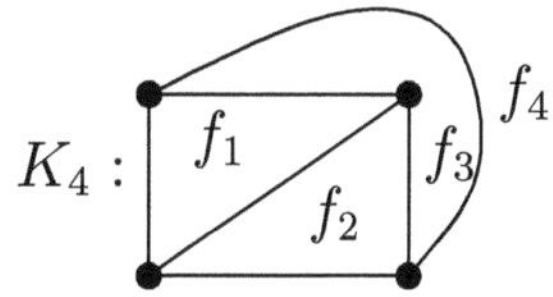

Proposition 20.16 *Every planar graph is k-vertex colorable if and only if every plane graph is k-face colorable.*

Proof. Suppose $G = (V, E)$ is a planar graph that is k-colorable. Let G be a connected graph that has no cut vertices and cut edges. Suppose G^* is the dual of G and $\overline{G}$ is the plane graph corresponding to G^*. Any two vertices in $\overline{G}$ are adjacent if and only if the corresponding faces of G are adjacent. So $\chi(\overline{G}) = \chi''(G)$.

Conversely let $G = (V, E)$ be a plane graph that is k-face colorable. Let G be connected and has no cut vertices or cut edges. Let G^* be the dual of G. G^* is loopless as G has no cut edges/vertices. Suppose the vertices f_1, f_2 in G^* have multiple edges, then subdivide these edges except one edge. Replace all such multi edges of G^* using substitution. Let the resultant graph be G'. G' is a plane graph. Clearly two vertices u, v are adjacent in G, then the corresponding regions/faces of G' are adjacent and conversely. So, $\chi''(G') = \chi(G)$. G is k-vertex colorable. □

Remark 20.17 Loops and multiple edges are irrelevant while vertex coloring a plane graph. So, for face coloring of a plane graph, we can assume that any two faces share atmost one edge in the common boundary.

Four Color Theorem (FCT)

Every planar graph is 4-colorable.

The proof of the above theorem was given by Appel and Haken (1977). There are some equivalent coloring concepts like FCT. We have the following one result.

Theorem 20.18 *A graph $G = (V, E)$ is 4-colorable if and only if every triangulation of G is 4-colorable.*

Proof. A plane graph $G = (V, E)$ with $|V| \geq 3$ is maximally plane if and only if every face of G is a boundary of a triangle. So every triangulation of a graph is a planar graph. As every planar graph being a subgraph of a triangulated graph, the result follows. □

Theorem 20.19 *The FCT is true if and only if every 3-regular plane graph without cut edges is 4-face colorable.*

Proof. A planar graph is 4-colorable if and only if every plane graph without cut edges is 4-colorable. Every plane graph G with a cut edge 'e' and let G' be the graph obtained by contracting e. So $\chi''(G) = \chi''(G')$ where $\chi''(G)$ is the face chromatic number of G.

Let G be a plane graph without cut edges. Let G' be the graph obtained from G as follows:

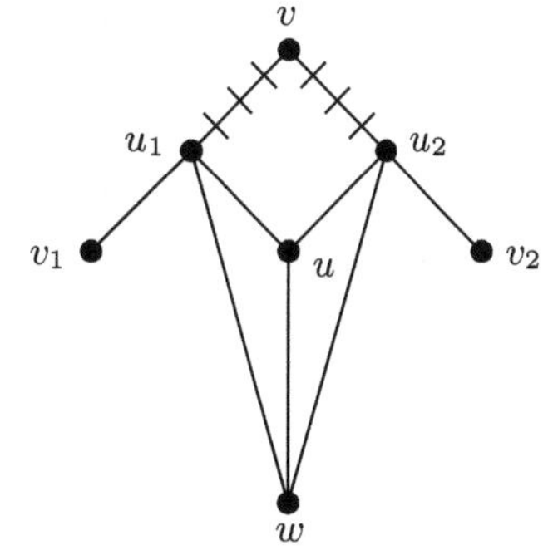

Fig. 20.3 A situation

Case i: If G has a vertex v of degree 2. Let v_1vv_2 be the edges at v. Subdivide v_1v and vv_2 by introducing two new vertices u_1, u_2. Now let u and w be two new vertices to be introduced such that edges uu_1, uu_2, wu_1, wu_2 and uw are the new edges to be drawn and delete the edges vu_1, vu_2. See the following part of the graph (Fig. 20.3)

So we can see that we replace 'v' by $K_4 - e$ which is a 3-regular maximal planar subgraph. Let G' be the new graph thus modified.

Case ii: Suppose G has a vertex 'v' of degree more than 4 i.e., $d(v) \geq 4$. Let $v_1, v_2, \ldots, v_s$ be the vertices incident at v. Let $u_1, u_2, \ldots, u_s$ be new vertices added to $G - v$ such that $v_1u_1, v_2u_2, \ldots, v_su_s$ are added as new edges (Fig. 20.4).

So the vertex 'v' is replaced by the above way leads to a 3-regular graph G'. G' will have no cut edges. Hence G' is 4-face colorable.

Now reversing or contracting $K_4 - e$ or the graph part of G modified as in Case (ii) above we get back G from G'. So G is 4-face colorable with maximum 4-colors. □

We state the following theorem whose proof is popular and widely available in the literature.

Theorem 20.20 *Every planar graph is 5-colorable.*

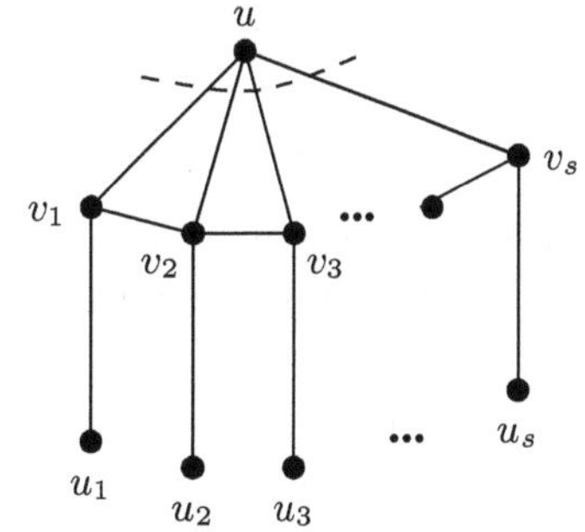

Fig. 20.4 A situation

20.4 Edge Coloring

Definition 20.21 Let $G = (V, E)$ be a graph. A k-edge coloring of G is a partition of the edge set E as $E = E_1 \cup E_2 \cup \cdots \cup E_k$ with $E_i \cap E_j = \emptyset, i \neq j$. For if the edges in E are colored with k-colors $\{1, 2, \ldots, k\}$ (say) such that no two adjacent edges are given same color. E_i's are called edge-color classes. If the edges E can be colored with k-colors and k being minimum, then G is said to be k-edge colorable. Here the number will be called as edge chromatic number and is denoted by $\chi'(G)$.

Example 20.22 For

(i) $K_5, \chi'(G) = 5$
(ii) $K_{3,2}, \chi'(G) = 3$
(iii) $C_6, \chi'(G) = 2$.

The following observations follow from definition.
Observations

1. For any graph G, $\chi'(G) \geq \Delta(G)$.
2. $\chi'(K_n) = \begin{cases} n & \text{if } n \text{ is odd} \\ n-1 & \text{if } n \text{ is even} \end{cases}$
3. $\chi'(C_n) = \begin{cases} 2 & \text{if } n \text{ is even} \\ 3 & \text{if } n \text{ is odd} \end{cases}$

Theorem 20.23 *For a bipartite graph G, $\chi'(G) = \Delta(G)$.*

Proof. The proof is by induction on the number of edges of G. Let $G = (V, E)$ be a graph.

For $|E| = 1$, the result is true. Assume the result to be true for all bipartite graphs with maximum $(s-1)$-edges.

Let G be a graph with s-edges. Let $e = uv$ be an edge of G. Consider $G - e$ has a proper Δ-coloring. Let this coloring be named as ζ.

Case i: In ζ, there is a color that is not represented at the vertices u and v, then this color can be used for e and then G has Δ-coloring.

Case ii: Suppose all the Δ colors are represented either at u or v in $G - e$. But the degree of u and v are atmost $\Delta - 1$ in $G - e$, there is a color, in ζ that is not represented at u in $G - e$. There will be a color for v as well from ζ in $G - e$. If 'i' is one such color not represented at u, then 'i' is represented at v and if 'i' is not represented at v then 'i' is represented at u in $G - e$. Note that u and v lie in different part as G is bipartite graph. But there exist no u-v path in G in which the colors alternate between i and j. Suppose P is a maximal path in

$G - e$ starting from 'u' and edges on P being alternately colored with i and j. Swapping the colors i and j in P, will retain the proper coloring of $G - e$ using Δ-colors. This swapping makes a color 'i' not represented at both u and v. Now color e with i and this yields a proper Δ-coloring of G. □

For all simple graphs, $\chi'(G) \geq \Delta(G)$. For example $\chi'(C_5) = 3$. So $\chi'(C_5) = \Delta + 1$.

Theorem 20.24 *For any simple graph $G = (V, E)$, $\Delta(G) \leq \chi'(G) \leq 1 + \Delta(G)$.*

Proof. The bound that $\chi'(G) \geq \Delta(G)$ is directly implied by definition and structure of $G = (V, E)$.

We now show that $\chi'(G)$ for G having maximum degree $\Delta(G) + 1$.

Suppose G is not $\Delta + 1$-edge colorable. But G definitely has a subgraph $H = (\overline{V}, \overline{E})$ that is $\Delta + 1$-colorable with maximum number of edges. Choose an edge $uv_1 \notin \overline{E}$ and $uv_1 \in E$. Clearly $d(u) \leq \Delta$ and as H is $\Delta + 1$-colorable, there exists a color 'c' not represented at u in H. The same argument applies to vertex $v_1 \in \overline{V}$ and a color c_1 is not represented at v_1 in H as in the following part graph of H (Fig. 20.5).

As per the above diagram and argument the color c_1 will be represented at u by an edge uv_2 (say). If not uv_1 can be given color c_1 and hence $H + uv_1$ has one more edge than H and will have proper $\Delta + 1$-coloring. Suppose the color c_2 is not represented at v_2, then an edge uv_3 can be assigned color c_2 and so on. By this we get a set of edges incident at u as $\{uv_1, uv_2, \ldots, uv_k\}$ (say) such that a color c_i for $1 \leq i \leq k$ is not represented at v_i and so edge uv_{j+1} can be assigned color c_j, $1 \leq j \leq k$ as seen in the above diagram.

While checking the colors represented at the vertices $v_1, v_2, \ldots, v_k$ that are adjacent to u, we notice at some stage 's' $1 \leq s \leq k$, the color 'c' is not represented at v_s, then one can reassign colors $c_1, c_2, \ldots, c_{s-1}$ from $uv_2, uv_3, \ldots, uv_s$ to $uv_1, uv_2, \ldots, uv_s$ to $uv_1, uv_2, \ldots, uv_{s-1}$ and assign color 'c' to u_s as 'c' is not represented at both u and v_s. This systematic checking and shifting fields $\Delta + 1$- edge coloring of $H + uv_1$ which is a contradiction to the edge coloring of H. So, we can assume that a color 'c' is represented at each of the vertices $v_1, v_2, \ldots, v_k$ that are adjacent to u. As $d(u)$ is finite the sequential checking will terminate. Also there may be an edge e not incident at u with color c_k assigned to ρ or the sequence reaches the assignment of same color c_ℓ such that $c_1 = c_s$, $\ell < s - 1$.

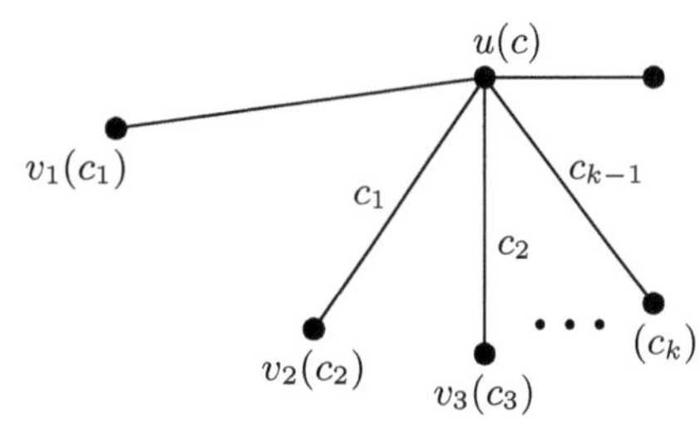

Fig. 20.5 A situation

As the color c_k is not represented at u in H, one can reassign the colors as before to get a $\Delta + 1$ coloring of $H + e$ which is a contradiction.

If $c_s = c_\ell$ for $\ell < s - 1$, assign colors $c_1, c_2, \ldots, c_\ell$ to $uv_1, uv_2, \ldots uv_\ell$ and let $uv_{\ell+1}$ be not colored. So the subgraph $\overline{H} = (H + uv_1) - uv_{\ell+1}$ and H have the same number of edges. Consider the subgraph T of $\overline{H}$ which is either a cycle or a path in which the colors 'c' and 'c_ℓ' alternate. Now the color 'c' is represented at every vertex $v_1, v_2, \ldots, v_s$ and in particular at $v_{\ell+1}$ and v_s, where as c_ℓ is not represented at $v_{\ell+1}$ and v_s because we shifted color c_ℓ to uv_ℓ and $c_\ell = c_s$ is not represented at v_s. In this case $d(v_{\ell+1}) = d(v_s) = 1$. Also the color c_ℓ is represented at 'u' but 'c' is not represented. It means $d(u) = 1$ in the subgraph T of $\overline{H}$ which is a path or an even cycle. So the vertices u, $v_{\ell+1}$ and v_s cannot be in the same component in the cycle or path subgraph T of $\overline{H}$ that is considered above.

If u, v_{s+1} are in different component of the subgraph T of $\overline{H}$, interchanging the colors c and $c_{\ell+1}$ in the components containing $v_{\ell+1}$, we get the color 'c' not represented at u and $v_{\ell+1}$. So the edge $uv_{\ell+1}$ can be given color 'c'. This leads to $\Delta + 1$-coloring of $\overline{H} + uv_{\ell+1}$.

Identically if $u, v_{\ell+1}$ are in the same component of T but v_s is not, interchanging c and c_ℓ in the component containing v_s we get $\Delta + 1$ coloring of $\overline{H} + uv_k$. So in all possible ways of coloring leads to $\Delta + 1$ coloring of $\overline{H} + e$, e an edge not in $\overline{H}$, which is a contradiction. So, $\chi'(G) \leq 1 + \Delta(G)$. □

The above theorem was given by Vizing and Gupta. Vizing also classifies graphs in terms of edge chromatic number.

Definition 20.25 (i) Let $G = (V, E)$ be a graph. G is called class-I graph if $\chi'(G) = \Delta$ and G is called class-II graph if $\chi'(G) = \Delta + 1$.

(ii) A graph $G = (V, E)$ is said to be edge-chromatic minimal if $\chi'(G - e) < \chi'(G)$ for every edge $e \in E$.

(iii) A graph G in class-II is said to be class-critical if $G - e$ is in class-I for every edge $e \in E(G)$. It is not easy to determine the class of a given graph as determining edge-chromatic number of a given graph is an NP-complete problem. It can be seen that $\chi'(K_{m,n}) = \Delta(K_{m,n})$ and hence any complete bipartite graph belongs to class-I. A cycles of odd length belong to class-II.

Theorem 20.26 *Any regular graph $G = (V, E)$ will be in class-I if and only if G has 1-factor.*

Proof. For a k-regular graph G, $\chi'(G) = k$ or $k + 1$. If $\chi'(G) = k$, then there will be k-colorings of the vertices of G such that $E_1 \cup E_2 \cdots \cup E_k$, $E_i \cap E_j = \emptyset$, is a color partition of the edges. Hence every vertex $v \in V$ is incident with exactly one edge in a set E_i, $1 \leq i \leq k$. This means each color class E_i forms a 1-factor, it is not difficult to show that $\chi'(G) = k$.
□

Corollary 20.27 *If $G = (V, E)$ is a k-regular graph with $|V|$ being odd, then G belongs to class-II.*

Exercises

1. Determine the vertex-chromatic number, edge-chromatic number and face-chromatic number of the following graphs.

   (i) Peterson graph.
   (ii) Cubic graph.
   (iii) Tripartite graph.

2. Show that if G is a non empty bipartite graph then G belongs to class-I.
3. Is it true that any complete graph is uniquely colorable? Justify your answer.
4. Let G be a uniquely k-colorable graph. Is it true that G is complete? Justify your answer.
5. Give 2 examples of graph G with

   (i) $\chi'(G) = \Delta(G)$
   (ii) $\chi'(G) > \Delta(G)$.

6. Is it true that every planar graph is 5-colorable? Justify your answer.
7. Let $G = (V, E)$ be a graph and let $S \subseteq V$ be an independent set of vertices of G. If G is k-critical then show that $\chi(G - S) = \chi(G) = 1$.
8. A vertex coloring of a graph $G = (V, E)$ is said to be complete if for any two coloring c_1 and c_2 in the set of colorings there is a pair of vertices $u, v \in V$ such that $uv \in E$ and u is colored with c_1 and v is colored with c_2. Give 2 examples of graphs G such that G has complete colorings.
9. The chromatic index $\psi(G)$ of a graph G is the maximum number k for which G has a complete k-coloring. Is it true that $\chi(G) < k < \psi(G)$? Justify your answer with some examples.

Ramsey Numbers and Ramsey Graphs 21

Theodore S. Motzlan, an Israeli-American mathematician commented that complete disorder is impossible. This also means that any irregular structure if large enough, will definitely contain some 'regular' substructure. Ramsey in 1929, proved some important results in set theory which gained attention due to their wide applicability. We will see how Ramsey theorems have graph theory formulations and generalizations. Ramsey theory is also called as mathematics of coloring.

21.1 Some Basic Theorems on Subsets and Coloring

In the following theorems on a set X, we use the notation $\mathcal{P}_n(X)$ to denote the subsets of X whose cardinality is always n. The elements of $\mathcal{P}_n(X)$ are said to be colored using a color set C, we mean that each of the n-element set in $\mathcal{P}(X)$ are assigned a color from C.

Definition 211 Let X be a set and C be a coloring of $\mathcal{P}_n(X)$. For a subset Y of X, if every element of $\mathcal{P}_n(X)$ get the same color, then the set Y is said to be monochromatic in C (Fig. 21.1).

Theorem 211 *Let n, k be positive integers. Let X be an infinite set and f be a coloring function that assigns colors from the set C, with $|C| = k$, to the elements of $\mathcal{P}_n(X)$. Then X contains an infinite monochromatic subset.*

In the case of X being finite, the above theorem can be stated as there definitely exists a subset of finite cardinality having assigned the same color as seen below.

R. Rama, *Topics in Combinatorics and Graph Theory*,
https://doi.org/10.1007/978-3-031-74252-1_21

Fig. 21.1 Subset mapping

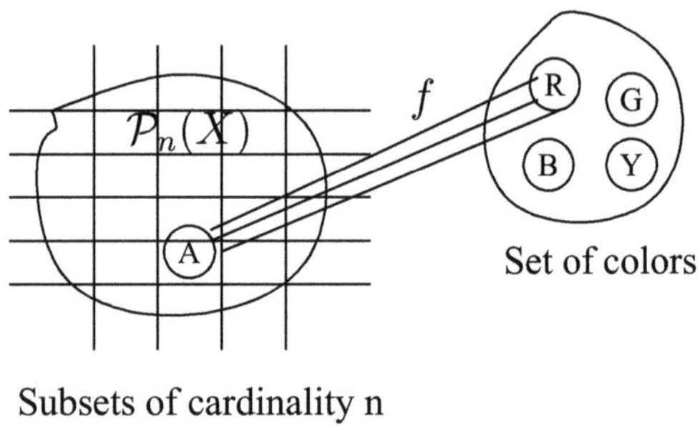

Theorem 212 *Let $m, k, s \in \mathbb{N}^+$. If there exists positive integer 'n' such that $\mathcal{P}_m(X)$ being subsets of X with $|X| = n$, then for any k-coloring of $\mathcal{P}_m(X)$, there exists a monochromatic subset Y of X with $|Y| = s$.*

The least integer 'n' for which Theorem 212 is valid is called the Ramsey number. It is denoted by $R(m, k, s)$ where m, k, s are the parameters.

The following theorem is a generalization of Theorem 212.

Theorem 213 *Let $m, k, s_1, s_2, \ldots, s_k$ be positive integers. If there exists a least number*

$$n = R(m, k, s_1, s_2, \ldots, s_k),$$

called the Ramsey number with parameters $m, k, s_1, s_2, \ldots, s_k$ and if $C = \{c_1, c_2, \ldots, c_k\}$, the coloring set for $\mathcal{P}_m(X)$ with $|X| = n$, then X has a subset Y with $|Y| = s_i$, which is monochromatic with assigned color being c_i.

21.2 Ramsey Numbers of Graphs

We first start with a puzzle on friends and strangers in a group.

A group has a set of n people. In this set there will always be either a subset of 3 people who are mutual friends or a subset of three people who are mutually strangers. Find minimum such 'n'. The value of 'n' in this puzzle is the Ramsey number.

Definition 212 1. The Ramsey number $R(s, t)$ is the least number n such that any graph on n vertices contains either an independent of size s or a clique of size t.
2. The Ramsey number $R_k(s_1, s_2, \ldots, s_k)$ is the least number 'n' such that any coloring of the complete graph K_n with colors $\{1, 2, \ldots, k\}$ contains a clique of size s_i, in color i for some i.

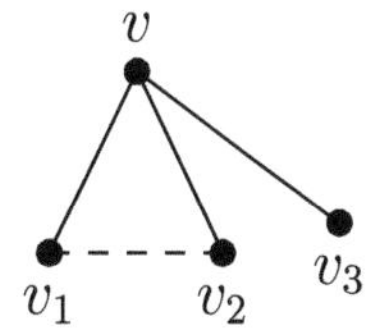

Fig. 21.2 A situation

Example 211 For any graph $G = (V, E)$ with $|V| = 6$. Show that G must contain either a K_3 or $\overline{K}_3$.

Solution: This solution gives the solution to the puzzle stated above on friends and strangers. Hence $R(3, 3) = 6$.

Let $v \in V$. This vertex 'v' must be adjacent to three vertices in $\overline{G}$ or G by stated conditions in the puzzle.

Suppose 'v' is adjacent to three vertices in G say v_1, v_2, v_3 (Fig. 21.2).

If two of these v_1, v_2, v_3 are adjacent in G, then G has K_3. If none of v_1, v_2, v_3 are adjacent in G, then $\overline{G}$ has K_3 on v_1, v_2, v_3.

Remark 211 For any graph G and its complement $\overline{G}$, $K_n = G \cup \overline{G}$. By coloring edges of G with a color 'red' and edges of $\overline{G}$ with color 'blue' say, we get a 2-coloring of edges of K_n. The converse is also true.

From the above remark we have the following lemma which can be proved by induction on number of vertices.

Lemma 211 *Let $G = (V, E)$ be a graph with $|V| = (s + t)C_s$. Then G contains either a clique of size $(s + 1)$ or an independent size of $(t + 1)$.*

We have the following simple observations on Ramsey numbers.

(i) $R(s, t) = R(t, s)$ as the two colors on the edges can be swapped.
(ii) $R(s, 1) = 1$, as K_1 has no edges. Any one color on K_1 contains a clique in that color.
(iii) $R(s, 2) = s$. If G contains a complete graph K_s with 'red' colored edges, then the result is true. If one of its edges is colored with 'blue' then G contains a K_2 with 'blue' color.

It is not easy to compute the values of Ramsey numbers. For some small numbers, $3 \leq s \leq 6, 3 \leq t \leq 7$, Ramsey numbers $R(s, t)$ are given in Table 21.1. In the table ? implies only upper and lower bounds are known.

Table 21.1 Upper and lower bounds

s t	3	4	5	6
3	6	9	14	18
4	9	18	25	?
5	14	25	?	?
6	18	?	?	?
7	23	?	?	?

Fig. 21.3 $R(3, 4) = 9$

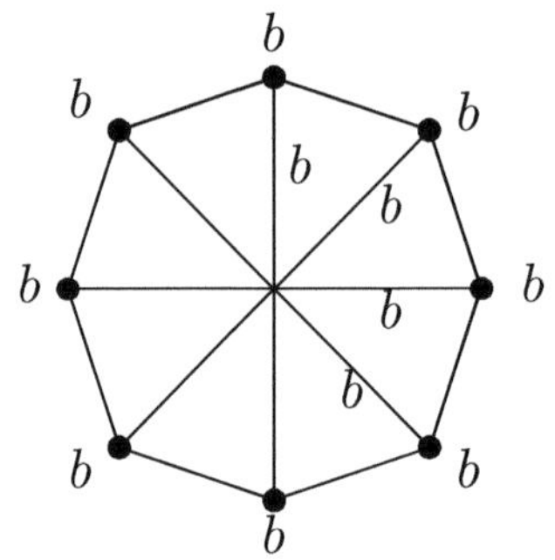

Example 212 $R(3, 4) = 9$.

Solution: We can first show that $R(3, 4) \geq 9$. The argument is based on coloring the edges of K_8 with 'red' and 'blue' colors (Fig. 21.3).

G is the graph of K_8 in which all the edges are given 'blue' color. $\overline{G}$ is the complement subgraph of G in K_8 whose edges are colored with 'red'. So K_8 has colored edges and does not contain a clique of size 4 or an independent set of size 3.

Now we have to show $R(3, 4) \leq 9$. For this consider K_9 and color its edges with 'red' and 'blue'.

There cannot be a 5-regular subgraph of K_9 whose edges are completely red. This is due to the fact that in any graph, number of odd degree vertices must be even. Then there cannot be a 3-regular subgraph of K_9 all whose edges are 'blue', as seen below for any vertex v of K_9 (Fig. 21.4).

So K_9, whose edges are 2-colored, must have a vertex 'v', with 6-edges incident with v are given 'red' color or 4-edges are given 'blue' color (Fig. 21.5).

These six red edges that are incident to six other vertices other than v must induce a monochromatic K_3 as $R(3, 3) = 6$. If such an induced K_3 is 'red' in color then we have the required clique. If this K_3 is 'blue' then its connecting edges with 'v' must induce a K_4 in color 'blue'. So we have the result.

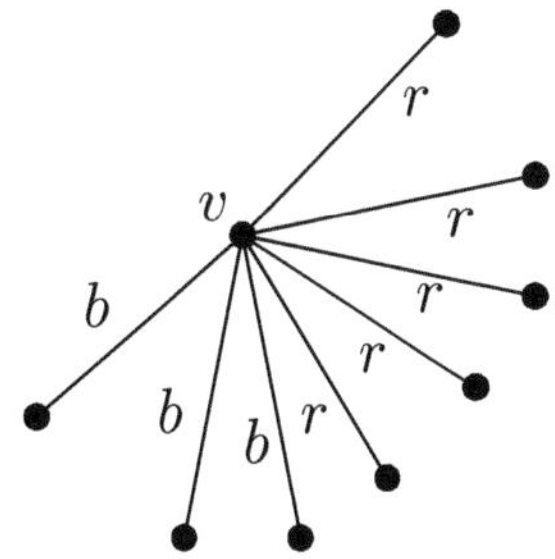

Fig. 21.4 A situation

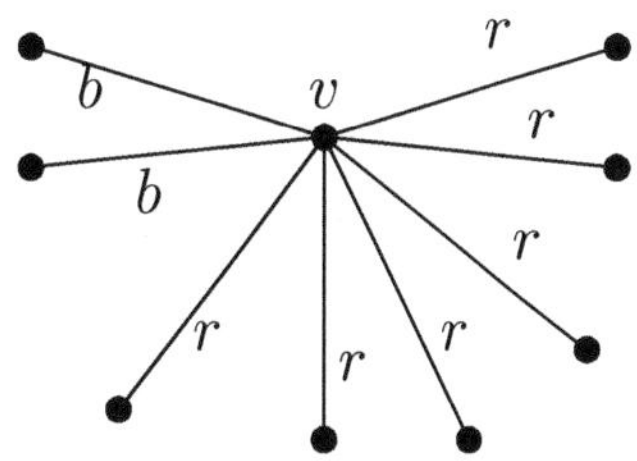

Fig. 21.5 A situation

The case with the edges incident at 'v', four of them being colored 'blue' and four of them colored 'red' can be argued identically. Hence $R(3, 4) = 9$.

21.3 Ramsey Theorem

We now have Ramsey theorem for 2-colored graphs.

Theorem 214 *Let $s, t \in \mathbb{N}$. There exists a natural number n such that any 2-colored complete graph on n vertices must contain a monochromatic K_s or K_t. Here 'n' is the least number and $R(s, t) = n$.*

Proof We know that $R(s, 2) = R(2, s) = s$ and $R(1, t) = R(t, 1) = 1$. The existence of $R(s, t)$ comes from the fact that $R(s, t)$ being bounded. We now show $R(s, t)$ is bounded. The proof is given by using method of induction on the number of vertices of the graph K_{s+t}.

Since the bounds is true for some base values, let us assume the result to be true for $R(s - 1, t)$ and $R(s, t - 1)$ that they exist.

We now claim that $R(s, t) \leq R(s - 1, t) + R(s, t - 1)$.

Let $n = R(s - 1, t) + R(s, t - 1)$ and consider edges of K_n being colored with colors red and blue. Let v be a vertex in K_n. Let R_v and B_v be two sets such that

$$R_v = \{u \mid uv \in E(K_n) \text{ and } uv \text{ is colored red}\}$$
$$B_v = \{u' \mid u'v \in E(K_n) \text{ and } u'v \text{ is colored blue}\}$$

Clearly $|R_v| + |B_v| = n - 1$. Also if $|R_v| < R(s-1,t)$ and $|B_v| < R(s,t-1)$, then

$$\begin{aligned} |R_v| + |B_v| &< R(s-1,t) + R(s,t-1) \\ &< n \\ \Rightarrow |R_v| + |B_v| &\le n-2 \end{aligned}$$

which is not true. So either $|R_v| \ge R(s-1,t)$ or $|B_v| \ge R(s,t-1)$.

Case (i): If $|B_v| \ge R(s,t-1)$ and B_v inducing a complete 'red' colored subgraph K_s, then we are done. If B_v induces a complete 'blue' colored subgraph K_{t-1}, then $B_v \cup \{v\}$ must induce a blue colored K_t. Actually each vu' is blue for all $u' \in B_v$. So $B_v \cup \{v\}$ induces a 'blue' colored K_t if B_v has 'blue' colored K_{t-1} being induced by it.

Case (ii): The argument for R_v is identical to case (i).

So we have seen that a 2-colored complete graph on 'n' must contain a 'red' K_s or blue K_t and hence $R(s,t)$ $(< n)$ is bounded.

Remark 212 When $K_{\overline{n}}$ is a complete graph whose vertex set $\overline{n}$ is countably infinite, every two colored $K_{\overline{n}}$ must contain $K_{\overline{s}}$, a countably infinite monochromatic complete subgraph. This is known as the infinite case of Ramsey theorem for two colors.

21.4 Bounds on Ramsey Numbers

The bounds of Ramsey number are not easy to determine. However in many cases, the lower bound for some Ramsey number $R(s,t)$ is found using 2-coloring of K_n containing a monochromatic subgraph K_s or K_t in all possible coloring so $R(s,t) > n$. Some bounds are determined using some combinatorial method. The following theorem is already proved in Sect. 21.3.

Theorem 215 *$R(s,t) \le R(s-1,t) + R(s,t-1) for s,t \ge 2$. Also we have a direct upper bound on $R(s,t)$.*

Theorem 216 *$R(s,t) \le (s+t-2)C_{s-1}$ for $s,t \ge 2$.*

Proof The proof is given using double induction on s, t as the inequality involves both the integers s, t.

Base case will be for $r = 2$, $s = 2$.

So $R(2, 2) \leq 2$ is true.
Let the bound be true for $R(s-1, t)$ and $R(s, t-1)$.
We know from the previous theorem that

$$\begin{aligned} R(s,t) &\leq R(s-1,t) + R(s,t-1) \\ &\leq ((s-1)+t-2)C_{(s-1)-1} + (s+(t-1)-2)C_{s-1} \\ &= (s+t-3)C_{s-2} + (s+t-3)C_{s-1} \\ &= (s+t-2)C_{s-1} \quad \text{[using Pascal's rule:} \\ &\qquad nC_k + nC_{k+1} = (n+1)C_{(k+1)}] \end{aligned}$$

Hence the theorem.

Remark 213 When $s = t$, we have

$$\begin{aligned} R(s,s) &\leq (2s-2)C_{s-1} \\ &= 2(s-1)C_{s-1}. \end{aligned}$$

Regarding the lower bound, we have the following special case theorem where $s = t$.

Theorem 217 *For an integer n with* $nC_s < 2^{sC_2 - 1}$*, we have* $R(s, s) > 2^{(s/2)}$.

Proof To find $R(s, s)$ with $R(s, s) > n$, we consider a 2-coloring of K_n such that edges of K_n are colored randomly with red or blue. We need to show that in all possibility, the 2-colored K_n neither contains a blue colored K_s or red colored K_s.

Let S be the set of s vertices of K_n such that the corresponding K_s being its induced subgraph. Also if χ_S is the set of 2-colorings of K_n where the induced subgraph K_s is either completely blue or red. This means there are $2^{nC_2 - sC_2}$ ways to color the remaining $(nC_2 - sC_2)$ edges of K_n by χ_S. Hence

$$|\chi_S| = 2.2^{(nC_2 - sC_2)} \tag{21.1}$$

For every possible subset 'S' of vertices of K_n. Consider

$$\left| \bigcup_{S:|S|=s} \chi_S \right|.$$

This union is the set all possible s-subset vertices of K_n satisfying the monochromatic condition. So,

$$\begin{aligned}\left|\bigcup_{S:|S|=s} \chi_S\right| &\leq \sum_{S:|S|=s} |\chi_S| \\ &= \sum_{S:|S|=s} 2.2^{(nC_2 - sC_2)} \text{ (using (21.1))} \\ &= nC_s.2.2^{(nC_2 - sC_2)} \qquad (21.2)\end{aligned}$$

Actually the number of all possible 2-coloring of K_n is $2^{(nC_2)}$.
But by the condition given in the theorem,

$$\begin{aligned} nC_s &< 2^{sC_2 - 1} \\ \Rightarrow nC_s.2.2^{(nC_2 - sC_2)} &< 2^{sC_2}.2^{(nC_2 - sC_2)} \\ &= 2^{nC_2}\end{aligned}$$

This means for the choice of 'n' there exists a 2-coloring of K_n containing neither a 'blue' colored K_s or red colored K_s. Hence $R(s, s) > n$.

We have seen in Table 21.1 some known Ramsey numbers. The table also contains some '?' marks. For some of these entries we know from the literature the lower and upper bounds of Ramsey numbers. We list some Ramsey number ranges in the Table 21.2.

Exercises

1. Let $r_1, r_2, r_3, \ldots$ be a sequence of distinct real numbers. Prove that it contains a monotonic subsequence.
2. Show that in a group of n students, $n > 2$, there exists atleast a set of 2 people who have the same number of friends.

Table 21.2 Ramsey number ranges

s	t	R(s, t)
3	10	[40, 42]
3	11	[46, 51]
4	6	[35, 41]
4	7	[49, 61]
5	5	[43, 48]
5	6	[58, 87]
6	6	[102, 165]
6	7	[113, 298]
7	7	[205, 540]

3. Consider all points (x, y) on a x-y plane such that both x, y are integers. By assigning colors red or blue to these points, show that there exists a rectangle all of whose corners have the same color.
4. Show that for any $r \geq 2$, there exists an integer $n \geq 3$ such that for any r-coloring of the set $\{1, 2, \ldots, n\}$, there always exists three numbers n_1, n_2, n_3 having the same color such that $n_1 + n_2 = n_3$.
5. Find (i) $R(4, 4)$ and (ii) $R(5, 5)$.
6. Show that $R(s, t) = R(t, s)$.
7. Find a tight upper bound for $R(3, s)$.

Spectral Properties of Graphs 22

Linear algebra is widely applied to several practical problems. Matrices, vector spaces, play vital role in the understanding of structural properties of graphs. Even today algebraic method to solve graph theory problem is a highly interesting topic. This area of graph theory is fitted under the title 'special graph theory'. We will see some concepts and techniques of relating matrices to graphs. The reader is expected to have only very basic knowledge of linear algebra to understand the approach.

22.1 Basics of Spectral Graph Theory

Let $G = (V, E)$ be a graph such that $V = \{v_1, v_2, \ldots, v_n\}$ and $E = \{v_i v_j / 1 \leq i, j \leq n\}$.

The definitions of adjacency matrix A_G and incidence matrix I_G of G have been defined in Chap. 2.

The matrix A_G is always symmetric and trace of A_G is zero. The matrix A_G is formed for an arbitrary labelling/arrangement of vertices in V. So the primary interest will be to see for those properties of A_G that are invariant under any permutation of vertices.

The entries of A_G are 0 and 1. In case we apply the linear algebra theory, the field of reference will be $\{0, 1\}$.

Suppose $G = (V, E)$ is a graph and the vertices in V are ordered. Let A_G be the corresponding adjacency matrix. Let D be a $\{0, 1\}$ matrix that represents the reordering of vertices or permutation of vertices in G. Then the new adjacency matrix $\overline{A}_G$ (say) can be obtained from this reordering and it will be DA_GD^{-1}. Because of this relation, we can ignore the mention on ordering of vertices and simply consider the adjacency matrix for discussions.

R. Rama, *Topics in Combinatorics and Graph Theory*,
https://doi.org/10.1007/978-3-031-74252-1_22

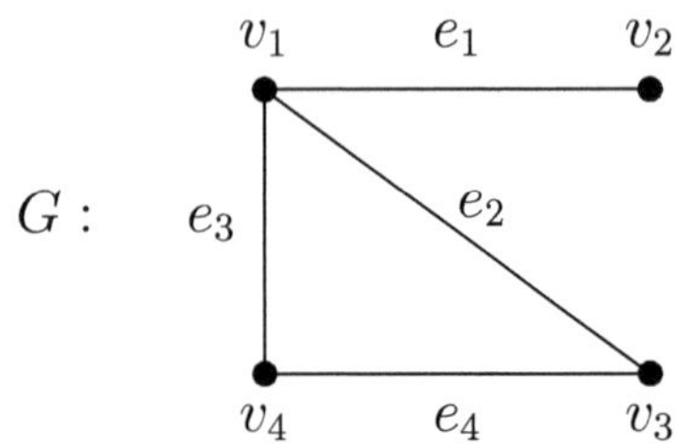

Fig. 22.1 Graph (1)

Example 22.1 Consider the graph (Fig. 22.1).

$$A_G = \begin{array}{c|cccc} & v_1 & v_2 & v_3 & v_4 \\ \hline v_1 & 0 & 1 & 1 & 1 \\ v_2 & 1 & 0 & 0 & 0 \\ v_3 & 1 & 0 & 0 & 1 \\ v_4 & 1 & 0 & 1 & 0 \end{array} \qquad I_G = \begin{array}{c|cccc} & v_1 & v_2 & v_3 & v_4 \\ \hline e_1 & 1 & 1 & 0 & 0 \\ e_2 & 1 & 0 & 1 & 0 \\ e_3 & 1 & 0 & 0 & 1 \\ e_4 & 0 & 0 & 1 & 1 \end{array}$$

Clearly A_G is a symmetric matrix. All the entries on the diagonal are '0'.

Proposition 22.2 *Let $G = (V, E)$ be a graph. Let A_G be its adjacency matrix. Then $(A_G^k)_{ij}$ gives the number of k-length paths between vertices $v_i v_j$.* □

Proof When $k = 0, 1$, the result is true. We use induction technique to complete the proof. Assume the result to be true for $(A_G^k)_{ij}$.
Consider

$$\begin{aligned} (A_G^{k+1})_{ij} &= A_G^k \cdot A_G \\ &= \sum_{t=1}^{n} (A_G^k)_{it} e_{tj} \end{aligned} \tag{22.1}$$

Here $e_{tj} = 1$ if there is an edge between v_t and v_j, otherwise $e_{tj} = 0$. $(A_G^k)_{it}$ gives the number of k-length paths beween v_i and v_t. Hence (22.1) enumerates all paths of length $(k + 1)$. □

Corollary 22.3 *G is a connected graph if and only if $(A + I)^{m-1}$ has no zero entry on all its non-diagonal positions. Here m denotes the number of vertices of G.*

Suppose λ is an eigen value of the adjacency matrix A_G, λ will be real because A_G is symmetric. Also the multiplicity of λ will be the dimension of the space of eigen vectors of λ.

The eigen values are determined from the equation $|A - \lambda I| = 0$ or $det(A - \lambda I) = 0$ where $|A - \lambda I|$ or $det(A - \lambda I)$ denotes the determinant of $A - \lambda I$.

Definition 22.4 The spectrum of a graph G is the set of eigenvalues of A_G including their multiplicities. That is if the distinct eigen values of A_G are $\lambda_1 > \lambda_2 > \cdots > \lambda_t$ and $m_i(\lambda_i)$ denote the multiplicity of λ_ℓ, $1 \leq i \leq t$. Then spectrum of G is denoted as $Spec(A_G)$ and is

$$Spec(A_G) = \begin{pmatrix} \lambda_1 & \lambda_2 & \lambda_3 & \cdots & \lambda_t \\ m_1(\lambda_1) & m_2(\lambda_2) & m_3(\lambda_3) & \cdots & m_t(\lambda_t) \end{pmatrix}$$

Example 22.5 Let

$$A_G = \begin{pmatrix} 0 & 1 & 1 & 1 \\ 1 & 0 & 1 & 0 \\ 1 & 1 & 0 & 1 \\ 1 & 0 & 1 & 0 \end{pmatrix}$$

$$Spec(A_G) = \begin{pmatrix} -\frac{\sqrt{17}+1}{2} & -1 & 0 & \frac{\sqrt{17}+1}{2} \\ 1 & 1 & 1 & 1 \end{pmatrix}$$

See Fig. 22.2.

Definition 22.6 The characteristic polynomial of a graph $G = (V, E)$ is nothing but the characteristic polynomial of its adjacency matrix A_G with some labeling of its vertices in V. This polynomial is denoted by $\chi(G : \lambda)$.
So,

$$\chi(G : \lambda) = |A - \lambda I| \text{ or } det(A - \lambda I)$$
$$= \sum_{i=0}^{n} (-1)^i c_i \lambda^{n-i}$$

Fig. 22.2 Graphs (2)

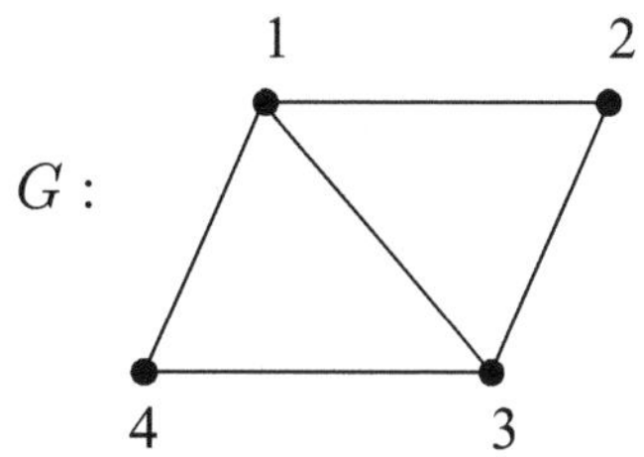

(i) The roots of the characteristic polynomial are the eigen values of the graph G.
(ii)

$$\chi(G:\lambda) = |A - \lambda I| = |P(A - \lambda I)P^{-1}|$$
$$= |PAP^{-1} - \lambda I|$$

for some permutation matrix P.

So, the characteristic polynomial $\chi(G:\lambda)$ is independent of the labelling of vertices. If we expand $\chi(G:\lambda)$,

$$\chi(G:\lambda) = c_0\lambda^n - c_1\lambda^{n-1} + c_2\lambda^{n-2} + \cdots + (-1)^n c_n$$

Here c_1 is the coefficient representing sum of zeros of the polynomial, which is sum of its eigen values. This is also the trace of the matrix A_G. So $c_1 = 0$. From the theory of matrices, we know that all the coefficients can be expressed in terms of principle minors of the adjacency matrix A_G. Hence we have the following theorem.

Theorem 22.7 *For a graph $G = (V, E)$, the coefficients of $\chi(G:\lambda)$ satisfy the following.*

(i) $c_1 = 0$.
(ii) $-c_2$ *is the number of edges in* G.
(iii) $-c_3$ *is twice the number of three cycles in* G.

Let $G = (V, E)$ be a graph and A_G be its adjacency matrix. Polynomial that are formed using $A_G^0, A_G^1, A_G^2, \ldots, A_G^n$ with coefficients either from $\mathbb{R}$ or $\mathbb{C}$.

Definition 22.8 The adjacency algebra $\mathscr{A}_G$ of a graph G comprises of polynomial in powers of A_G with coefficients either from $\mathbb{R}$ or $\mathbb{C}$.

Clearly every element of $\mathscr{A}_G$ is a linear combination of $A_G^0, A_G^1, \ldots, A_G^n$. Earlier we had a theorem for the count of walks from a vertex v_i to a vertex v_j in a graph G. We have the next proposition, which mentions about the dimension of adjacency algebra of a connected graph G.

Proposition 22.9 *Let $G = (V, E)$ be a connected graph and let $\mathscr{A}_G$ be its adjacency algebra and 'd' be its diameter. Then dimension of $\mathscr{A}_G$ is atleast $d + 1$.*

Proof Let u and v be two vertices such that $d(u, v) = d$ and suppose

$$u = u_1, u_2, \ldots, u_{d+1} = v$$

is a walk of length d. Consider the vertex u_i and the walk $u = u_1, u_2, \ldots, u_i$ is of length $i-1$ and there is no shorter walk from $u = u_1$ to u_i. This means A_G^i will have a non zero entry at this position and $A_G^{i-1}, A_G^{i-2}, \ldots, A_G, I$ will have zero at this position. This means A_G^i is not linearly dependent on $A_G^{i-1}, A_G^{i-2}, \ldots, A_G, I$. This argument is true for each i, $1 \leq i \leq d$. This means $\{I, A_G, A_G^2, \ldots, A_G^d\}$ are linearly independent in $\mathscr{A}_G$. So the dimension of $\mathscr{A}_G$ is $d+1$. □

Corollary 22.10 *Let G be a connected graph with diameter 'd'. Then A_G will have atleast $(d+1)$ distinct eigen values.*

Let us see the spectrum of some special graphs.

Theorem 22.11 *Let $G = (V, E)$ be a k-regular graph with $|V| = n$. Then every eigen value λ of A_G is such that $|\lambda| \leq k$.*

Proof Let λ be an eigen value of A_G and x be the corresponding eigen vector. Then $A_G x = \lambda x$, $x \neq 0$. Let $x = (x_1, x_2, \ldots, x_n)^T$ and let x_i be the largest positive entry in the eigen vector x. Let $u_1, u_2, \ldots, u_k$ be the vertices adjacent to the vertex v_i in G. Consider the inner product of ith row of A_G with the vector x. We get

$$\begin{aligned} x_{j_1} + x_{j_2} + \cdots + x_{j_k} &= \lambda x_i \\ \Rightarrow |\lambda||x_i| &= |\lambda x_i| \\ &= |x_{j_1} + x_{j_2} + \cdots + x_{j_k}| \\ &\leq k|x_i| \\ \Rightarrow |\lambda| &\leq k. \end{aligned}$$

□

Corollary 22.12 *Let $G = (V, E)$ be any graph. Then $|\lambda| \leq \Delta(G)$ where λ is an eigen value of A_G.*

Definition 22.13 A linear subgraph of a graph G is a subgraph of G whose components are single edges (K_2) or cycles ($C_n, n \geq 3$).

We state the following theorem without proof. This theorem evaluates the value of the determinant of the adjacency matrix using linear subgraphs.

Theorem 22.14 *Let $G = (V, E)$ be a graph. For the adjacency matrix A_G,*

$$det(A_G) = \sum (-1)^{e(H)} 2^{c(H)}$$

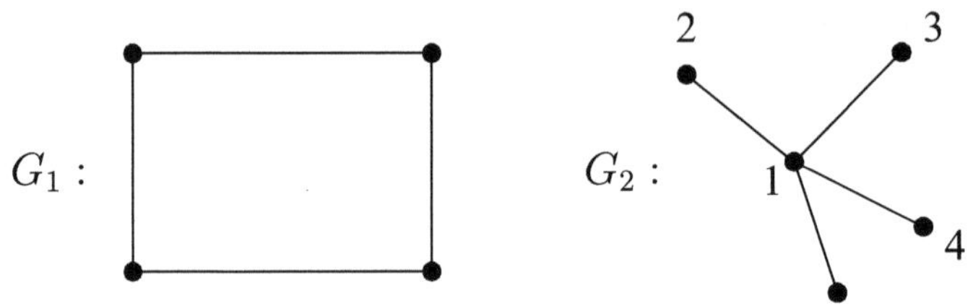

Fig. 22.3 Graphs (3)

where the summation is taken over all linear subgraphs H of G. Here $e(H)$ is the number of even components and $c(H)$ the number of cycles of H.

Example 22.15 Consider the following two graphs (Fig. 22.3).

$$A_{G_1} = \begin{bmatrix} 0\,1\,0\,1\,0 \\ 1\,0\,1\,0\,0 \\ 0\,1\,0\,1\,0 \\ 1\,0\,1\,0\,0 \\ 0\,0\,0\,0\,0 \end{bmatrix}$$

$$A_{G_2} = \begin{bmatrix} 0\,1\,1\,1\,1 \\ 1\,0\,0\,0\,0 \\ 1\,0\,0\,0\,0 \\ 1\,0\,0\,0\,0 \\ 1\,0\,0\,0\,0 \end{bmatrix}$$

Characteristic polynomial of A_{G_1} and A_{G_2} will be the same and it is $\lambda^5 - 4\lambda^3$. Hence G_1 and G_2 have same eigen values.

From the above example one can see that the spectrum does not characterize the structure of a graph. We can use the concept only for understanding other invariant parameters of a graph.

Definition 22.16 Two graphs that have the same characteristic polynomial are said to be cospectral graphs.

Finding a family of graphs having the same characteristic polynomial is in general a very hard problem.

22.2 Laplacian Matrix of Graphs

We will define another type of matrix associated with a graph G known as Laplacian matrix $L(G)$. This matrix is directly related to the already defined adjacency matrix A_G and the degrees of the vertices of the graph G.

Definition 22.17 Let $G = (V, E)$ be a simple graph. Consider the matrix L for G and it will be a matrix such that

$$L_{ij} = \begin{cases} d(v_i) & \text{if } v_i = v_j \\ -1 & \text{if } v_i v_j \in E \\ 0 & \text{otherwise} \end{cases}$$

for $1 \le i, j \le n$, $V = \{v_1, v_2, \dots, v_n\}$.

Consider the diagonal matrix D

$$D_{ij} = \begin{cases} d(v_i) & \text{if } v_i = v_j \\ 0 & \text{if } v_i \neq v_j \end{cases}$$

for $1 \le i, j \le n$.

The Laplacian matrix L defined above can be seen as

$$L = D - A_G.$$

Example 22.18 1. Consider the graph G (Fig. 22.4).

Its Laplacian is given below:

$$L = \begin{array}{c} \\ 1 \\ 2 \\ 3 \\ 4 \\ 5 \end{array} \begin{array}{c} \begin{array}{ccccc} 1 & 2 & 3 & 4 & 5 \end{array} \\ \begin{bmatrix} 2 & -1 & 0 & 0 & -1 \\ -1 & 4 & -1 & -1 & -1 \\ 0 & -1 & 2 & -1 & 0 \\ 0 & -1 & -1 & 3 & -1 \\ -1 & -1 & 0 & -1 & 3 \end{bmatrix} \end{array}$$

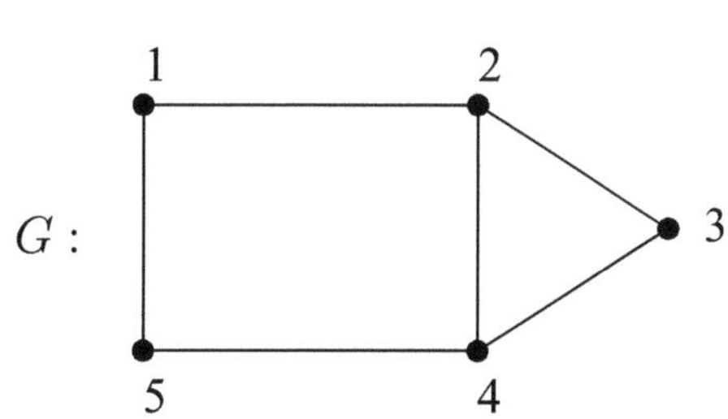

Fig. 22.4 Graphs (4-1)

Fig. 22.5 P_5

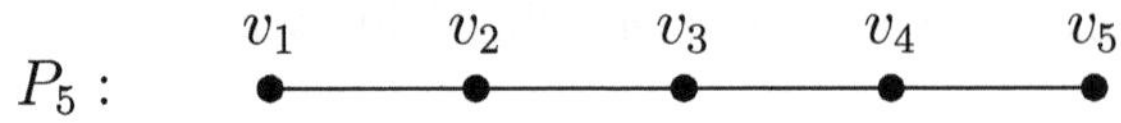

2. Consider the path graph (Fig. 22.5).

$$L(P_5) = \begin{array}{c} \\ v_1 \\ v_2 \\ v_3 \\ v_4 \\ v_5 \end{array} \begin{array}{c} \begin{array}{ccccc} v_1 & v_2 & v_3 & v_4 & v_5 \end{array} \\ \begin{bmatrix} 1 & -1 & 0 & 0 & 0 \\ -1 & 2 & -1 & 0 & 0 \\ 0 & -1 & 2 & -1 & 0 \\ 0 & 0 & -1 & 2 & -1 \\ 0 & 0 & 0 & -1 & 1 \end{bmatrix} \end{array}$$

3. Consider the complete graph K_4 (Fig. 22.6).

$$L(K_4) = \begin{bmatrix} 3 & -1 & -1 & -1 \\ -1 & 3 & -1 & -1 \\ -1 & -1 & 3 & -1 \\ -1 & -1 & -1 & 3 \end{bmatrix}$$

4. Consider the cycle C_5 (Fig. 22.7).:

$$L(C_5) = \begin{bmatrix} 2 & -1 & 0 & 0 & -1 \\ -1 & 2 & -1 & 0 & 0 \\ 0 & -1 & 2 & -1 & 0 \\ 0 & 0 & -1 & 2 & -1 \\ -1 & 0 & 0 & -1 & 2 \end{bmatrix}$$

In all the above examples one can see that all the Laplacian matrices are real symmetric. We will now see some basic properties. Since the eigen values of real symmetric matrix are real, any Laplacian matrix will have all its eigen values real.

Fig. 22.6 K_4

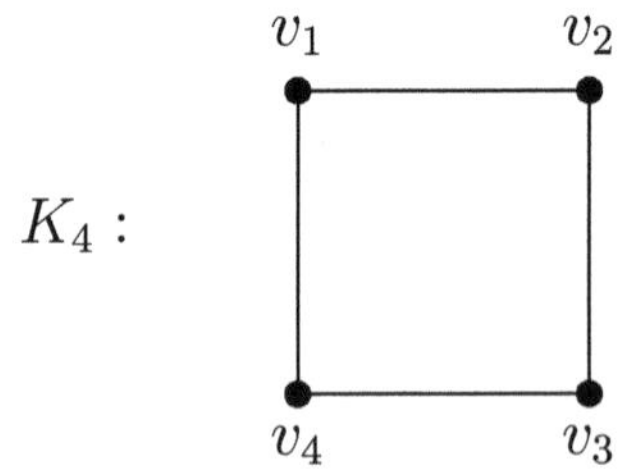

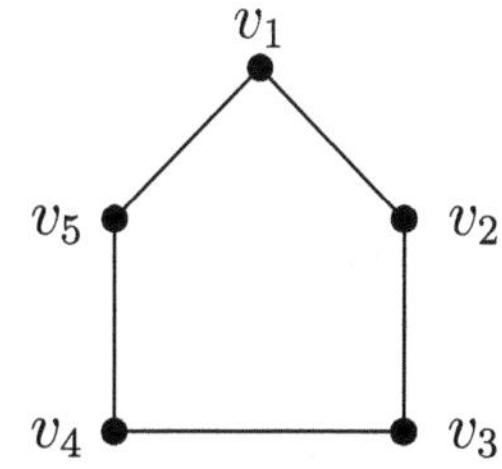

Fig. 22.7 C_5

Definition 22.19 Let M be a $n \times n$ real matrix. M is called positive semi definite if

$$\overline{x}^T M\overline{x} \geq 0 \text{ for all } \overline{x} \in \mathbb{R}^n.$$

Now for a Laplacian matrix $L(G)$ of a graph, $\overline{x}^T L(G)\overline{x} \geq 0$. For any edge $e = uv \in E$ of $G = (V, E)$, we have,

$$\overline{x}^T L_e\overline{x} = (\overline{x}_u - \overline{x}_v)^2, \ \forall \, \overline{x} \in \mathbb{R}^n$$

where

$$L_e = \begin{cases} 1 & \text{if } i = j, i = u, v \\ -1 & \text{if } i = u, j = v \\ 0 & \text{otherwise} \end{cases}$$

Since

$$\overline{x}^T L(G)\overline{x} = \sum_{e=uv\in E} \overline{x}^T L_e\overline{x} \geq 0, \text{ we have.}$$

Theorem 22.20 *Let $G = (V, E)$ be any graph. Then the Laplacian $L(G)$ is symmetric positive semi definite.*

Lemma 22.21 *The Laplacian of the complete graph K_n has the eigen value '0' and eigen value 'n' occurring $(n-1)$-times.*

Proof For any graph G, $\lambda = 0$ is an eigen value of $L(G)$. Consider $\overline{x} = (1, 1, \ldots, 1) \in \mathbb{R}^n$ and $L(G)\overline{x}$. $L(G)$ is a matrix with row sum zero and column sum zero. Hence $L(G)\overline{x}$ will be a zero matrix. So '0' is an eigen value of $L(K_n)$.

The Laplacian matrix $L(K_n)$ has entries

$$a_{ij} = \begin{cases} -1 & \text{if } i \neq j \\ n-1 & \text{if } i = j. \end{cases}$$

$L(K_n)$ has no entries with '0'.

Consider the matrix

$$L(K_n) - nI = M.$$

This M matrix has all its entries as '−1', M is not invertible and has rank 1. So 'n' must be an eigen value of $L(K_n)$ and nullity of M must be $(n-1)$. So the eigen value 'n' has multiplicity $(n-1)$. □

Lemma 22.22 *Let $G = (V, E)$ be a graph and let $\lambda_1, \lambda_2, \ldots, \lambda_n$ be the eigen values of the Laplacian matrix $L(G)$ such that*

$$0 = \lambda_1 \leq \lambda_2 \leq \cdots \leq \lambda_n.$$

Then G is connected if and only if $\lambda_2 \neq 0$.

Proof Suppose G is disconnected. Then there are atleast two components in G and G is the union of such components. Let G_1, G_2 be the components of G. Then the Laplacian of G can be put in the following term.

$$L(G) = \begin{bmatrix} L(G_1) & 0 \\ 0 & L(G_2) \end{bmatrix}$$

So, the orthogonal eigen vectors of $\lambda_1 = 0$ are $(1, 0)^T$ and $(0, 1)^T$ which are linearly independent. Hence the multiplicity of $\lambda = 0 > 1$.

Conversely let $\overline{x}$ be an eigen vector of $\lambda_1 = 0$ of the matrix $L(G)$ and, let G be connected. Then, $L(G)\overline{x} = 0$.

$$\Rightarrow \overline{x}^T L(G)\overline{x} = \sum_{e=uv\in E} (\overline{x}_u - \overline{x}_v)^2 = 0.$$

This means $\overline{x}_u = \overline{x}_v$ for $e = uv \in E$. As G is connected, every pair of vertices u, v are connected by a path, $\overline{x}_u = \overline{x}_v$ is true for any two vertices u, v. This means $\overline{x}$ is a constant vector. So the eigen space of the eigen value $\lambda_1 = 0$ is 1. □

The eigen values of star graphs and path graphs are as below.

(i) The star graph S_n on n-vertices has eigen value 0 with multiplicity 1, eigen value 1 with multiplicity $(n-2)$ and eigen value n with multiplicity 1.
(ii) The Laplacian matrix of C_n has eigen values

$$2 - 2\cos\left(\frac{2\pi k}{n}\right)$$

The eigen values of $L(P_n)$ are same as the eigen values of $L(C_{2n})$.

We know what a spanning tree of a graph is. We also saw the definition of a Laplacian matrix of a graph. We now state matrix-tree theorem which gives a nice connection between graph theory and matrix theory. The matrix-tree theorem is an important theorem in many ways. The theorem simply modifies the given Laplacian matrix L of a graph G by eliminating a row i and column i from L to form a new matrix $\hat{L}_i$ (say). The number of spanning trees of G is simply the value of $det(\hat{L}_i)$. Here the vertices of G are assumed to be distinguishable and hence $det(\hat{L}_i)$ counts some isomorphic spanning trees also.

Example 22.23 Consider the graph (Fig. 22.8).

$$L = \begin{bmatrix} 2 & -1 & -1 & 0 \\ -1 & 3 & -1 & -1 \\ -1 & -1 & 2 & 0 \\ 0 & -1 & 0 & 1 \end{bmatrix}$$

The spanning trees of G are Fig. 22.9.

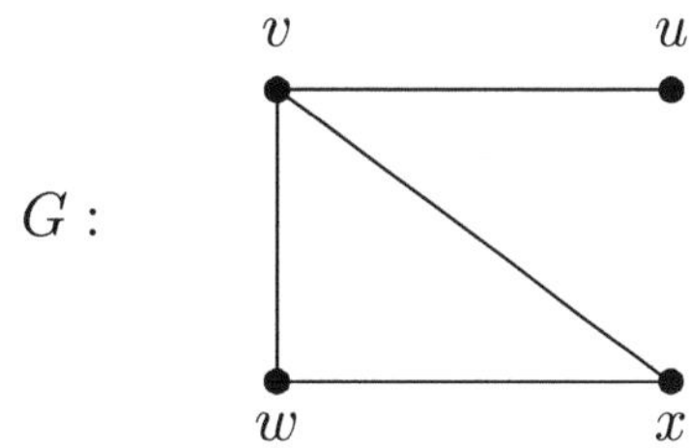

Fig. 22.8 Graph (5)

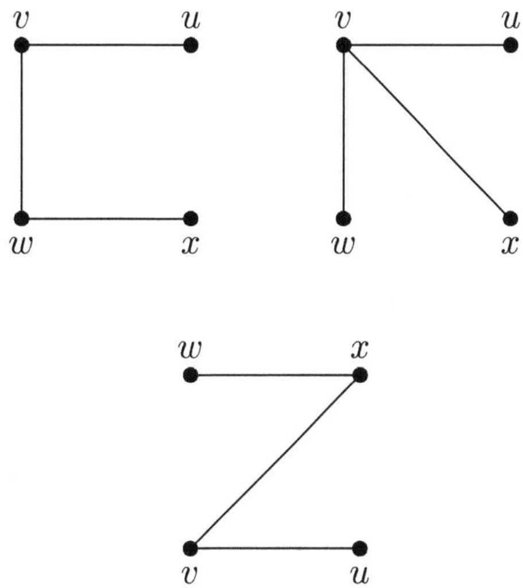

Fig. 22.9 Spanning trees

Now let

$$\hat{L}_1 = \begin{bmatrix} 3 & -1 & -1 \\ -1 & 2 & 0 \\ -1 & 0 & 1 \end{bmatrix}$$

Hence the count on the number of spanning trees of G is 3.

Remark 22.24 The calculation of $det(\hat{L}_j)$ that gives the number of spanning trees is independent of the choice of 'j'. i.e., any vertex of G occurring in L can be choosen.

Example 22.25 Consider K_4 and

$$L(K_4) = \begin{bmatrix} 3 & -1 & -1 & -1 \\ -1 & 3 & -1 & -1 \\ -1 & -1 & 3 & -1 \\ -1 & -1 & -1 & 3 \end{bmatrix}$$

Number of spanning trees of K_4 are 16.

$$\begin{aligned} det(\hat{L}_3) &= \begin{bmatrix} 3 & -1 & -1 \\ -1 & 3 & -1 \\ -1 & -1 & 3 \end{bmatrix} \\ &= 16. \end{aligned}$$

We state the Matrix-Tree Theorem without proving it. The proof can be seen in D.B. West.

Theorem 22.26 *Let $G = (V, E)$ be a graph and let L be its Laplacian. The number of spanning trees n_T contained in G is given by the following method.*

Choose a vertex v_i and eliminate the ith row, ith column from L and the determinant value of the resultant matrix will be n_T.

22.3 Eigen Values and Eigen Vectors of Graphs

We have seen two types of matrices associated with any simple undirected graphs. They are adjacency matrix and Laplacian matrix. We had a brief and basic ideas discussed in the previous two sections. In this section we can see more concepts on eigen values and eigen vectors of these two matrices.

Note that both adjacency matrix and Laplacian matrix are real and symmetric. In these matrices, there will be an orthoganal basis comprising of the eigen vectors. Note that the

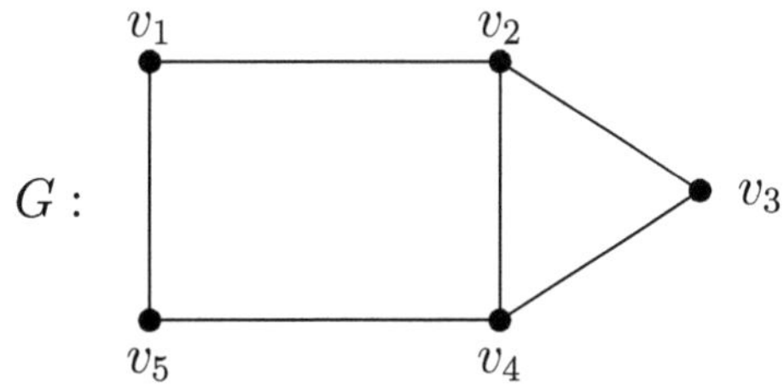

Fig. 22.10 Graph (7)

eigen values are counted along with their multiplicities. Both A and L of G are $n \times n$ matrices where n is the number of vertices of the graph G. For further discussion and understanding let us use the following notation.

Let $\lambda_1 \leq \lambda_2 \leq \cdots \leq \lambda_n$ be the arrangement of eigen values of the adjacency matrix and let $\underline{\lambda}_{n_1} \leq \underline{\lambda}_{n-1} \leq \cdots \leq \underline{\lambda}_1$ be the arrangement of eigen values of $L(G)$. Also note that $A + L = dI$ where

$$D = \begin{cases} d & d = deg(v_i) \text{ if } i = j \\ 0 & \text{otherwise.} \end{cases}$$

Hence we see that $\lambda_k \underline{\lambda}_k = d$ for regular graphs.

Example 22.27 Consider the graph (Fig. 22.10).

$$A_G = \begin{bmatrix} 0\,1\,0\,0\,1 \\ 1\,0\,1\,1\,0 \\ 0\,1\,0\,1\,0 \\ 0\,1\,1\,0\,1 \\ 1\,0\,0\,1\,0 \end{bmatrix}$$

$$L(G) = \begin{bmatrix} 2 & -1 & 0 & 0 & -1 \\ -1 & 3 & -1 & -1 & 0 \\ 0 & -1 & 2 & -1 & 0 \\ 0 & -1 & -1 & 3 & -1 \\ -1 & 0 & 0 & -1 & 2 \end{bmatrix}$$

$$A_G + L(G) = \begin{bmatrix} 2\,0\,0\,0\,0 \\ 0\,3\,0\,0\,0 \\ 0\,0\,2\,0\,0 \\ 0\,0\,0\,3\,0 \\ 0\,0\,0\,0\,2 \end{bmatrix}$$

Fig. 22.11 Graphs (8)

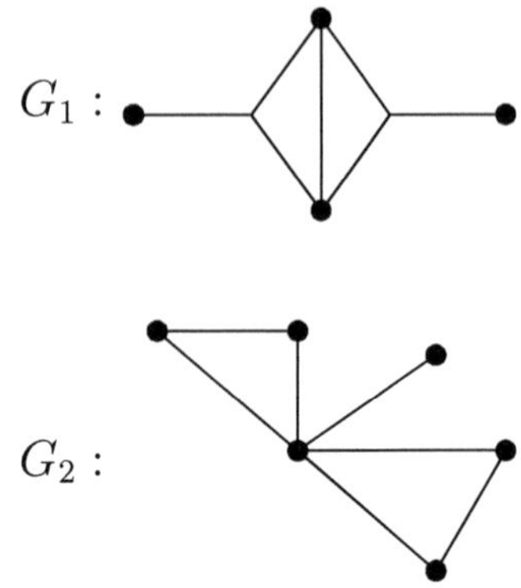

Eigen values are of A_G are

$$-2, -1.170, 0, 0.689, 2.481$$

Eigen values of $L(G)$ are

$$4.618, 3.618, 2.382, 1.382, 0.$$

Example 22.28 Consider the following graphs (Fig. 22.11).

Clearly G_1 and G_2 are not isomorphic. The characteric polynomial of G_1 is

$$\chi(G_1 : \lambda) = \lambda^6 - 7\lambda^4 - 4\lambda^3 + 7\lambda^2 + 4t - 1$$

One can check that the characteric polynomial of G_2 is also same as that of G_1. Hence G_1 and G_2 are cospectral graphs.

Exercises

1. Does there exist two trees that are cospectral? Justify your answer.
2. Find two cospectral regular connected graphs.
3. If G is a connected graph that has 4 distinct eigen values, then show that $diam(G) \leq 3$.
4. Find the spectrum of the following graphs (Fig. 22.12).
5. Consider the graph (Fig. 22.13).
   Find the characteristic polynomial of G. Is it same as the product of characteristic polynomials of the two components of G? Justify your answer.

Fig. 22.12 Graphs (4)

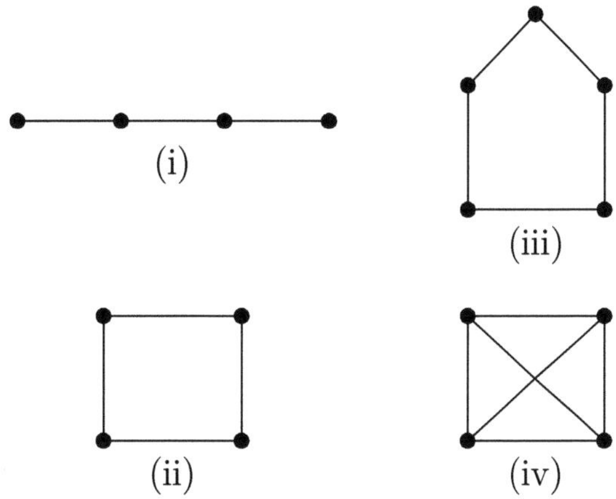

Fig. 22.13 Graphs (5)

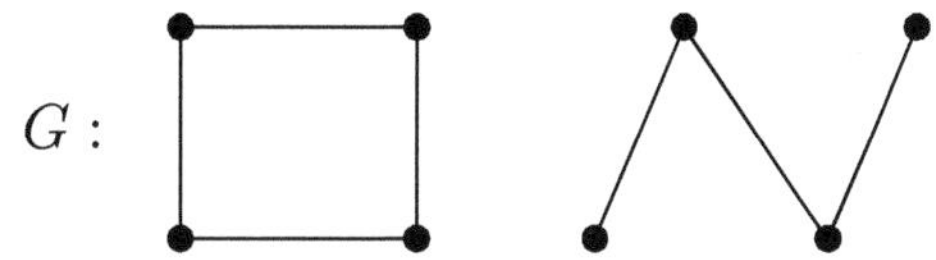

6. Let $G = (V, E)$ be a graph and let

$$ave(G) = \frac{\sum_{v \in V} deg(v)}{n}$$

where $|V| = n$. Then show that

$$ave(G) \leq \overline{\lambda}(G) \leq \Delta(G)$$

where $\overline{\lambda}(G)$ is the largest eigen value.

7. Find the largest eigen values of K_n and $K_{m,n}$.
8. Let $G = K_{m,n}$, a complete bipartite graph. Show that it has exactly one positive eigen value.

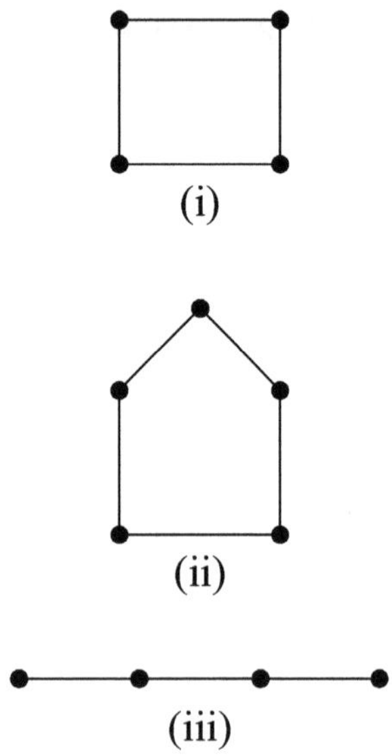

Fig. 22.14 Graphs (10)

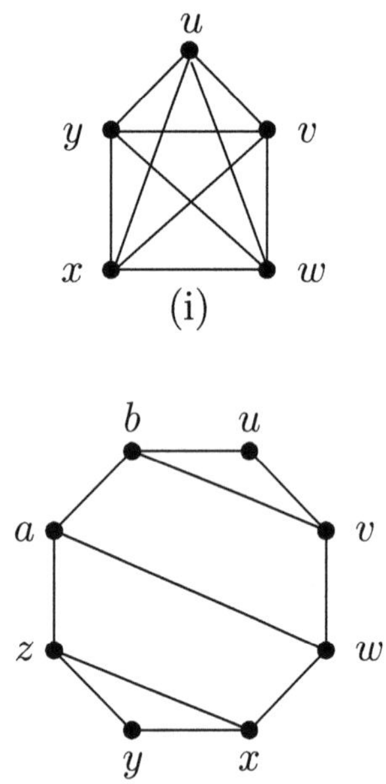

Fig. 22.15 Graphs (11)

9. Is the statement in question 8 true for any complete t-partite graph? Justify your answer.
10. Find the value of determinant of A_G of the following graphs (Fig. 22.14).
11. Find the number of spanning trees of the following graphs (Fig. 22.15).

Directed Graphs and Graph Algorithms 23

23.1 Directed Graphs

Definition 23.1 A directed graph or digraph D is a pair $(V, \vec{E})$ where V is a finite non empty set of vertices and $\vec{E} = \{e = (u, v) | u, v \in V, u \neq v\}$. Here $(u, v) \neq (v, u) \; \forall \, u, v \in V$. The elements of $\vec{E}$ are called as directed edges or arcs. For $(u, v) \in \vec{E}$, u will be called as the initial vertex or tail vertex and v will be called as terminal vertex or head vertex. The edge $e = (u, v)$ is understood as an arc from u to v.

Remark 23.2 A digraph can be drawn as a diagram with the vertices being represented as 'dots' and edges being represented as 'arrows' to indicate the head and tail vertices.

Example 23.3 Consider

$$D = (\{u, v, w, x\}, \{(u, v), (v, w), (w, x), (x, u), (u, x)\})$$

The digraph diagram will be as in Fig. 23.1.

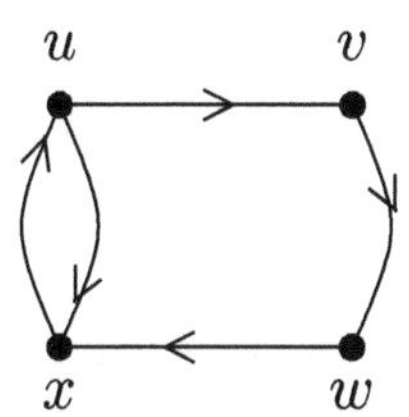

Fig. 23.1 Digraph

Note that we often refer the diagram of a digraph as the digraph itself. We have the following definitions of a digraph which are similar to those of graphs.

R. Rama, *Topics in Combinatorics and Graph Theory*,
https://doi.org/10.1007/978-3-031-74252-1_23

Definition 23.4 Let $D = (V, \vec{E})$ be a digraph.

(1) The indegree of a vertex $v \in V$ in D is the number of arcs that have vertex 'v' as their head or terminal vertex. It is denoted by $d^-(v)$. Outdegree of a vertex $v \in V$ denoted as $d^+(v)$ will be the number of arcs that have the vertex 'v' as their tail.
(2) A directed walk in a digraph D will be a sequence of alternating vertices and arcs $\vec{w} : v_0, e_1, v_1, e_2, \ldots, e_t, v_t$ such that $v_i \in V, 0 \le i \le t, e_j \in \vec{E}, 1 \le j \le t$. Note that the arc e_j will be an arc with v_{j-1} as its tail and v_j as its head, for $1 \le j \le t$. We can simply write $\vec{w}$ as a directed walk from v_0 to v_t. $\vec{w}$ will be a closed walk when $v_0 = v_t$.
(3) A directed path $\vec{P}$ is a directed walk with all its vertices are distinct. A directed cycle is a non trivial closed directed walk having head and tail being the same.
(4) A digraph in which $(u, v) \in \vec{E}$ for every $u, v \in V$ is said to be strongly connected. That is there is a directed walk between every pair of vertices. It is said to be weakly connected if the underlying graph is connected. If the underlying graph is disconnected, then the digraph is disconnected.
(5) A component of a digraph is said to be a strong component if it is a maximal strongly connected component (or subgraph) of D.
(6) A digraph D is said to be acyclic if it has no directed cycle.

Example 23.5 $w : v_1 e_1 v_2 e_5 v_4$ is a walk.
There is no walk from v_3 to v_1.
There is no closed walk in D.
The graph D is weakly connected.
D has no strong component (Fig. 23.2).

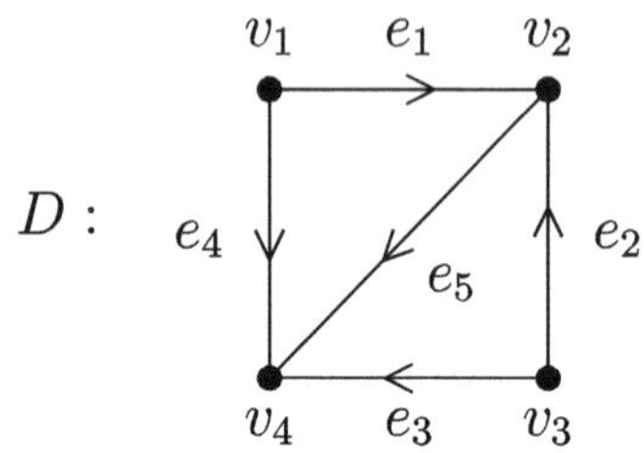

Fig. 23.2 Graphs (2)

We state a few basic results which are not difficult to prove.

Theorem 23.6 *1. For any digraph $D = (V, \vec{E})$*

$$\sum_{v \in V} d^+(v) = \sum_{v \in V} d^-(v)$$

2. *If the digraph D has a directed walk from a vertex v to a vertex u then D has a directed path from v to u.*
3. *Let* H_1 *and* H_2 *be two subgraphs of D that are strongly connected and* $V(H_1) \cap V(H_2) \neq \phi$. *Then* $H_1 \cup H_2$ *is strongly connected.*
4. *If in a digraph D every vertex v is such that* $d^+(v) \geq 1$, *then D contains a directed cycle.*
5. *The necessary and sufficient condition for a digraph D to be acyclic is that its vertex set can be ordered as* $v_1, v_2, \ldots, v_n$ *such that* $(v_i, v_j) \in \vec{E}$ *with* $i \ngtr j$.

23.2 Minimal Connector Problem

Suppose there are 'n' buildings to be connected for internet facility. The designer is also given the cost for every pair of buildings. The problem is to establish a network at a minimum cost irrespective of the other influencing parameters. For this we first define a weighted graph.

Definition 23.7 A graph $G = (V, E)$ is called a weighted graph if there is a real function $\phi : E \to \mathbb{R}$. The function ϕ is called the weight function or cost function.

So the connector problem can be seen to be a graph G where vertices represent buildings, and the edges represent the possible connections that can be made. The 'ϕ' function assigns the cost against each edge indicating the cost of construction. The problem is to construct a network of minimal cost. One can see that the selection will be a spanning tree of G whose cost is minimal.

Theorem 23.8 *Let* (G, ϕ) *be a network representation. The minimal cost network* T_ϕ *will be a minimal spanning tree of G whose total weight will be the least.*

The following is an algorithm which computes the minimal weight spanning tree.

Kruskal's Algorithm

Input: G and ϕ.

1. Select an edge $e_1 \in E$ such that $\phi(e_1)$ being the smallest.
   (If there two edges with least cost select any one).
2. Continue to select in sequence edges $e_2, e_3, \ldots, e_i$ all having costs with

$$\phi(e_1) \leq \phi(e_2) \leq \cdots \leq \phi(e_i).$$

   Then choose $\phi(e_{i+1})$ such that it is the least weight edge among leftover edges, such that $e_1, e_2, \ldots, e_{i+1}$ do not form a cycle.

3. Stop when step 2 cannot be done any more.

Remark 23.9 The correctness of the algorithm follows from the above theorem and the steps of the algorithm being direct computation.

Example 23.10 Consider the graph (Fig. 23.3)

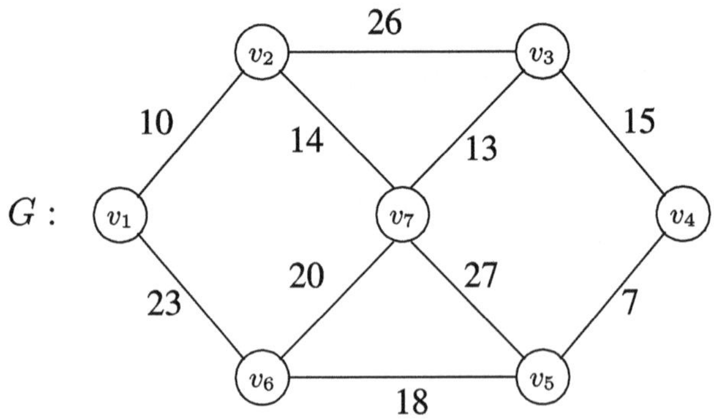

Fig. 23.3 Graphs (3)

First draw all the vertices without edges.

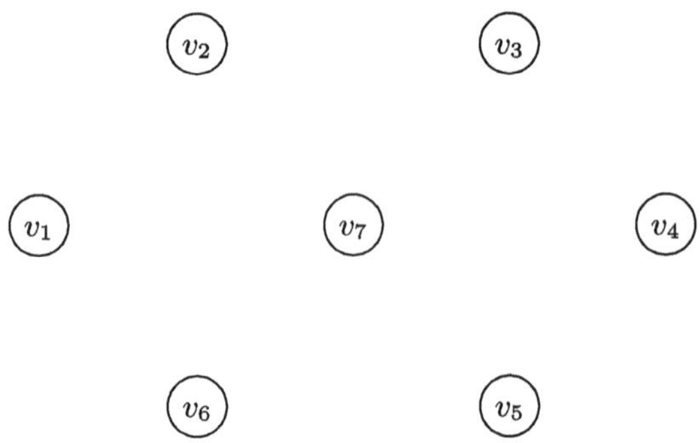

Fig. 23.4 Step 1

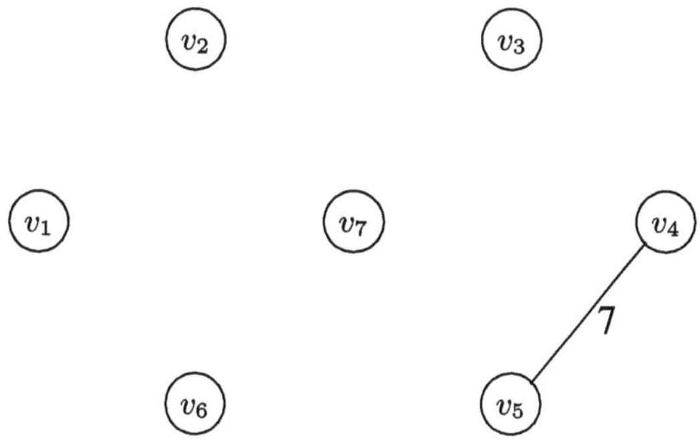

Fig. 23.5 Step 2

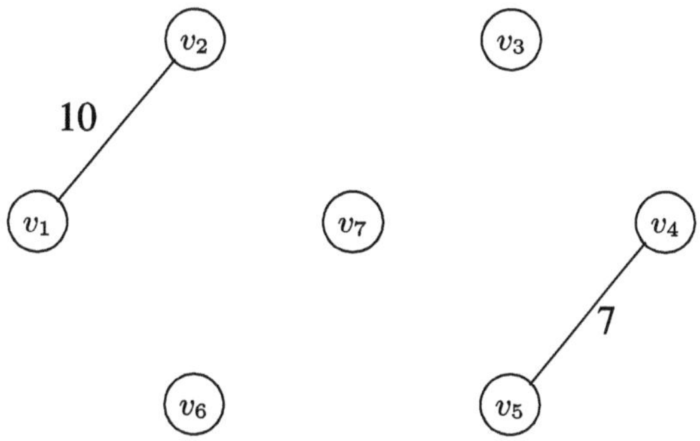

Fig. 23.6 Step 3

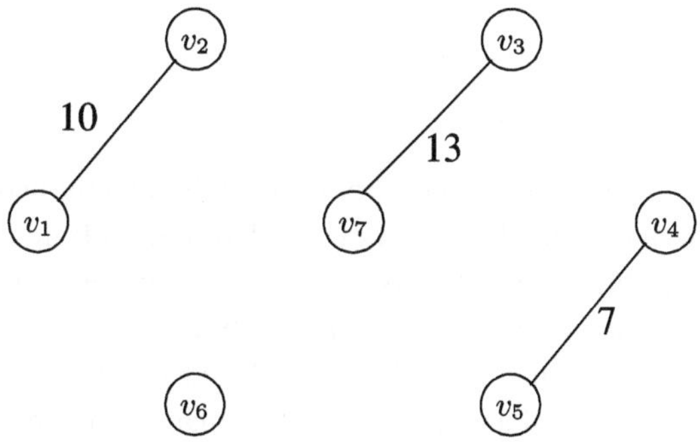

Fig. 23.7 Step 4

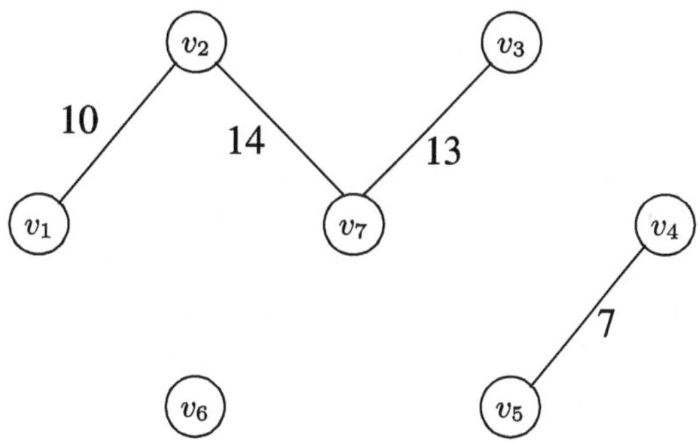

Fig. 23.8 Step 5

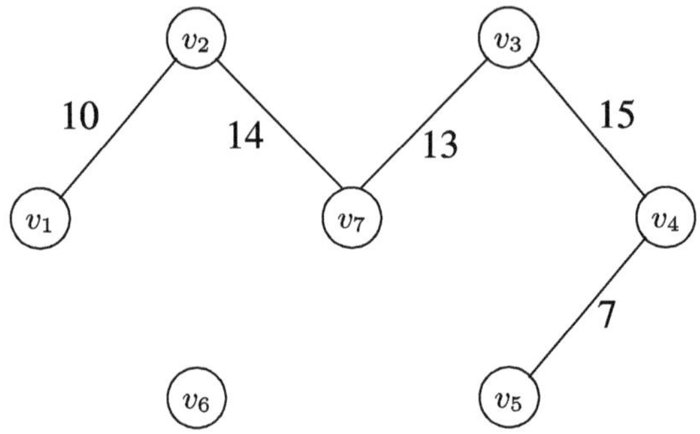

Fig. 23.9 Step 6

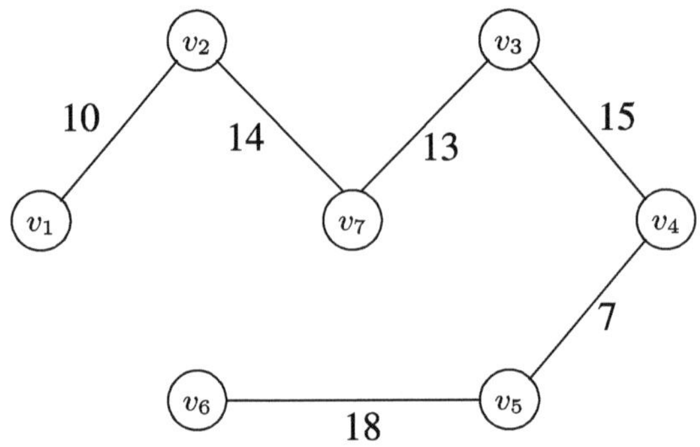

Fig. 23.10 Step 7

Total cost = 77

The minimal connector problem is actually finding a minimal spanning tree in a given weighted graph. The following algorithm by Prim finds a minimum spanning tree of a given weighted graph. This is also a greedy algorithm. To form a spanning tree having minimum weight, actually a tree T is constructed in steps by adding edges of least weight. Initial choice of an edge to T will be the edges or edge with the least weight. Initially T will be disconnected and have edges less than the number of vertices of the underlying weighted graph (G, ϕ). The algorithm is as below.

Prim's Algorithm

Input: (G, ϕ) with n vertices, m edges $n > 1$.
Output: T, a minimum weight spanning tree.
Step 1: $T \leftarrow \{u\}$. u, any vertex in G.
Step 2: Add to T those edges uv such that $\phi(uv)$ is minimum such that $u \in T, v \notin T$. If there are more than one edge with minimum weight, choose one arbitrarily.
Step 3: If T has $(n-1)$ edges, then stop else repeat Step 2 until adding edges to T, with minimum weight edges become impossible. If T has less than $(n-1)$ edges and adding edges become impossible means G is not a connected graph.

Example 23.11 Consider the following weighted graph (G, ϕ)

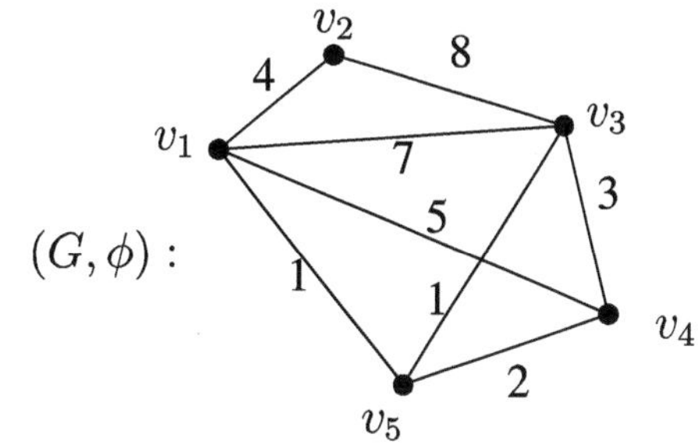

Fig. 23.11 Weighted graph

Initially T is a tree with $\{v_1\}$ say. Now edge v_1v_5 will be added to T as it is the edge with least weight which is incident with v_1. In the next iteration consider vertices in G other than v_1, v_5 which are incident either with v_1 or v_5 having the least weight edge. This iteration adds v_5v_3 to T. Now the vertices of T are $\{v_1, v_3, v_5\}$. In the next iteration all edges incident with v_1, v_3, v_5 that are not examined so far are considered. Now T is updated with vertices $\{v_1, v_3, v_4, v_5\}$ and the edge v_4v_5 is added.

At this stage the tree T looks as below

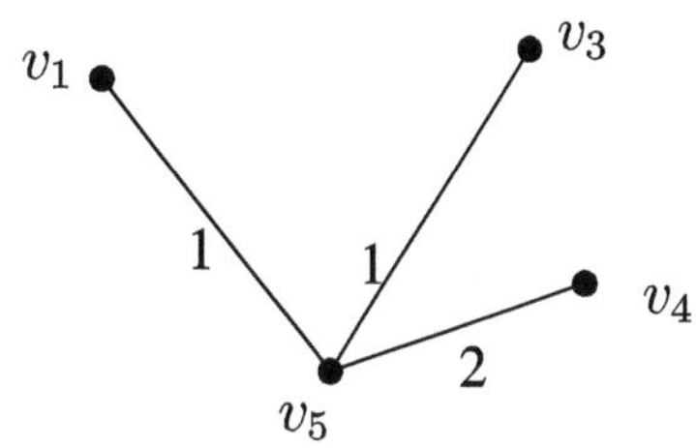

Fig. 23.12 Weighted tree

Finally the spanning tree with minimum weight will be

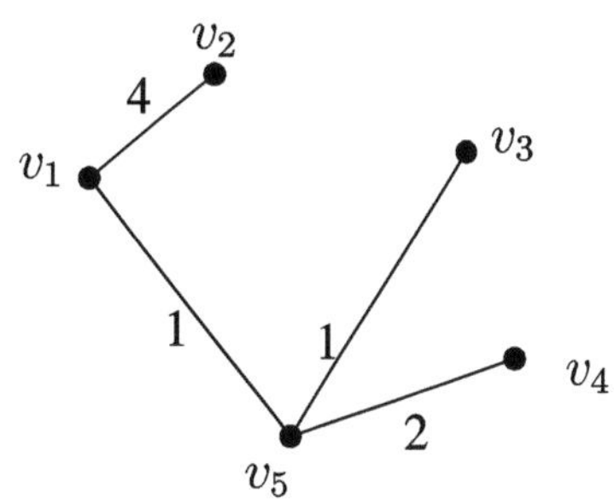

Fig. 23.13 Weighted spanning tree

Remark 23.12 If n is the number of vertices and m is the number of edges in (G, ϕ), then the time complexity of Prim's algorithm will be $O(m + n \log_2 n)$ steps.

23.3 Shortest Path Problem

Given a weighted digraph, the shortest path problem is the problem of finding a path between two vertices in the graph such that the sum of the weights on the edges of the path is minimum. This problem can be defined for both directed and undirected graphs. This is a model for describing layout of road map between n cities (say) and one wishes to find the shortest route between any two given cities. Here the distance between any two cities are described by weighted edges, if they have a direct connection. We give a simple algorithm called.

Dijkstra's Algorithm

Problem: Let G be the weighted digraph and let u_0, v_0 be any two vertices of G.
Input: $\vec{G} = (V, E, \phi)$, $u_0, v_0 \in V$.

Let $S_0 = \{u_0\}$. Let $u_1 \in V \backslash S_0$ such that the path P_0 from u_0 to u_1 is the shortest one. In the next step identity $u_2 \in V - \{u_0, u_1\}$ such that the path P_1 from u_0 to u_2 which includes P_1 as a concatenated path from u_0 to u_1 and an edge u_1u_2 such that the path P_2 from u_0 to u_2 being the shortest. Continue the above process to find a path from u_0 to v_0 which is shortest.

Note that the process computes the distance from u_0 in steps exploiting the minimum distance paths in sequence. We illustrate the algorithm with an example and then give the actual algorithm.

Example 23.13 Let $(\vec{G}, \phi)$ be the given weighted graph as in the following diagram.

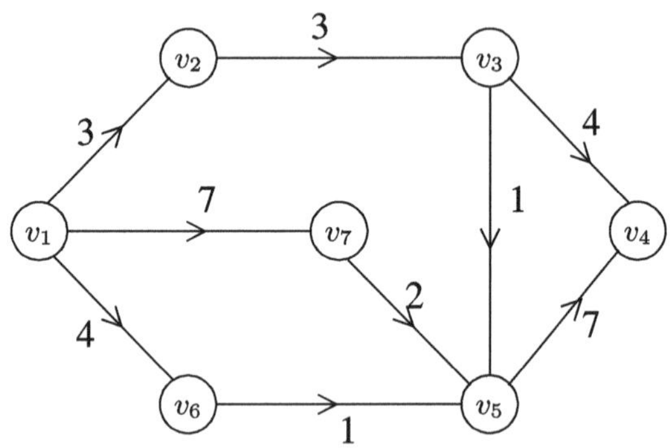

Fig. 23.14 Weighted graph (5)

Let us find the shortest path between v_1 and v_4.

In the first iteration, we get the source vertex $v_1 \in S$, where S is the set updated iteratively for identifying vertices on the path from v_1 to v_4 which is the shortest. So $v_1 \in S$ initially.

In the next iteration we see vertices v_2, v_7, v_6 reachable by an edge from v_1 directly. The weights of $\phi(v_1v_2) = 3$, $\phi(v_1v_7) = 7$, $\phi(v_1v_6) = 4$. Since v_1v_2 has the least weight. We add $v_2 \in S$. So $S = \{v_1, v_2\}$ now.

In the third iteration we compute the paths from v_1 to vertices v_3, v_6 and v_7 with their weights respectively. Hence

$$W(v_2v_3) = W(v_1, v_2) + \phi(v_2v_3) = 3 = 3 = 6$$

where $W(u, v)$ means the sum of weights of the edges of paths from u to v which is the shortest. Similarly

$$W(v_1, v_6) = 4 \text{ and } W(v_1, v_7) = 7$$

and so $v_6 \in S$.

Fourth iteration checks for

$$W(v_1, v_7), W(v_2, v_3), W(v_6v_5)$$

and adds $v_5 \in S$.

In the fifth iteration the weights

$$W(v_1, v_7), W(v_2v_5), W(v_5v_4)$$

are computed and their values are

$$\begin{aligned} W(v_1, v_7) &= 0 + 7 = 7 \\ W(v_2v_3) &= 3 + 3 = 6 \\ W(v_5, v_4) &= 5 + 7 = 12 \end{aligned}$$

Hence add $v_3 \in S$.

Sixth iteration finds v_7 to be put in S because

$$\begin{aligned} W(v_1, v_7) &= 0 + 7 = 7 \\ W(v_5v_4) &= 5 + 7 = 12 \\ W(v_3v_4) &= 6 + 4 = 10 \end{aligned}$$

Finally we compute

$$W(v_5v_4) \text{ and } W(v_3v_4).$$

They are 12 and 10 respectively. Hence we have $v_4 \in S$. Now

$$S = \{v_1, v_2, v_6, v_5, v_3, v_7, v_4\}.$$

The shortest path $P_1 : v_1v_2v_3v_4$ which can be identified by backtracking the path from v_4 to v_3, v_3 to v_2 and v_2 to v_1. This is done by collecting the respective arcs leading to minimum distance from v_1.

Dijkstra's Algorithm

Input $(\bar{G}, \phi)$ with source vertex v_0 and end vertex v_t. $\bar{G}(V, \bar{E})$, $|V| = n$.

1. For $i = 0$ $W(v_i, v_i) \leftarrow 0$, $W(u, v) \leftarrow \infty$ if $uv \notin E$, $S_i \leftarrow \{v_i\}$.
2. For $i \neq 0$, for each

$$x \in V - S_i W(u_i, x) \leftarrow \min\{W(v_0, u_i) + \phi(u_i x)\}.$$

   Let u_{i+1} be the vertex of minimality not in S_i.
3. If $i = n - 1$ stop. If $i < n - 1$,
4. Output $W(v_0v_i)$ then $i \leftarrow i + 1$ and go to step 2.

When the algorithm ends the value of shortest distance will be the output.

The complexity of the algorithm will be $O(n^2)$. As we see step 2 will be using $\frac{n(n-1)}{2}$ additions which is the major step in this algorithm.

Exercises

1. Find the minimum weight spanning tree in the following graphs (Figs. 23.15 and 23.16).

   (i)

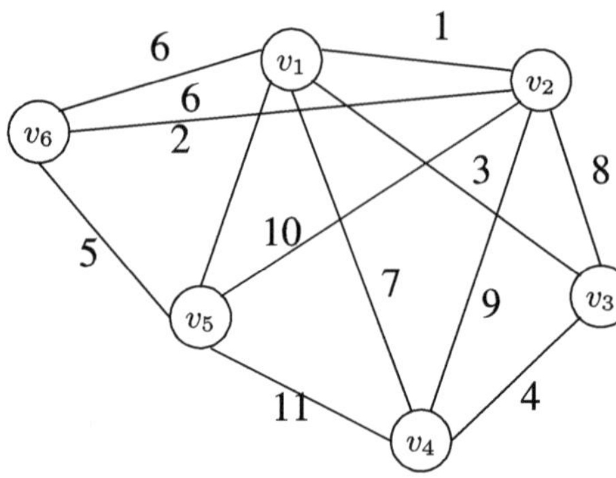

Fig. 23.15 Weighted graph (1)

   (ii)

2. Find the shortest path in the following digraphs.

   (i) Source vertex v_1 and end vertex v_6 (Fig. 23.17).

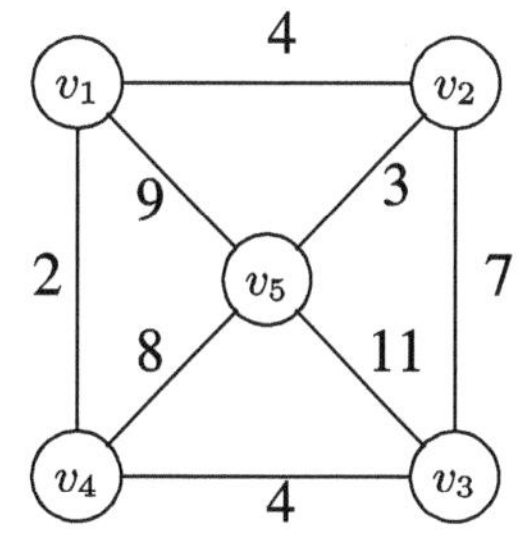

Fig. 23.16 Weighted graph (2)

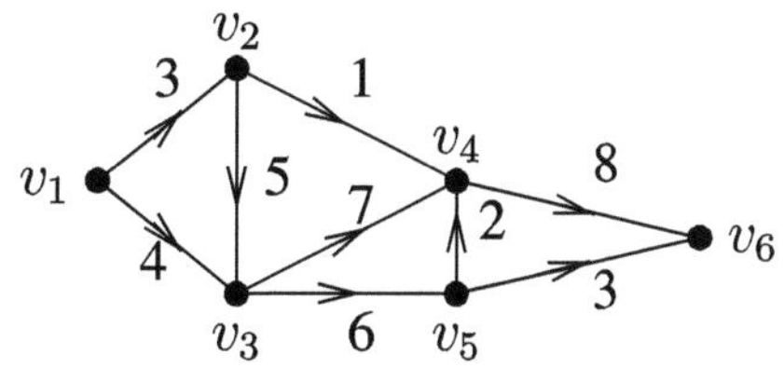

Fig. 23.17 Weighted graph (2-1)

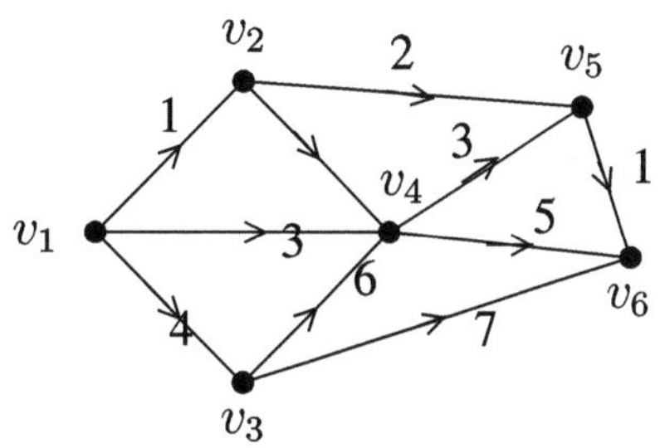

Fig. 23.18 Weighted graph (2-2)

(ii) Source vertex : v_2 and end vertex : v_6 (Fig. 23.18).

3. For each of the weighted graphs given below find the minimum spanning tree using Prim's algorithm (Figs. 23.19 and 23.20).

(i)

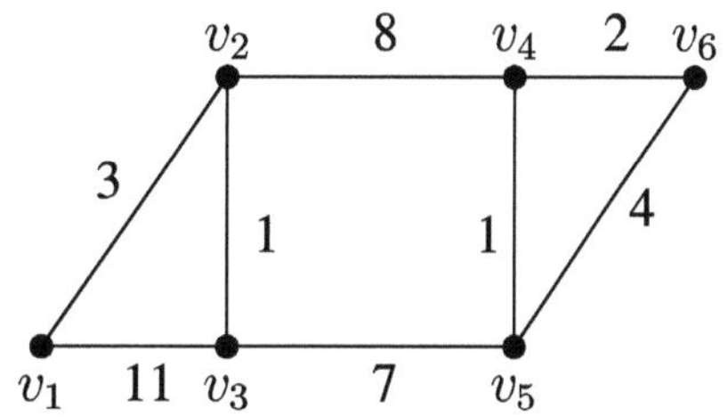

Fig. 23.19 Weighted graph (3-1)

(ii)

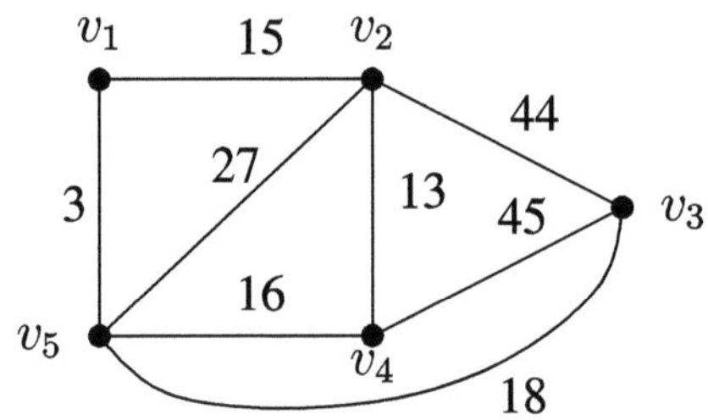

Fig. 23.20 Weighted graph (3-2)

4. If $\vec{G} = (V, \vec{E})$ is a strongly connected digraph, then show that there is a subset $\bar{V}$ of V, with $\bar{V} \neq \phi$ such that outdegree of every vertex $v \in \bar{V}$ is atleast 1.

5. A Directed Acyclic Graph (DAG) is a digraph with no directed cycles.
Check whether the following graph is a DAG. Justify your answer (Fig. 23.21).

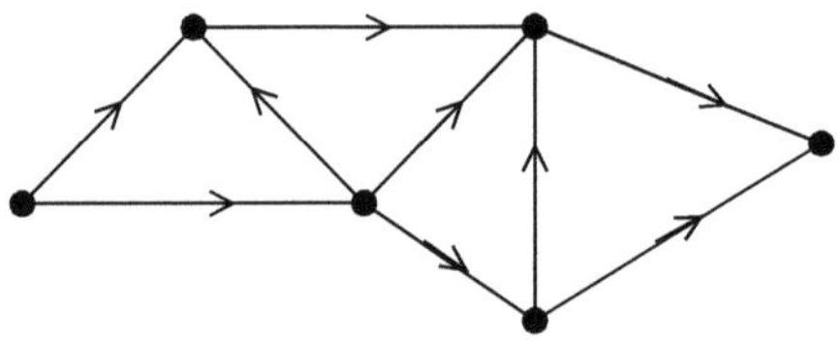

Fig. 23.21 DAG

6. Let $G = (V, E)$ be an undirected graph and let $E = E_1 \cup E_2$ with a condition that every vertex $v \in V$ is incident with atmost one edge from E_1 and E_2. Show that G is 2-colorable.

7. A directed walk in a digraph $\vec{G} = (V, \vec{E})$ is a sequence $v_0e_1v_2e_2 \ldots e_nv_n$ such that $v_i \in V$, $0 \leq i \leq n$ and e_i is an arc from v_{i-1} to v_i, $1 \leq i \leq n$. If $v_0 = v_n$, then the directed walk is said to be a closed one and if $v_0, v_1, \ldots, v_n$ are distinct, then the walk is called a directed path.
Find two distinct directed walks, two distinct directed paths and two distinct closed walks in the following digraph, if they exists (Fig. 23.22).

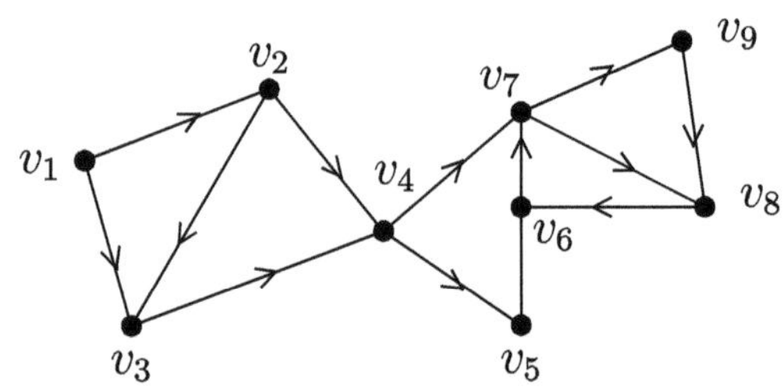

Fig. 23.22 Graph (7)

8. Let $\vec{G} = (V, \vec{E})$ be a digraph and let $\bar{H}_1, \bar{H}_2$ be two subgraphs of $\vec{G}$ which are strongly connected. If vertices of $\bar{H}_1$ and $\bar{H}_2$ are disjoint, then show that $\bar{H}_1 \cup \bar{H}_2$ is strongly connected.

Problems and Solutions—Graph Theory

1. Definition: A graph whose vertices are named distinctly is called a labelled graph.

Question: A graph G has 4 vertices and 5 edges. Its vertices are labelled as $\{a, b, c, d\}$. Draw all such possible graphs.

Solution: The graphs are in Fig. 23.23.

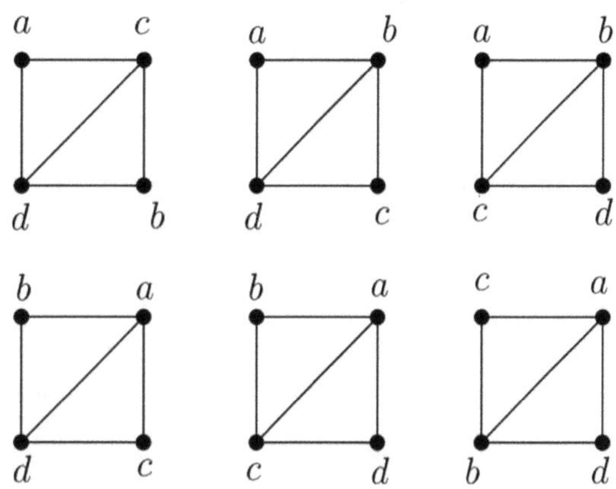

Fig. 23.23 Graphs (1)

2. Are the following graphs bipartite? (Fig. 23.24).

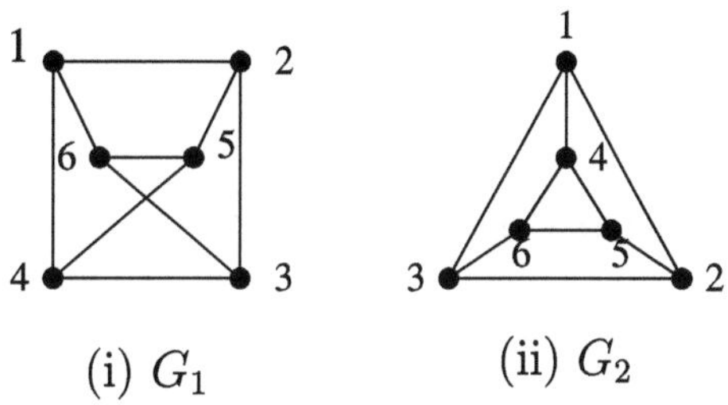

Fig. 23.24 Graphs (2)

Solution: (i) G_1 contains no odd cycles. Its 4-cylces are in Fig. 23.25.

Fig. 23.25 Graphs (2-1)

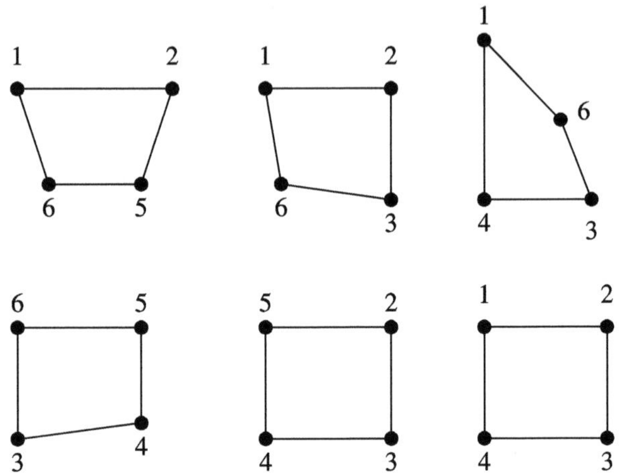

The 6-cycles are in Figs. 23.26 and 23.27.

Fig. 23.26 Graphs (2-2)

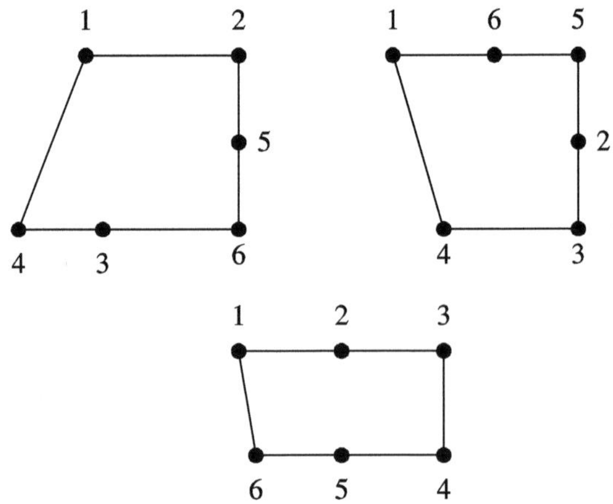

Fig. 23.27 Graphs (2-3)

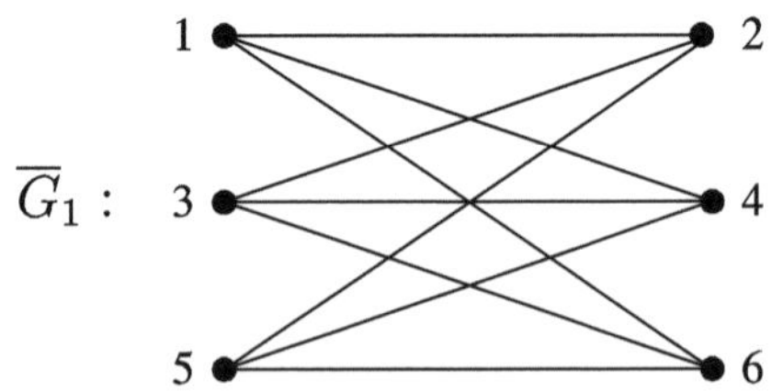

(ii) G_2 contains an odd cycle 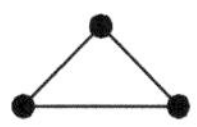

Hence G_2 is not bipartite.

(Try yourself) Are the following graphs bipartite? Justify your answer (Fig. 23.28).

Fig. 23.28 Graphs (2-4)

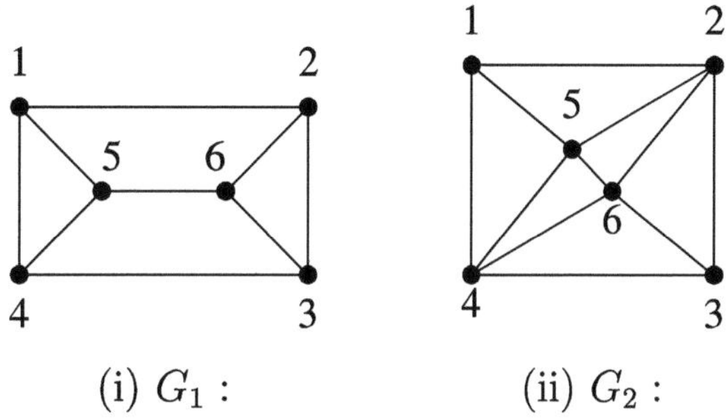

3. Definition: Let G be a graph on 'n' vertices. Let $d_1, d_2, \ldots, d_n$ be the degrees of the vertices of G. Degree vector $D(G) = (d_1, d_2, \ldots, d_n)$ such that $d_1 \leq d_2 \leq \cdots \leq d_n$ and $d_i \geq 0 \ \forall \ i, 1 \leq i \neq n$.

Find the degree vector of the following graphs (Fig. 23.29).

Fig. 23.29 Graphs (3)

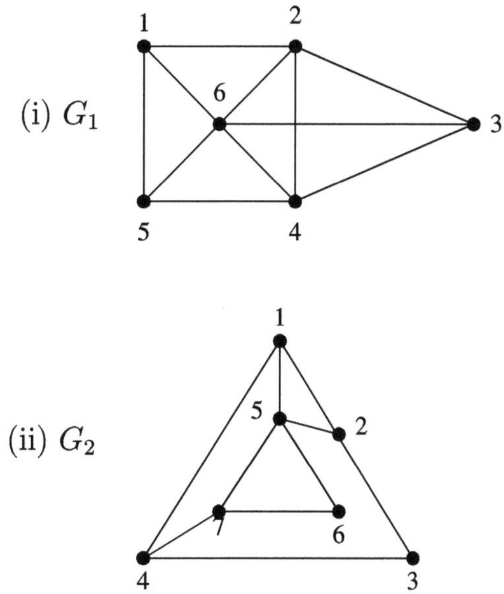

Solution:
(i) $D(G_1) = (3, 3, 3, 4, 4, 5)$
(ii) $D(G_2) = (2, 2, 3, 3, 3, 3, 4)$.

(Try yourself) Draw a graph 'G' for the degree vector (i) $D(G) = (1, 3, 3, 3)$ (ii) $D(G) = (3, 3, 3, 3, 3, 3)$

4. Draw the complement graph of the following graphs and write down their degree vectors (Figs. 23.30 and 23.31).

Fig. 23.30 Graphs (4)

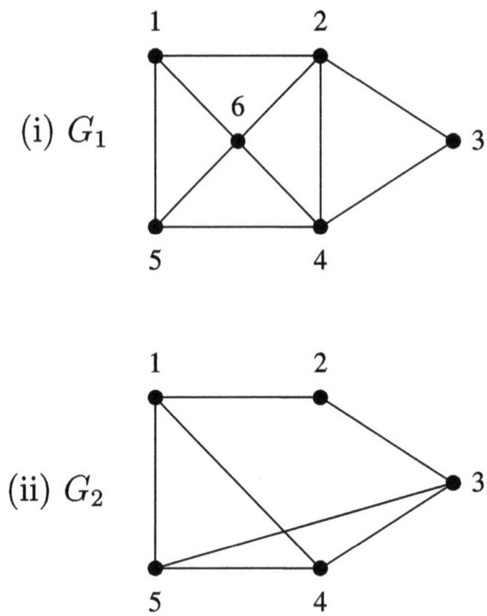

Solution:
(i) $D(\overline{G}_1) = (1, 1, 1, 2, 2, 3)$
(ii) $D(\overline{G}_2) = (1, 1, 2, 2, 2)$
(Try yourself) Is it true that a complement of a bipartite graph is bipartite? Justify your answer.

5. Let G be a graph with 'n' vertices and m edges. If every vertex of G is degree either t or $t + 1$, then find the number of vertices of degree't' in G.

Solution: Let 's' be the number of vertices of degree 't' in G. Then we can write the equation

$$s.t + (n - s)(t + 1) = 2m$$
$$\Rightarrow s = n(t + 1) - 2m.$$

6. Let G be a simple graph with 9 vertices and let the sum of the degrees of the vertices of G be atleast 27. Show that G has a vertex of degree at least 4.

Fig. 23.31 Graphs (4-1)

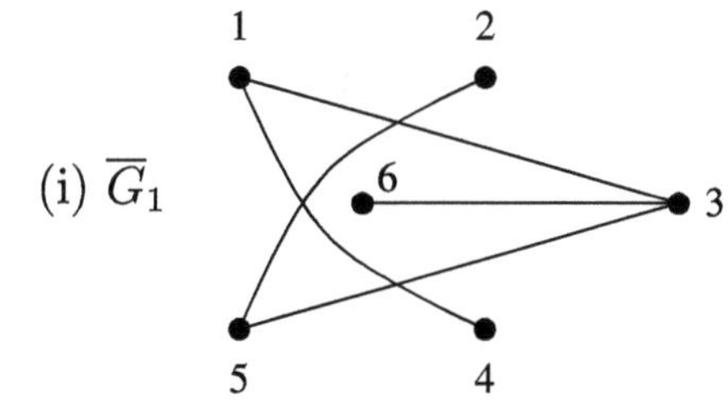

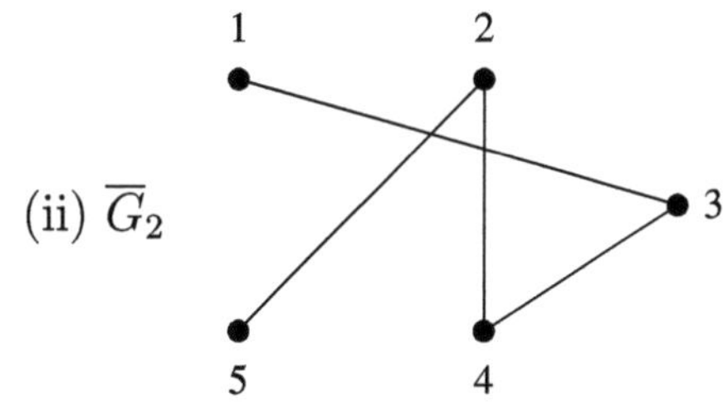

Solution: We know that the sum of the degrees cannot be 27. The first even number after 27 will be 28. If every vertex of G is of degree 3, then the sum of the degrees will be 27. But this is not possible for two reasons.

Reason 1: Sum of the degrees must be even.

Reason 2: Number of odd degree vertices in G must be even.

Hence the sum of the degrees will be atleast 28. By pigeon hole principle there is a vertex of degree atleast 4 in G.
(Try yourself) Does there exist a simple graph G with degree vector $D(G) = (1, 1, 2, 4, 4)$? Justify your answer.

7. Definition: Let G be a graph. A subgraph $G - \{v\}$ is a subgraph of G obtained from G by deleting the vertex 'v' and all the edges incident at 'v'. For the following graph G find $G - \{v_1\}$, $G - \{v_1, v_2\}$, $G - \{v_1, v_2, v_3\}$ (Figs. 23.32, 23.33, 23.34 and 23.35).

Fig. 23.32 Graphs (7)

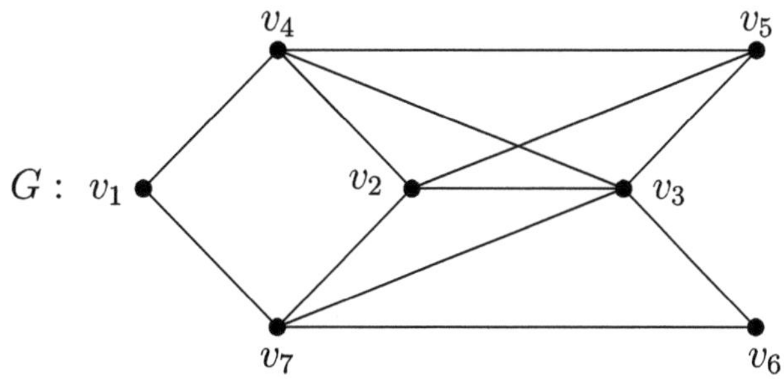

Solution:

$G - \{v_1\}$:

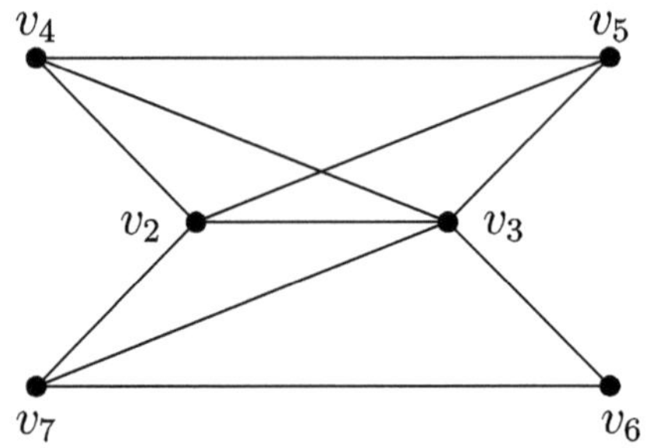

Fig. 23.33 Graphs (7-1)

8. In the following graph G find all the cut-vertices and cut-edges (Fig. 23.36). Find the cut-vertices and cut-edges of G.

Solution: Cut-vertices are $\{v_7, v_9, v_{11}\}$ cut-edges are $\{v_9v_{11}, v_{11}v_{12}, v_7v_{13}\}$.

9. Does there exist a graph with the following conditions?

Conditions: G is a simple graph on 7 vertices and has exactly two cycles.

$G - \{v_1, v_2\}$:

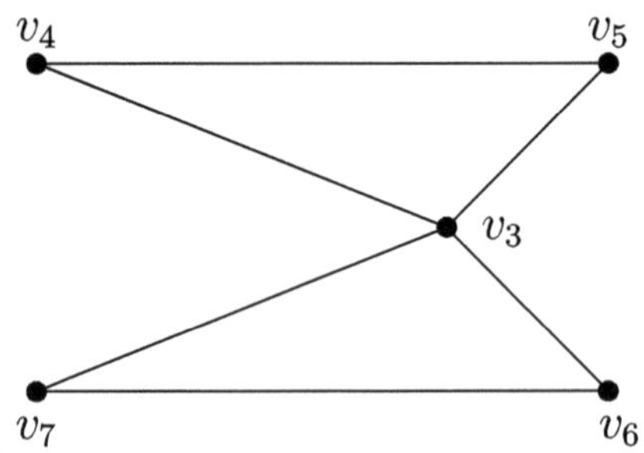

Fig. 23.34 Graphs (7-2)

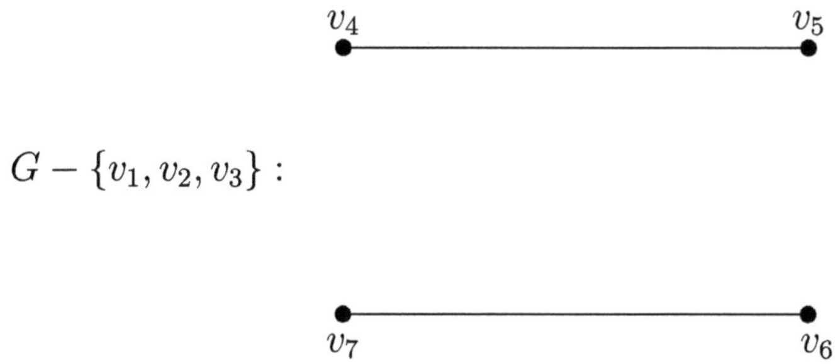

Fig. 23.35 Graphs (7-3)

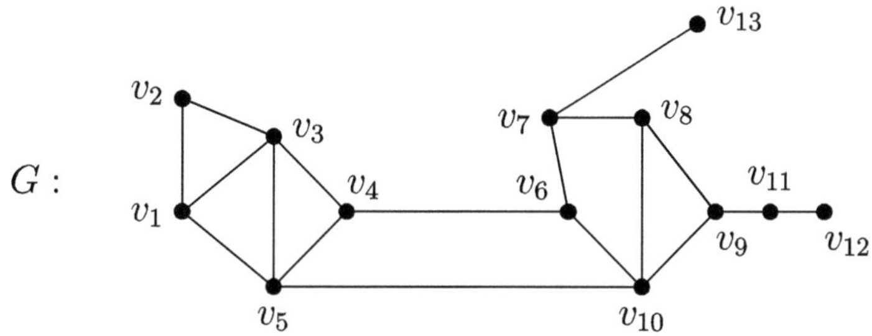

Fig. 23.36 Graphs (8)

Solution: The answer is 'Yes'.

(i) If G is connected, then G will be the graph
(ii) If G is not connected, then G will be the graph (Figs. 23.37 and 23.38).

10. Show that there always exists a connected graph on n-vertices and $(n + 1)$-edges containing exactly two cycles, for $n \geq 4$.

Solution: Let us see how to draw such a graph. A connected graph on n-vertices and $(n - 1)$-edges will be a tree. In particular we can consider a path P_n on n vertices and $(n - 1)$ edges. By adding an edge between the two leaf nodes will make $(P_n + e)$ as a cycle C_n on n vertices and n-edges. Now an edge drawn inside C_n by connecting two non adjacent vertices on C_n will give the desired graph.

Fig. 23.37 Graphs (9-1)

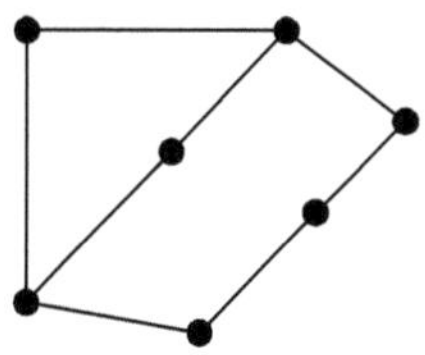

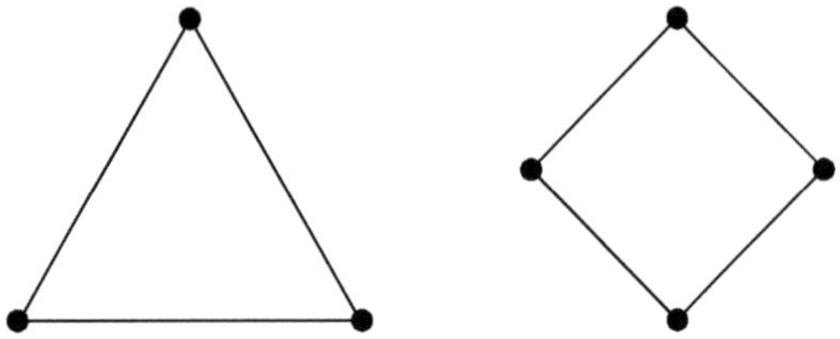

Fig. 23.38 Graphs (9-2)

11. A graph G has 24 edges 10 vertices. Each vertex has either degree 3 or degree 5. Find the number of vertices of G that are of degree 3.

Solution: Suppose $G = (V, E)$, then $|V| = 10$, $|E| = 24$. By handshaking lemma we have the sum of the degrees of the vertices being equal to 48. i.e.,

$$\sum_{v \in V} d(v) = 48. \tag{23.1}$$

Suppose the number of vertices of degree 3 is x, then $(10 - x)$ vertices will be of degree 4. Hence

$$3x + 5(10 - x) = 48 \Rightarrow x = 1.$$

Hence there is only one vertex of degrees and a vertices of degree 5.

12. Definition: Let S be a set with $|S| = n$. Consider a subset I of $\mathcal{P}(S)$, the power set of S. Consider the subsets in I to represent vertices and for any $A, B \in I$ there will be an edge between A and B if their intersection is non-empty. A simple graph thus constructed will be called as the 'intersection graph'.

Question: Draw the intersection graph for the subsets. $A = \{1, 2, 3\}$, $B = \{2, 3, 4\}$, $C = \{4, 5, 6\}$, $D = \{2, 3, 4, 5\}$, $E = \{1, 3, 5, 6\}$ (Fig. 23.39).

Solution: Here $I = \{A, B, C, D, E\}$.

13. Show that given a simple graph G, there always exists an intersection graph $I(G)$ of a set of subsets of a finite set such that G is isomorphic to $I(G)$.

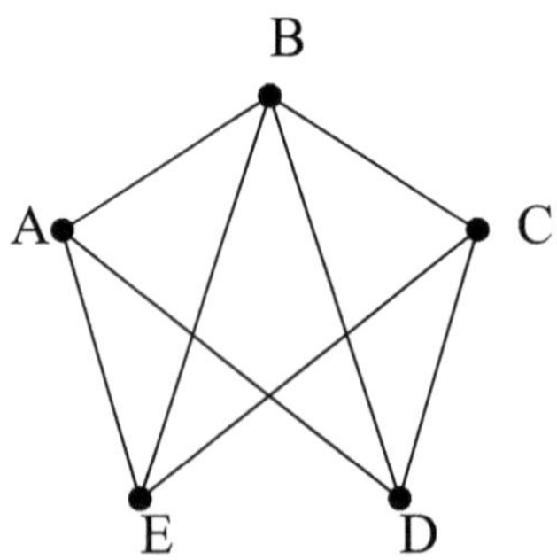

Fig. 23.39 Graphs (12)

Solution: Let $G = (V, E)$ and let $V = \{1, 2, \ldots, n\}$. Now we construct the intersection graph $I(G)$ from G such that $G \cong I(G)$. For $I(G)$, the vertices are subsets of V. Let $X_1, X_2, \ldots, X_n$ be the subsets such that $X_i = \{j/ij \in E\} \cup \{i\}$. Clearly for $I(G)$, the edges will be supplied based on whether $X_i \cap X_j$ being empty or not. This is same as saying X_i, X_j are adjacent in $I(G)$ iff $ij \in E$. Clearly $I(G)$ is unique and isomorphic to G.

14. Definition: In the intersection graph, instead of considering subsets as vertices, we consider open interals on real line as vertices what we obtain is known as 'Interval graph'. That is the edges are between two open interals, if they overlap.

Question: Find the value of n, such that C_3 is isomorphic to an interval graph IC_3 (say) and $C_4 \ncong IC_4$.

Solution: (i) Consider C_3. Let I_1, I_2, I_3 be three intervals with

$$I_1 = (a_1, a_2) \tag{23.2}$$

$$I_2 = (b_1, b_2) \tag{23.3}$$

$$I_3 = (c_1, c_2). \tag{23.4}$$

One can construct an interval graph IC_3 with the following conditions on I_1, I_2 and I_3 (Fig. 23.40).

Fig. 23.40 Graphs (14)

$$a_1 < b_1 < c_1 < a_2 < b_2 < c_2.$$

Clearly $IC_3 \cong C_3$.

(ii) Such an assignment of intervals to the vertices of C_4 is not possible. Let the vertices of IC_4 be I_1, I_2, I_3 and I_4 say. Clearly each I_j is adjacent to I_ℓ, I_m and I_ℓ, I_m are nonoverlapping intervals. Suppose two edges are drawn satisfying this condition with $I_jI_\ell, I_jI_m \in E(IC_4)$, $I_\ell \cap I_m = \emptyset$. One can argue that the further supply of edges becomes impossible to make $IC_4 \cong C_3$.

15. Definition: A graph G is said to be a chordal graph if in every cycle subgraph C of length ≥ 4, there is an edge of G joining two nonadjacent vertices in C.

Question: Are the following graphs chordal? (Figs. 23.41, 23.42, 23.43 and 23.44).

Fig. 23.41 Graphs (15-1)

Fig. 23.42 Graphs (15-2)

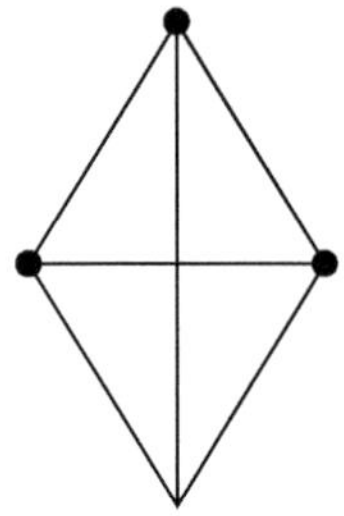

Fig. 23.43 Graphs (15-3)

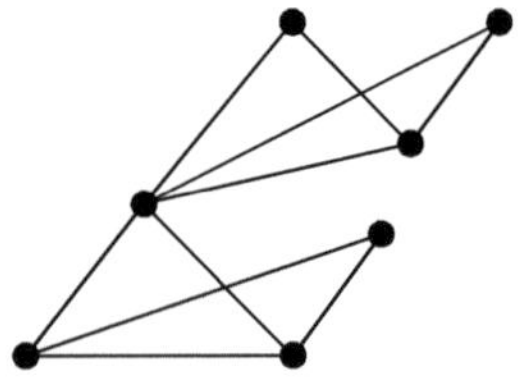

Fig. 23.44 Graphs (15-4)

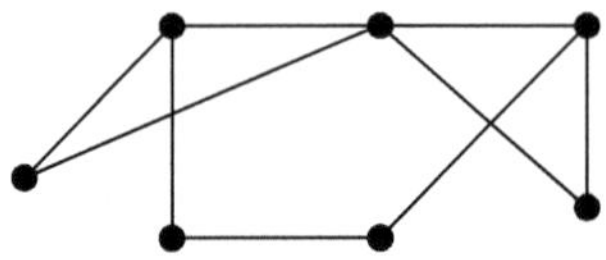

Solution: Graphs in (i), (ii) are chordal graphs (iii) and (iv) are non chordal graphs.
In graph (i), there are no cycles.
In graph (ii) the cycle C_4 contains a chord.
In graph (iii) all the cycles are of length 3. There are no cycles of length ≥ 4.
In graph (iv), C_3, C_5 are present. But C_5 does not contain a chord.

16. Let $G = (V, E)$ be a simple connected graph. Show that G has two vertices of the same degree.

Solution: The graph is given to be a simple graph and let $|V| = n$.

If a vertex $v \in V$ is of degree $(n-1)$, then 'v' is incident with all the remaining $(n-1)$ vertices. In this case there will be no vertex $v \in V$ whose degree 0. All the remaining $(n-1)$ vertices must be with degree $t_i, t_i \in \{1, 2, \ldots, n-1\}$. Hence by Pigeon-hole principal as the 'n' vertices can have $(n-1)$ different degrees, there must be two vertices of same degree.

17. Draw all graphs on 6 vertices in which each vertex is of degree less than or equal to 2. Also find the number of connected graphs in this drawing (Fig. 23.45).

Solution:

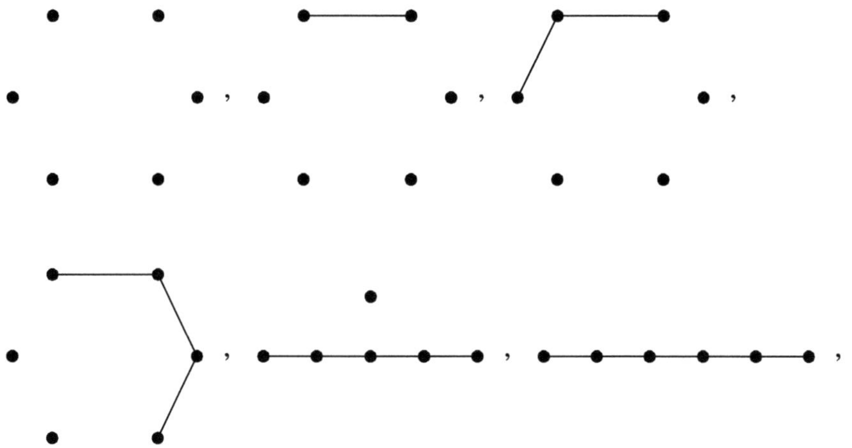

There are only two graphs that are connected.

18. Definition: A rooted tree is a tree in which a vertex is identified as a root. With respect to this root, we define depth of a vertex to be the number of edges in the path from root to this vertex.

Question: Find the depth of the vertices

$$\{u_2, u_3, u_4, u_5, u_6, u_7, u_8, u_9, u_{10}, u_{11}\}$$

in the following rooted tree whose root is 'u_1' (Fig. 23.46).

Solution:

Depth of $\{u_2, u_3, u_4\}$ is 1.
Depth of $\{u_5, u_6, u_7\}$ is 2.
Depth of $\{u_8, u_9, u_{10}, u_{11}\}$ is 3.

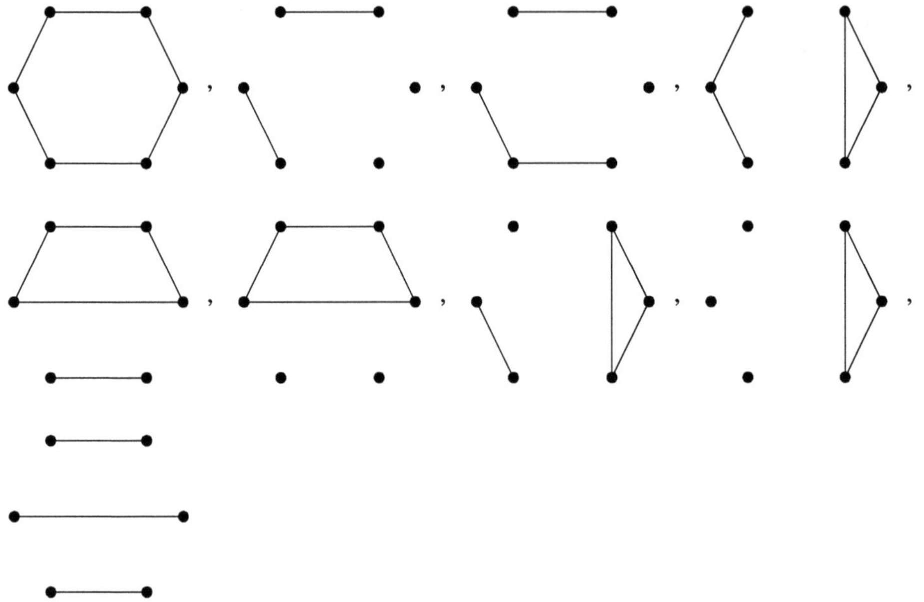

Fig. 23.45 Graphs (17)

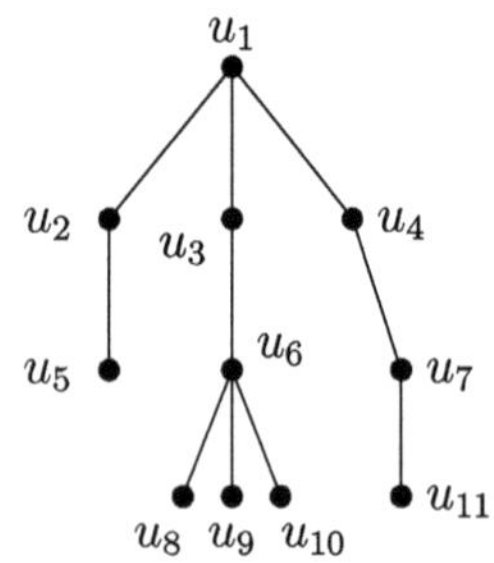

Fig. 23.46 Graphs (18)

19. Definition: A rooted binary tree is one in which evey vertex that is not a leaf has degree 3. That is each non-leaf vertex has two children. To be full if every vertex has degree 3 or degree 1 except the root vertex which of degree 2.

Question: Let T be a rooted binary tree that is full having 49 vertices. What is the maximum height of T and what is the minimum height of T. Justify your answer.

Solution: Note that maximum depth of T is called as height of the tree. As T is a full binary tree, one can actually draw a binary tree T with maximum height is as Fig. 23.47.

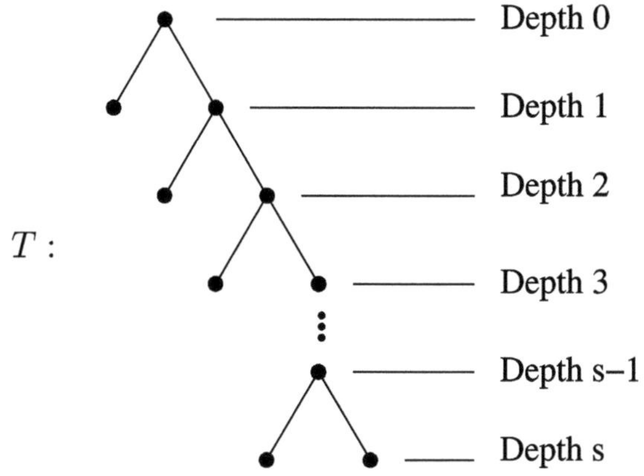

Fig. 23.47 Graphs (19-1)

Number of nodes of T is

$$\begin{aligned} 49 &= 1 + \underbrace{2 + 2 + \cdots + 2 + 2}_{s \text{ times}} \\ &= 2s + 1 \\ \Rightarrow s &= 24. \end{aligned}$$

Hence maximum height of such a tree will be 24.

One can observe from the above construction that there has to be maximum number of vertices at each level satistying the given conditions. We draw such a tree as Fig, 23.48.

Hence $1 + 2 + 2^2 + 2^+ \cdots + 2^k = 49$.
$\Rightarrow 2^5 < 49 < 2^6$
Hence the minimum height of the tree will be 6.

20. Does there exist a rooted binary tree with

(i) depth 3 and 9 leaf nodes?
(ii) a full binary tree with 8 internal vertices or non-leaf vertices and 7 leaf vertices? Justify your claim.

Solution:

(i) Let T be the binary tree with n vertices. Let T be of height 3 and let the leaf vertices of T be 9. Then

$$9 \le 2^3 = 8.$$

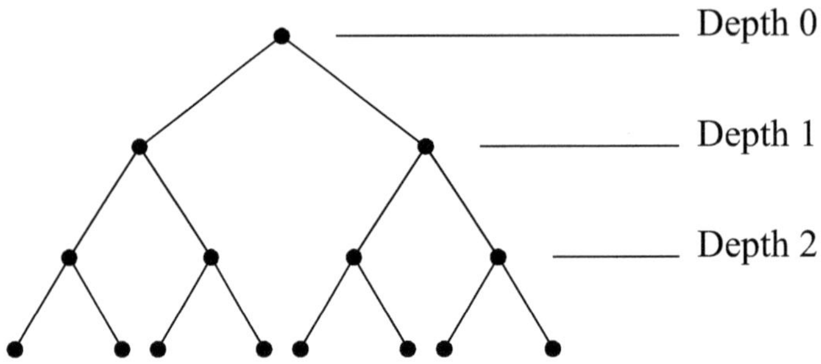

Fig. 23.48 Graphs (19-2)

Hence such a tree is not possible.

(ii) A full binary tree with 't' internal vertices has a total of $2t + 1$ vertices. Such a tree will have $(t + 1)$ leaf vertices. Hence $t + 1 = 8 + 1 = 9$. So such a tree is not possible.

21. Let $G = (V, E)$ be a graph on even number of vertices.

(i) Does there exist a graph G that is not a tree such that G has exactly two vertices with degree 't' for $t \in \{1, 2, \ldots, n\}$, $|V| = 2n$?
(ii) Show that number of trees on n vertices with the property that such trees have exactly two vertices of degree $t \in \{1, 2, \ldots, n\}$ is $n(n + 1)$, with $|V| = 2n$.

Solution:

(i) The following graph is a graph G on 8 vertices.

Fig. 23.49 Graphs (21-1)

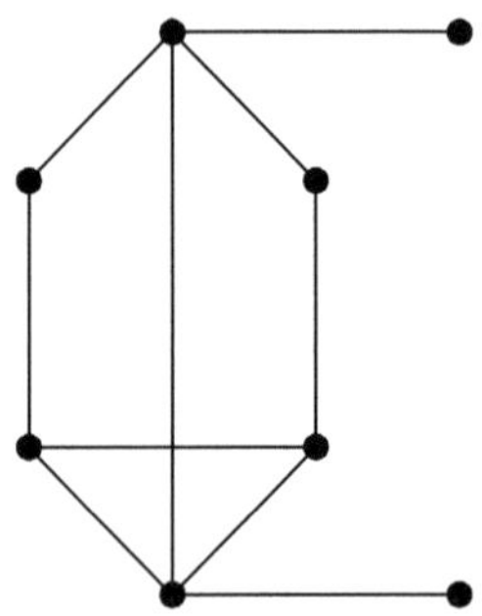

(ii) Let $T = (V, E)$ be a tree on $2n$-vertices satisfying the given property that there are exactly 2 vertices of degree $t \in \{1, 2, \ldots, n\}$. Hence sum of the degrees will be $2(1 + 2 + \cdots + n) = n(n + 1)$. So, $|E| = \frac{n(n+1)}{n}$. Since T is a tree, T must have $(2n - 1)$-edges. Hence

$$\frac{n(n+1)}{n} = 2n - 1$$
$$\Rightarrow n^2 - 3n + 1 = 0$$
$$\Rightarrow n = 1 \text{ or } 2.$$

This means T must be a tree on either 2 vertices or 4 vertices. Such trees are drawn in Fig. 23.50.

Fig. 23.50 Graphs (21-2)

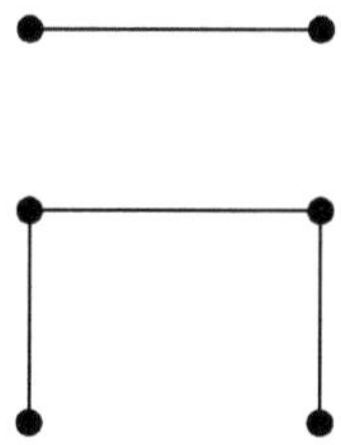

22. Let G be a graph with maximum matching of size $2m$. Show that the size of a maximal matching in G is atleast m.

Solution: One can understand such a graph by looking at G having m components, with each component has a path P_3 of length 3. Then the maximum matching will use two of these edges in P_3. This gives a maximum matching of size $2m$. Clearly in a matching M of size $2m$, one can identity a maximal matching $\overline{M}$ of size atleast m. Each of the edges in $\overline{M}$ will cover one end vertex of an edge in M. i.e. Fig. 23.51,

Fig. 23.51 Graphs (22)

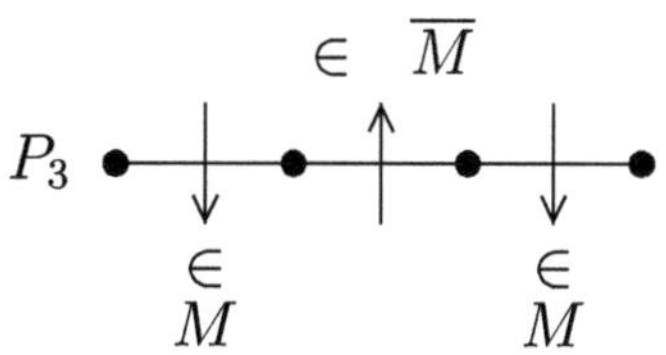

If this is not true then one can add an edge to $\overline{M}$ contradicting the fact that $\overline{M}$ being maximal. The number of vertices covered by M will be $2m$ and $\overline{M}$ is of size atleast m.

23. Let G be a graph having $2m$ vertices and let degree of every vertex of G be atleast $(m+1)$. Then show that G has a perfect matching.

Solution: The graph G described in the question will contain a Hamilton cycle. Taking alternate edges on this cycle will produce a perfect matching.

24. Let S be a finite set. Let $S_1, S_2, \ldots, S_n$ be subsets of S. Prove the following using the "matching" concepts of graph theory.

(i) There exists a subset F of $\{1, 2, 3, 4, \ldots, n\}$ such that $|S_1 \cup S_2 \cup \ldots S_n| < |F|$.
(ii) There exists elements $x_1, x_2, \ldots, x_n \in S$ with $x_i \neq x_j, i \neq j$ such that $x_i \in S_i$ for each $1 \leq i \leq n$.

Solution: Let us define a graph $G = (V, E)$ such that the set of vertices will be $\{1, 2, \ldots, n\} \cup S$. $E = \{e = iv/i \in \{1, 2, \ldots, n\}, v \in S$ and $v \in S_i\}$

Clearly G is a bipartite graph.

G being bipartite, if it has a matching every element of $\{1, 2, \ldots, n\}$, then there will be an element $v_i \in S$ such that $iv_i \in E$ and $iv_i \in M$. Clearly such v_i, $1 \leq i \leq n$ are distinct as $iv_i \in M$

If there is no such matching M, then by Hall's theorem there will be a subset F of $\{1, 2, \ldots, n\}$ covered by edges in E. This means

$$|S_1 \cup S_2 \cdots \cup S_n| < |F|.$$

But we have

$$|I| < |S_1 \cup S_2 \cdots \cup S_n|.$$

25. For what values of m, does the complete graph K_m contain

(i) a Euler trail?
(ii) a Hamilton path?
(iii) a Hamilton cycle?

Solution:

(i) K_1, K_2 contain Euler trail, $K_{2n+1}, n \geq 3$ contain Euler trial.
(ii) K_m for $m \geq 1$ contain Hamilton path.
(iii) $K_m, m \geq 3$ contain Hamilton cycle.

26. Let G be a simple connected graph having an Euler tour—Are the following statements true? Justify your answer.

(i) If the number of vertices in G is even, then G must have even number of edges.
(ii) If G is bipartite, then the number of edges of G must be odd.

Solution:

(i) This statement is false. Consider the following graph as an example.
Number of vertices is 6, but the number of edges is 7. The graph definitely contains an Euler tour.

Fig. 23.52 Graphs (26)

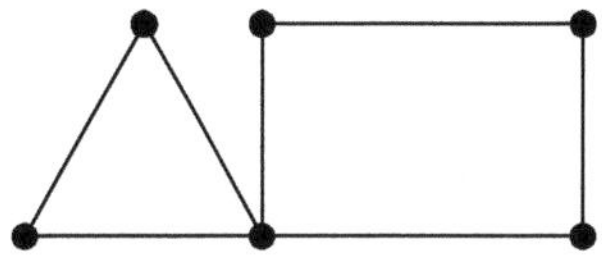

(ii) The statement will be false for the following reason.
Total number of edges in any bipartite graph will be sum of the degrees of the vertices in any one of the parts. Also the graph is given to contain an Euler tour. So every vertex of G is of even degree. So the graph must contain even number of edges (Fig. 23.52).

27. Let G be a connected graph having even number of edges. Also assume that the degree of every vertex of G is even. Every edge of G being colored using two colors say red or green. Is it possible to color the edges in such a way that every vertex of G has the same number of red and green edges incident with it? Justiy your answer.

Solution: Since G is connected and every vertex of G is even, G has an Euler tour say C. Now color edges of C alternately using the colors red and green. One can understand that every pair of edges occurring as adjacent ones in C get different colors. Since every vertex has even degree, coloring the edges satisfying the conditions of the question is possible.

28. Let G be a graph on atleast 'n' vertices such that $\delta(G) = t, n \geq 2$. Show that for every such 't', there exists a graph G containing no cycle of length $t + 1$.

Solution: Let $k = \lceil \frac{n}{t} \lceil$. Consider k different copies of K_{t+1}. Here the total number of vertices will be $k(t + 1)$ and $k(t + 1) > n$. i.e., for $t = 3, n = 4$. We have the following graph that has no cycle of length greater than 4 (Fig. 23.53).

Fig. 23.53 Graphs (28)

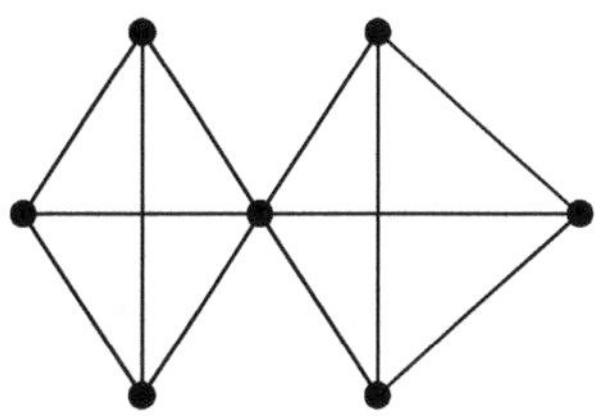

This procedure produces a graph satisfying the given conditions.

29. Does there exist a non-Hamiltonian graph G on n vertices $n \geq 1$ such that

(i) the number of edges is $(n - 1)C_2 + 1$.
(ii) $\delta(G) = \lceil \frac{n}{2} \lceil +1$.
Justify your answer.

Solution:

(i) One can construct a non-Hamilton graph as below.
Consider two copies of the complete graph $K_{\lceil \frac{n}{2} \lceil}$. Let it be disjoint copies. So we can call them as G_1 and G_2. Let $v \in V(G_1)$, a specific one. Supply edges from v to all vertices of G_2. Let the resultant graph be denoted as G (say). Clearly minimum degree of G is $\lceil \frac{n}{2} \lceil - 1$. G is non-Hamilton because every such cycle passing through all the vertices of G will meet the specified vertex 'v' atleast twice.

(ii) Consider K_{n-1}. Now add a new vertex 'u' such that u is adjacent to one of the vertices of K_{n-1}. Clearly the number of edges in the resulting graph will be $(n-1)C_2 + 1$. Also the constructed graph is non-Hamiltonian.

30. Definition: A tournament is a directed graph tk_n obtained by assigning direction or orientation to every edge of a complete graph K_n.

Question: Show that every tournament tK_n has a Hamilton path.

Solution: We prove the result using mathematical induction on the number of vertices of tK_n.

For $n = 2$, tK_2 : 

Hence there is an Hamilton path in tK_2.

Assume the result to be true for all tournaments tK_n on n vertices that they have a Hamilton path.

Consider tK_{n+1}. Let v be a vertex of tK_{n+1} and consider $tK_{n+1} - \{v\}$, the graph obtained by removing 'v' from tK_{n+1}. Note that the resultant graph is a tournament tK_n. By assumption tK_n has a Hamilton path say $v_1 \to v_2 \to \cdots \to v_n$, say.

Case i: If $v \to v_1$ in tK_{n+1}, then $v \to v_1 \to v_2 \cdots \to v_n$ is a Hamilton path in tK_{n+1}.

Case ii: If $v_j \to v$ is in tK_{n+1} and 'j' be the maximum indexed vertex of tK_n such that $v_j \to v$. Here for $j = n$,

$$v_1 \to v_2 \to \cdots \to v_n \to v$$

is a Hamilton path in tK_{n+1}.

$$v_1 \to v_2 \to \cdots \to v_j \to v \to v_{j+1} \to \cdots \to v_n$$

is a Hamilton path in tK_{n+1}.

31. Consider the complete graph $K_n, n \geq 4$. Show that K_{2n} can be partitioned into n Hamilton paths. Note that no two Hamilton paths in this collection share an edge.

Solution: K_{2n} for $n \geq 4$ has a Hamilton path and a Hamilton cycle as well. For example the graph K_4 has two paths (Figs. 23.54 and 23.55).
which are edge-disjoint Hamilton paths.

Let the vertices of the graph K_{2n} be denoted by $\{0, 1, 2, \ldots, 2n-1\}$ and let them be arranged in a circular fashion. One can write the Hamilton paths as Fig. 23.56.

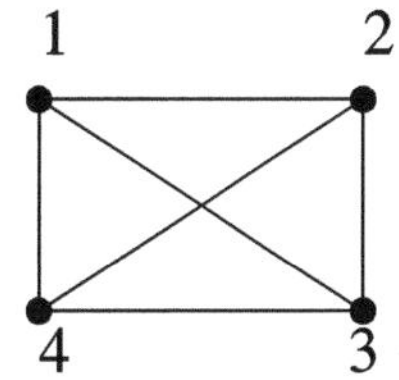

Fig. 23.54 Graphs (31-1)

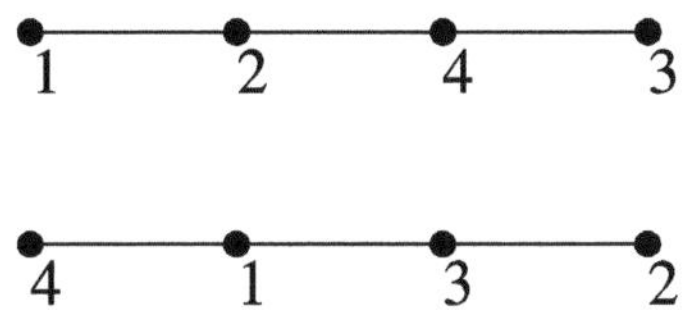

Fig. 23.55 Graphs (31-2)

kth Hamilton path:

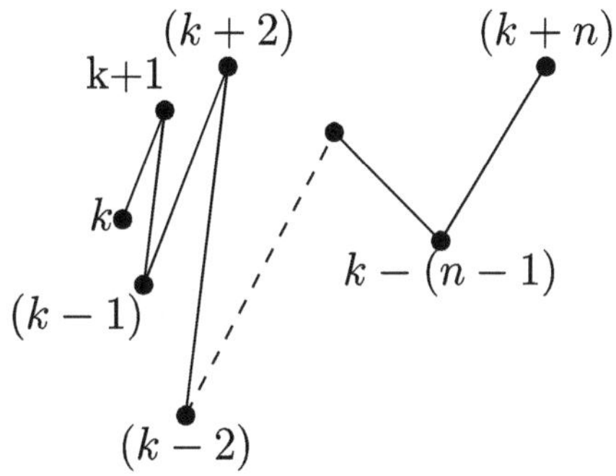

Fig. 23.56 Graphs (31-3)

Here use the indices as (mod $2n$) where $t = 0, 1, \ldots, 2n-1$. This gives all the n distinct Hamilton paths. One has to the paths do not share any edge.

32. Show that if a plane graph has different faces having the same set of edges as boundary, then the graph must be a cycle.

Solution: Let G be the plane graph and let f_1, f_2 be two different faces of G. Let $\overline{G}$ be the subgraph of G and also bounding edges of the faces f_1, f_2. Then $\overline{G}$ must contain a cycle say C. This means that f_1 and f_2 are in different faces of C. This is possible only when $\overline{G}$ itself is C. Also the vertices and edges of $\overline{G}$ will be in one of the faces of C. This imply that f_1, f_2 are different faces of C. C clearly has only two faces. So $f_1 \cup f_2 \cup C$ will be the whole plane $\mathbb{R}^2$. Hence $G \cong C$.

33. Prove the following.

(i) Let G be a triangulated plane graph. Show that G has even number of faces.
(ii) Does there exist a planar graph having the property that each vertex of degree atleast 5 and atleast one vertex of degree 8?

Solution:

(i) Every face of $G = (V, E)$ is now a triangle except the exterior face. Sum of the edges bounding the faces will be $2|E|$. This means $3|F| = 2|E|$, F is the set of all faces. Hence $|F|$ must be even.
(ii) We know

$$|E| \leq 3\,V| - 6 \tag{23.5}$$

for $G = (V, E)$, G being planar.

Every vertex being of degree atleast 5 and the vertex being of degree atleast 8 means,

$$\sum_{\substack{v\in V,\\ v\neq w}} 6 + 8 \geq 14, \text{ where } w \text{ is a vertex of degree 8.}$$

which is true as (23.5) imply that

$$6\,V| - \sum_{v\in V} deg(v) \geq 12$$

$$\Rightarrow \sum_{\substack{v\in V\\ v\neq w}} (6 - deg(v)) + (6 - deg(w)) \geq 12$$

$$\Rightarrow \sum_{\substack{v\in V\\ v\neq w}} (6 - deg(v)) \geq 14.$$

This is possible for a graph G with 15 vertices.

34. Prove or disprove the following statements.

(i) If $G = (V, E)$ is a planar graph having no subgraph isomorphic to K_3, then $\chi(G) \leq 4$.
(ii) If $G = (V, E)$ is a planar graph having no subgraph isomorphic to K_4, then $\chi(G) \leq 3$.

Solution:

(i) This statement is a direct consequence of Four Color theorem.
It can be shown directly as below as well.

Since G has no subgraph isomorphic to K_3, G must contain a vertex of degree atmost three. We can proceed by induction on the number of vertices. For $|V| = 1, 2$, the result is true that $\chi(G) \leq 4$. Assume the result to be true for graphs with $|V| < n$. Let G be a graph on n vertices as with the property of the statement that it has no subgraph isomorphic to K_3. Let v be a vertex of degree ≤ 3. Clearly $G - \{v\}$ is such that $\chi(G - \{v\}) \leq 4$ by assumption, 'v' can be colored with one of the colors not used for its neighbours in $G - \{v\}$. This shows $\chi(G) \leq 4$.

(ii) This statement is not true. Consider the following graph.

Fig. 23.57 Graphs (34)

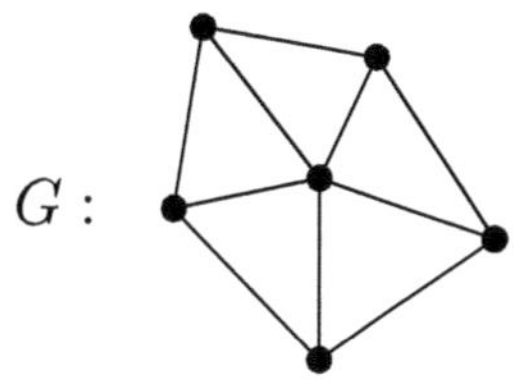

$\chi(G) \neq 3$ and G has no subgraph isomorphic to K_4 (Fig. 23.57).

35. Show that a plane graph $G = (V, E)$ is 2-connected if and only if every face of G is bounded by a cycle.

Solution: If every face of G is bounded by a cycle, then obviously G is 2-connected.

Assume G is 2-connected. To show that every face of G is a cycle. We adopt the method of contraction. That is let there be a face 'f' of G that is not bounded by a cycle. As G is given to be 2-connected, G must have atleast one of its faces bounded by a cycle. Let $\overline{G}$ be a maximal subgraph of G made up of faces bounded by cycles. None of the faces in $\overline{G}$ will have two non adjacent vertices that are being adjacent in G. This is becuse $\overline{G}$ is a maximal subgraph of G containing all the cycle faces of G. In fact $\overline{G} = (\overline{V}, \overline{E})$ is 2-connected. For any two vertices $u, v \in \overline{V}$. $x \in V \backslash \overline{V}$ there is a path between u and v passing through x (Harary's characterization). So there are two paths between any pair of vertices $u, v \in V$, one being a path present in $\overline{G}$ and the other being as seen in the previous statement. Hence the boundary of every face of G is a cycle.

36. If $G = (V, E)$ is a 3-connected graph, $|V| \geq 5$, then show that there is an edge $uv(= e) \in E$ such that $G.e$ is 3-connected. Here $G.e$ is the graph obtained from G by contracting the edge $e(= uv)$.

Solution: Suppose $G.e$ is not 3-connected for each edge $e \in E$. Consider, $e = uv$. Let w be the vertex in $G.e$ representing the contraction of e. Clearly $G.e$ is 2 connected as G is 3-connected. Also the subgraph $G = \{u, v\}$ must be connected, for any $u, v \in V$.

Let $u_1, u_2 \in V$ such that $G.e - \{u_1, u_2\}$ is not connected and u_1, u_2, w be distinct. The graph $G.e - \{u_1, u_2\}$ can be obtained from $G - \{u_1, u_2\}$ by contracting the edge $uv = e$

as $u \neq u_1$, $v \neq v_1$. This means $G.e - \{u_1, u_2\}$ is connected. This is a contradiction as $G.e - \{u_1, u_2\}$ is disconnected. Hence our assumption that u_1, u_2, w being distinct is not a valid one. Hence the contracted vertex w must be in any separting set of $G.e$ as either $u_1 = w$ or $v_1 = w$. So $G - \{u_1, v_1, w\}$ is not connected.

Now let $\overline{G}$ be component of $G - \{u_1, v_1, x\}$ with least number of vertices. The vertices u_1, v_1 and x must be adjacent with three distinct vertices of $\overline{G}$. Let p, q, r be vertices in $\overline{G}$ such that $\rho_1 = pu_2$, $\rho_2 = qv_1$, $\rho_3 = rx$. $\{u_1, u_2, x\}$ has to be a minimum disconnecing set of G. If one considers $G - \{p, q, x\}$ having G' as the minimum vertex component G' must be a proper subgraph of $\overline{G}$. Continuing like this one can reach a component with a single vertex and an edge joining to one of the three vertices whose removal disconnected the graph. Let one of the edges be say $\overline{e}$ and $G.\overline{e}$ is 3 connected, which is not possible.

37. Definition: A maximal outerplanar graph is one that loses its outer-planarity property if any two non-adjacent vertices are joined by an edge.

Question: Let $G = (V, E)$ be an outerplanar graph with $|V| = n$, $|E| = m$ and with 'f' number of faces. Prove the following.

(i) $m = 2n - 3$, $f = n - 1$
(ii) there is atleast 3 vertices whose degrees are <4
(iii) there is atleast 2 vertices of degree 2
(iv) $\kappa(G) = 2$.

Solution:

(i) Degree of outer face will be 'n' and degree of each interior face is 3. Hence

$$3(f - 1) + n = 2m \tag{23.6}$$
$$n - m + f = 2 \tag{23.7}$$
$$\Rightarrow f = n - 1, m = 2n - 3$$

(ii) Let d_i be the number of vertices of degree i.

$$\begin{aligned} d_1 &= 0 \\ 2d_2 + 3d_3 + \cdots = 2d_2 + 2d_3 + d_3 + 4d_4 + 4d_5 + d_5 + \cdots &= 2m \\ &= 4n - 6 \\ \Rightarrow 2d_2 + 2d_3 + 4(d_4 + d_5 + \cdots &\leq 4n - 6 \\ \text{But } d_1 + d_2 + d_3 + \cdots + d_n &= n \\ d_4 + d_5 + \cdots &= n - d_2 - d_3 \end{aligned}$$

So

$$2d_2 + 2d_3 + 4(d_4 + d_5 + \cdots) \leq 4n - 6$$
$$\Rightarrow 2d_2 + 2d_3 + 4(n - d_2 - d_3) \leq 4n - 6$$
$$\Rightarrow d_2 + d_3 \geq 3$$

(iii) If $d_2 = 0$, then

$$3d_3 + 4d_4 + \cdots = 4n - 6$$
$$\Rightarrow 3d_3 + 4(n - d_3) \leq 4n - 6$$
$$\Rightarrow d_3 \geq 6, \quad \text{which is a contraction.}$$
$$\text{Hence } d_2 \neq 0$$
$$\text{can } d_2 = 1$$

(iv) Clearly all the vertices are lying in the outer face. So every vertex is adjacent to two of its neighbouring vertices. This means $\kappa(G) > 1$. But there is atleast one vertex of degree 2. So $\kappa(G) < 3$. This means $\kappa(G) = 2$.

38. Show that, if $G = (V, E)$ is outer planar, then G contains atleast 2 vertices of degree <3.

Solution: Let $|V| \geq 5$ as the statement is true for all outer planar graphs with $|V| \leq 4$. If G is not maximal outer planar, add edges to G such that G becomes maximal outer planar graph say $G' = (V, E')$. This makes the outer face of G being bounded by a Hamilton cycle C of G. Let $u, v \in V$ such that $uv \in E'$. Let this chord uv of C along with edges on the path between u and v lying on C will be a face bounded by a cycle and this cycle will include minimum number of interior faces of G'. This number must be 1. Every vertex on this cycle will be of degree 2. Let $x \in V$, $x \neq u, v$ such that degree of x is 2. The other path between u, v on C will identify a vertex x' of degree 2 in a similar manner.

39. Definition: The crossing number of a non planar graph $G = (V, E)$ is the minimum number of crossover of its edges for all possible drawings of G in a plane. It is denoted by $cr(G)$.

Question: Let $G = (V, E)$ be a graph with $|V| \geq 3$. Then

$$cr(G) \geq |E| - 3\,V| + 6.$$

Solution: Let the drawing of G on the plane have t crossings such that 't' being minimum. Suppose we take each crossing point be a vertex then the drawing now becomes a plane graph H (say) such that H has $|V| + t$ vertices and $|E| + 2t$ edges. This is because each new vertex divides the edge into two new edges. We know, for a planar graph H,

$$|E| + 2t \le 3(|V| + t) - 6$$
$$\Rightarrow t \ge |E| - 3\,V| + 6.$$

40. Let $G = (V, E)$ be a graph. Show that

$$\chi(G) \ge \frac{|V|^2}{|V|^2 - 2|E|}.$$

Solution: Let $\chi(G) = k$ (say). Coloring the vertices of G will produce a partition of the vertices V of G such that $V = V_1 \cup V_2 \cup V_3 \cup \cdots \cup V_k$ such that the vertices in V_i are colored with same color i (say). Clearly one can see G as a k-partite graph where there are edges between the partitions preserving its chromatic number. The edges of G will be maximum when G is a complete k-partite graph with each V_i being equal to $\frac{|V|}{k}$, $1 \le i \le |V|$. Hence the edges between the partitions V_i and V_j will be $\left(\frac{|V|}{k}\right)^2$ for $i \ne j$. This means

$$|E| \le kC_2 \cdot \frac{|V|^2}{k^2} \quad \text{and}$$
$$2|E| \le \frac{(k-1)|V|^2}{k}$$

Also

$$|V|^2 - 2|E| \ge |V|^2\left(1 - \left(\frac{k-1}{k}\right)\right)$$
$$\Rightarrow \chi(G) \ge \frac{|V|^2}{|V|^2 - 2|E|}.$$

41. Prove that a graph $G = (V, E)$ is t-colorable if and only if every block of G is t-colorable.

Solution: Suppose G is t-colorable. Then it is true that every block of G is t-colorable.

Suppose every block of G is t-colorable. We prove that G is also t-colorable using the induction on the number of blocks. For one block, the result is true. Let G be with n blocks and let the result that G being t-colorable be true. Now consider G to have $(n + 1)$ blocks and that each block being t-colorable. Let the graph be with one pendant block 'B' (say) and let the remaining part H of G be with n blocks. So for H, the result that H being t-colorable is true. Also as a block, B is also t-colorable. Clearly B shares a unique vertex 'v' with H. If 'v' is assigned the same color in H and B, the result is true. Otherwise a recoloring of the vertices of B can be done to assign 'v' the same color as that in H. Hence G is t-colorable.

42. Show that the subgraph induced by any two color classes in a k-coloring of a graph G that is uniquely k-colorable must be connected.

Solution: A graph G is said to be uniquely k colorable if there is exactly one partition of its vertices into k-color classes.

Let K_1, K_2 be two color classes in the k-coloring of G that are colored with colors $1, 2$. Consider the induced subgraph G_{12} induced by the vertices in K_1 and K_2. Suppose G_{12} is not connected. Then every component of G_{12} must contain vertices that use both color 1 and 2. Suppose there is a component F in G_{12}, then interchange the colors 1, 2 in the vertices of F, we get a different coloring of G. This is not possible. Hence G_{12} is connected.

43. Let G be an undirected simple graph. Let A_G be the adjacency matrix of $G = (V, E)$. Then show that

(i) $\sum_{i=1}^{n} \lambda_i = 0$

(ii) $\sum_{i=1}^{n} \lambda_i^2 = 2|E|$

(iii) $\sum_{i=1}^{n} \lambda_i^3 = 6.t$
where t is the number of cycles of length 3 in G.

Solution:

(i) By the definition of trace of a matrix. We know that it will be zero as A_G has all its diagonal entries zero. This is always true for any simple graph. Also A_G is a symmetric matrix and hence it is diagonalizable. Then '$P^{-1}A_G P = D$' with the diagonal matrix D has all the eigen values of A_G. Hence

$$\sum_{i=1}^{n} \lambda_i = 0$$

as the property 'trace' being invariant under similarity transformation. Also in the characteristic polynomial of A_G, the coefficient of t^{n-1} will be $\sum_{i=1}^{n} \lambda_i$ where the graph G has 'n' vertices. Hence $\sum_{i=1}^{n} \lambda_i = 0$.

(ii) In the matrix A_G^2, any coefficient a_{ij}^2 represents the number of paths of length 2 between the vertex v_i and v_j for $V = \{v_1, v_2, \ldots, v_n\}$. This result can be proved by direct argument on constructing A_G^2 from A_G. So if one considers the trace of A_G^2, it will be sum of the vertex degrees of G. Hence

$$\sum_{i=1}^{n} \lambda_i^2 = 2|E|.$$

(iii) This result can be proved using the result that if $\overline{a}_{ij}$ is the ijth entry of A_G^H, then it represents number of paths between v_i and v_j of length n. One has to enumerate all $\overline{a}_{ii}$ in A_G^3 that occur on the diagonal.

44. Find the spectrum of Figs. 23.58 and 23.59.

Solution:

Fig. 23.58 Graphs (44-1)

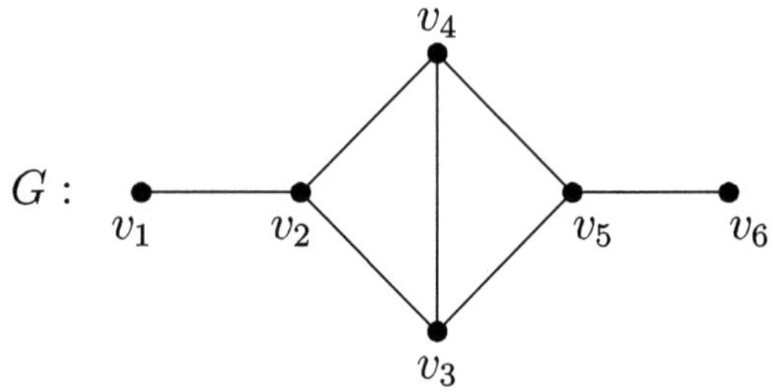

Fig. 23.59 Graphs (44-2)

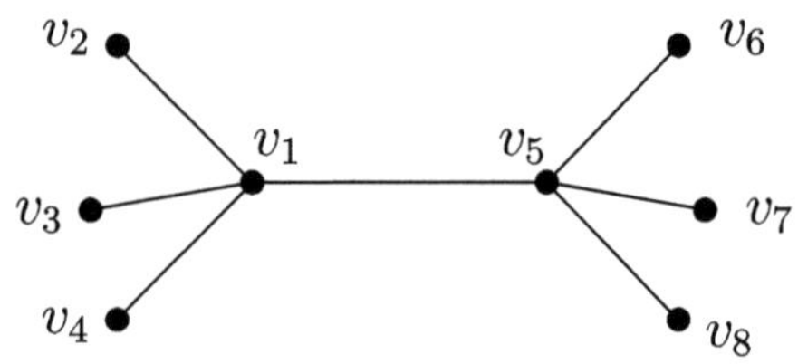

(i) $A_G = \begin{bmatrix} 0&1&0&0&0&0 \\ 1&0&1&1&0&0 \\ 0&1&0&1&1&0 \\ 0&1&1&0&1&0 \\ 0&0&1&1&0&1 \\ 0&0&0&0&1&0 \end{bmatrix}$

The characteristic polynomial will be

$$(\lambda^2 - 1)(\lambda^4 - 6\lambda^2 + 4\lambda + 1)$$

The spectrum will be

$$\begin{pmatrix} 1.90 & 1 & -0.19 & -1 & -2.70 \\ 1 & 2 & 1 & 1 & 1 \end{pmatrix}$$

(ii) The characteristic poly of the tree will be

$$\lambda^8 - 7\lambda^6 + 9\lambda^4.$$

The spectrum will be

$$\begin{pmatrix} 2.30 & 1.30 & 0 & -1.30 & -2.30 \\ 1 & 1 & 4 & 1 & 1 \end{pmatrix}$$

45. Definition: A graph G is said to be orientable if each edge of G can be given direction such that the resultant graph is strongly connected.

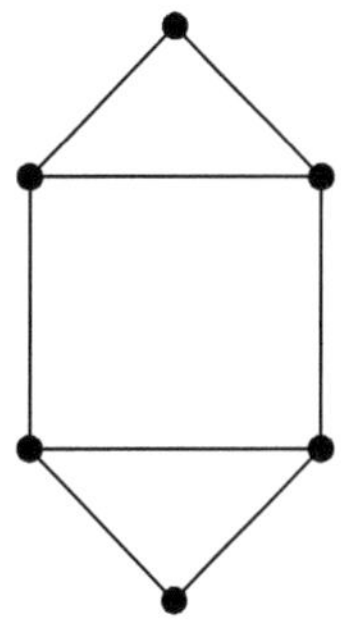

Fig. 23.60 Graphs (45-1)

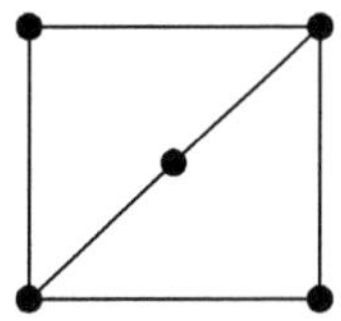

Fig. 23.61 Graphs (45-2)

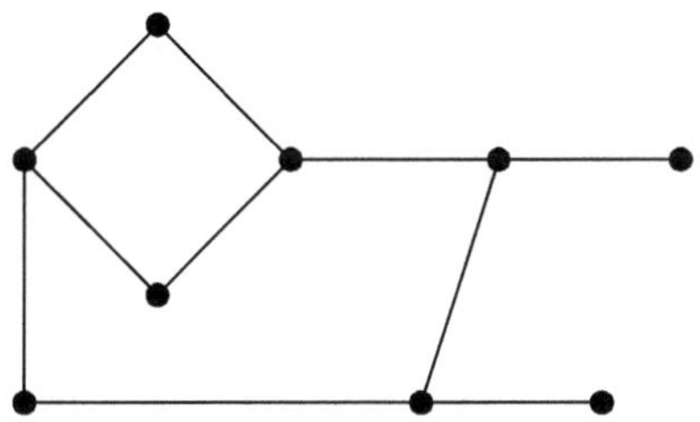

Fig. 23.62 Graphs (45-3)

Check whether the following graphs are orientable or not. If so provide a orientation (Fig. 23.60, 23.61 and 23.62).

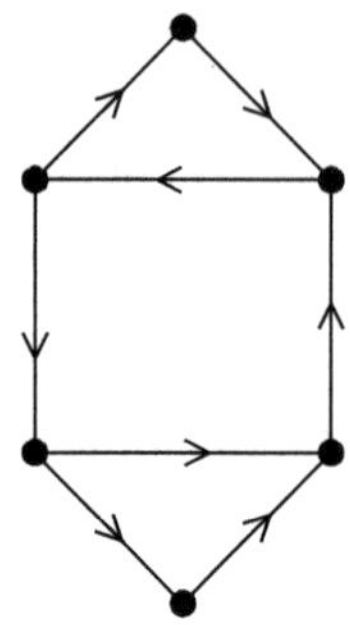

Fig. 23.63 Graphs (45-4)

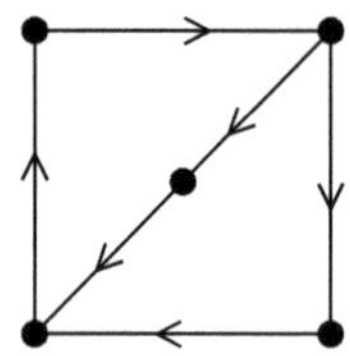

Fig. 23.64 Graphs (45-5)

Solution:

(iii) It is not possible to orient this graph. The necessary condition for a graph to be orientable is that every edge of it must be contained in a cycle (Fig. 23.63 and 23.64).

46. Show that a graph G is 2-connected if and only if it is strongly connected.

Solution: Let G be strongly connected. Suppose G has a cut edge and the edge be $e = uv$ i.e., Figure. It means there is only one way to move from u to v. That is G is not strongly connected. Hence every edge of G must lie on a cycle. So G is 2-connected.

Conversely let G be 2-connected. So every edge of G must lie on a cycle C. Direct all edges of C cyclically. If every every edge of G is lying on C then G is strongly connected. Suppose there is an edge e of G that does not lie on C, but e being adjacent to an edge lying on C. Since G is 2-connected e must lie on a cycle C' (say). Now provide orientation to the edges of C' cyclicly except those edges that already received orientation. Now it is clear to see that the edges of C and C' have orientation that makes the graph strongly connected. Continue the procedure until every edge of G is oriented.

Bibliography

M. H. Albert, M. D. Atkinson and V. Vatter, Counting 1324, 4231-avoiding permutations. Electronic J. Combin 16 (2009), # R136.

M. H. Albert, M. Elder, A. Rechnitzer, P. Westcott and M. Zabrocki, On the Stanley-Wilf limit of 4231-avoiding permutations and a conjecture of Arratia. Adv. in Appl. Math. 36 (2006), no. 2, 96–105.

R. Alter, Some remarks and results on Catalan numbers, in Proc. Second Louisiana Conference on Combinatorics, Graph Theory and Computing (1971) (R. C. Mullin et al., eds.) Congressum Numeratium III, Utilitas Mathematica Publishing Co., Winnipeg, 1971, 109–132.

N. Alon and E. Friedgut, On the number of permutations avoiding a given pattern. J. Combin. Theory Ser. A. 89 (2000), no. 1, 133–140.

I. Anderson, Perfect matchings of a graph. J. Combin. Theory Ser. B 10 (1971) 183–186.

K. Appel and W. Haken, Every planar map is four colorable. Bull. Amer. Math. Soc. 82 (1976) 711-712.

R. Arratia, On the Stanley-Wilf conjecture for the number of permutations avoiding a given pattern. Electronic J. Combin., 6 (1999), no. 1, N1.

V. Bafna and P. Pevzner, Sorting by transpositions. SIAM J. Discrete Math. 11 (1998), no. 2, 224–240 (electronic).

R. B. Bapat, Graphs and Matrices, Springer, University Text (2010).

C. Berge, Two theorems in graph theory. Proc. Nat. Acad. Sci. U.S.A. 43 (1957) 842–844.

C. Berge, Some classes of perfect graphs. Six Papers on Graph Theory. Indian Statistical Institute, Calcutta (1963) 1–21.

N. Biggs, Algebraic Graph Theory, Second edition, Cambridge University Press, Cambridge (1993).

N. L. Biggs, E. K. Lloyd and R. J. Wilson, Graph Theory, 1736-1936. Oxford University Press, London (1976).

M. Bóna, Introduction to Enumerative Combinatorics. McGraw-Hill, Boston, MA, 2007.

M. Bóna, A Walk Through Combinatorics, 3rd Edition. World Scientific, River Edge, NJ, 2011.

J. A. Bondy and V. Chvátal, A method in graph theory. Discrete Math. 15 (1976) 111–136.

J. A. Bondy and U. S. R. Murty, Graph Theory. Springer, New York (2008).

W. G. Brown and J. W. Tutte, On the enumeration of rooted non-separable planar maps, Canad. J. Math., 16 (1964) 572–577.

R. Rama, *Topics in Combinatorics and Graph Theory*,
https://doi.org/10.1007/978-3-031-74252-1

A. Burstein, A short proof for the number of permutations containing the pattern 321 exactly once. Elec. J. Combin. 18 (2001), no. 2, p21.

R. Chapman, An involution on derangements. 17th British combinatorial Conference (Canterbury, 1999). Discrete Math. 231 (2001), no. 1–3, 121–122.

C. Charalambides, Enumerative Combinatorics. Chapman & Hall / CRC, Boca Raton, FL, 2002.

G. Chartrand and F. Harary, Graphs with prescribed connectivities, Theory of Graphs. Academic Press, New York (1968) 61–63.

G. Chartrand and J. Mitchem, Graphical theorems of the Nordhaus Gaddum class. Recent Trends in Graph Theory. Springer, Berlin (1971) 55–61.

G. Chartrand and P. Zhang, Chromatic Graph Theory. Chapman & Hall/CRC Press, Boca Raton (2009).

V. Chvátal, Tree-complete graph Ramsey numbers. J. Graph Theory 1 (1977) 93.

V. Chvátal and P. Erdös, A note on Hamiltonian circuits. Discrete Math. 2 (1972) 111–113.

Claude Berge, Graphs and Hypergraphs. New York: Elsevier, 1973.

J. Cofman, Catalan numbers for the classroom? Elemente der Mathematik, 52 (1997), 108–117.

C. W. Curtis, Frobenius, Burnside, Schur and Brauer: Pioneers of Representation Theory, American Mathematical Society, Providence, to appear.

D. M. Cvetković and M. Doob, Recent results in the theory of graph spectra, Annals of Discrete Math., Elsevier, Amsterdam/New York (1988).

N. G. de Bruijn, Pólya's theory of counting. in Applied Combinatorial Mathematics (E. F. Beckenbach, ed.), Wiley, New York (1964); reprinted by Krieger, Malabar, Florida (1981).

J. Désarmenien and M. Wachs, Descent classes on permutations. J. Combin. Theory A, 64 (1993), no. 2, 311–328.

B. Descartes, A three colour problem. Eureka 9 (1947) 21.

R. Diestel, Graph Theory, Third Edition. Springer-Verlag, Berlin (2005).

G. A. Dirac, Some theorems on abstract graphs. Proc. London Math. Soc. 2 (1952) 69–81.

G. A. Dirac, A property of 4-chromatic graphs and some remarks on critical graphs. J. London Math. Soc. 27 (1952) 85-92.

S. Doty and G. Walker, Modular symmetric functions and irreducible modular representations of general linear groups, J. Pure Appl. Algebra 82 (1992) 1–26.

S. Dulueq. S. Gire and J. West, Permutations with forbidden subsequences and nonseparable planar maps. Discrete Math., 153 (1996), no. 1–3, 85–103.

P. H. Edelman, Inversions and cycles in permutations. Europ. J. Combinatorics, 8 (1987), 269–279.

E. Egerv́ary, On combinatorial properties of matrices (Hungarian). Mat. Lapok 38 (1931) 16–28.

S. Elizalde, Asymptotic enumeration of permutations avoiding generalized patterns. Adv. Appl. Math. 26 (2006), 138–155.

P. Erdös, Some remarks on the theory of graphs. Bull. Amer. Math. Soc. 53 (1947) 292–294.

P. Erdös, Extremal problems in graph theory. A Seminar on Graph Theory. Holt, Rinehart and Winston, New York (1967) 54–59.

P. Erdös and T. Gallai, Graphs with prescribed degrees of vertices (Hungarian). Mat. Lapok 11 (1960) 264–274.

Fede S. Roberts and Barry Tesman, Applied Combinatorics (second edition), CRC Press (2009).

H. J. Finck, On the chromatic numbers of a graph and its complement. Theory of Graphs. Academic Press, New York (1968) 99–113.

Frank Harary, Graph Theory. Reading, MA: Addison-Wesley, 1969.

P. Franklin, The four color problem. Amer. J. Math. 44 (1922) 225–236.

Gary Chartand, Linda Lesniak and Ping Zhang, Graphs and Digraphs (6th edition), CRC Press (2016).

George E. Andrews and Kimmo Eriksson, Integer Partitions, Cambridge, England: Cambridge University Press, 2004.

S. W. Golomb, How to number a graph. Graph Theory and Computing. Academic Press, New York (1972) 23–37.

J. Gross and J. Yellen (editors), Handbook of Graph Theory. CRC Press, Boca Raton, FL (2004).

R. P. Gupta, The chromatic index and the degree of a graph. Notices Amer. Math. Soc. 13 (1966) 719.

R. K Guy, Crossing numbers of graphs. Graph Theory and Applications. Springer-Verlag, New York (1972) 111–124.

F. Harary, The maximum connectivity of a graph. Proc. Nat. Acad. Sci. U.S.A. 48 (1962) 1142–1146.

F. Harary and C. St. J. A. Nash-Williams, On Eulerian and Hamiltonian graphs and line graphs. Canad. Math. Bull. 8 (1965) 701–709.

P. J. Heawood, On the four-colour map theorem. Quart. J. Pure Appl. Math. 29 (1898) 270–285.

P.J. Hilton and J. Pedersen, Catalan numbers, their generalization, and their uses, Math. Intelligencer, 13 (Spring 1991), 64–75.

B. Jackson, Hamilton cycles in regular 2-connected graphs. J. Combin. Theory Ser. B 29 (1980) 27–46.

J. B. Kelly and L. M. Kelly, Paths and circuits in critical graphs. Amer. J. Math. 76 (1954) 786–792.

T. P. Kirkman, On a problem in combinatorics. Cambridge and Dublin Math. J. 2 (1847) 191–204.

D. Kremer, Permutations with forbidden subsequences and a generalized Schröder number. Discrete Mathematics, 218 (2000), no. 1–3, 121–130.

L. Lesniak, Results on the edge-connectivity of graphs. Discrete Math. 8 (1974) 351–354.

C. L. Liu, Topics jn Combinatorial Mathematics. Washington, DC: Mathematical Association of America, 1972.

Louis Comtet, Advanced Combinattorics. Boston: Reidel, 1974.

L. Lovász, A characterization of perfect graphs. J. Combin. Theory Ser. B 13 (1972) 95–98.

Luo Jian-Jin, Catalan numbers in the history of mathematics in China, in Combinatorics and Graph Theory, Proc. Spring School and International Conference on Combinatorics, Hefei, 6–27 April 1992 (H. P. Yap et al., eds.), World Scientific, (1992) 68–70.

M. M. Matthews and D. P. Sumner, Hamiltonian results in $K_{1,3}$-free graphs. J. Graph Theory 8 (1984) 139–146.

Miklős Bona, Combinatorics of Permutations. Boca Raton, FL: Chapman & Hall/CRC 2004.

M. Molloy and B. Reed, A bound on the total chromatic number. Combinatorica 18 (1998) 241–280.

J. Noonan and D. Zeilberger, The enumeration of permutations with a prescribed number of forbidden patterns. Adv. in Appl. Math., 17 (1996), no. 4, 381–407.

O. Ore, Note on Hamilton circuits. Amer. Math. Monthly 67 (1960) 55.

O. Ore, Theory of Graphs. Amer. Math. Soc. Colloq. Pub., Providence, RI (1962).

O. Ore, Hamilton connected graphs. J. Math. Pures Appl. 42 (1963) 21–27.

S. Pan and R. B. Richter, The crossing number of K_{11} is 100. J. Graph Theory 56 (2007) 128–134.

K. R. Parthasarathy, Basic Graph Theory. Tata McGraw Hill Pub. (1994).

F. P. Ramsey, The Foundations of Mathematics, and Other Logical Essays. Harcourt, Brace and Company, New York (1931).

R. C. Read, An introduction to chromatic polynomials. J. Combin. Theory 4 (1968) 52–71.

B. Reed, ω, Δ, and χ. J. Graph Theory 27 (1998) 177–212.

A. Reifegerste, On the diagram of Schröder permutations. Electronic J. Combin., 9 (2003), no. 2, R8.

A. Reifegerste, On the diagram of 132-avoiding permutations. European J. Combin., 24 (2003), no. 6, 759–776.

J. Remmel, A note on a recursion for the number of derangements. European J. Combin., 4 (1984), no. 4, 371–374.

Richard A. Brualdi and Herbert J. Ryser, Combinatorial Matrix Theory. New York: Cambridge University Press, 1991.

R. D. Ringeisen and L. W Beineke, The crossing number of $C_3 \times C_n$. J. Combin. Theory Ser. B 24 (1978) 134–136.

G. Ringel, Problem 25. Theory of Graphs and its Applications. Nakl. CSAV, Prague (1964) 162.

D. G. Rogers, A Schröder triangle: Three combinatorial problems, in Combinatorial Mathematics V: Proceedings of the Fifth Australian Conference, Lecture Notes in Math. 622, Springer-Verlag, Berlin/Heidelberg/New York (1977) 175–196.

D. P. Sanders and Y. Zhao, Planar graphs of maximum degree seven are class I. J. Combin. Theory Ser. B 83 (2001) 201–212.

M. Sekanina, On an ordering of the set of vertices of a connected graph. Spisy Přír Fak. Univ. Brno (1960) 137–141.

R. Stanley, Binomial posets, Möbius inversion, and permutation enumeration, J. Combinatorial Theory Ser. A, 20 (1976), 336–356.

R. Stanley, Enumerative Combinatorics, Volume 1, Third Edition, Cambridge University Press, Cambridge UK, 2011.

R. Stanley, Enumerative Combinatorics, Volume 2, Cambridge University Press, Cambridge UK, 1999.

G. Szekeres and H. S. Wilf, An inequality for the chromatic number of a graph. J. Combin. Theory 4 (1968) 1–3.

P. G. Tait, Remarks on the colouring of maps. Proc. Royal Soc. Edinburgh 10 (1880) 729.

C. Thomassen, Some remarks on Hajós' conjecture. J. Combin. Theory Ser. B 93 (2005) 95–105.

S. Toida, Properties of an Euler graph. J. Franklin Inst. 295 (1973) 343–345.

P. Turán, Eine Extremalaufgabe aus der Graphentheorie. Mat. Fiz. Lapok 48 (1941) 436–452.

P. Turán, A note of welcome. J. Graph Theory 1 (1977) 7–9.

W. T. Tutte, On Hamiltonian circuits. J. London Math. Soc. 21 (1946) 98–101.

W. T. Tutte, A theorem of planar graphs. Trans. Amer. Math. Soc. 82 (1956) 99–116.

V. G. Vizing, On an estimate of the chromatic class of a p-graph. (Russian) Diskret. Analiz. 3 (1964) 25–30.

V. G. Vizing, Critical graphs with given chromatic class. Metody Diskret. Analiz. 5 (1965) 9–17.

I. Vun and P. Belcher, Catalan numbers, Math. Spectrum, 30 (1997/98), 3–5.

D. B. West, Introduction to Graph Theory, Second Edition. Prentice-Hall, Upper Saddle River, NJ (2001).

A. T. White, Graphs of Groups on Surfaces, Interactions and Models. North-Holland, Amsterdam (2001).

H. Whitney, The coloring of graphs. Ann. of Math. 33 (1932) 688–718.

J. E. Williamson, Panconnected graphs. II. Period. Math. Hungar. 8 (1977) 105–116.

R. M. Wilson, Decompositions of complete graphs into subgraphs isomorphic to a given graph. Congr. Numer. 15 (1976) 647–659.

R. Wilson, Four Colors Suffice: How the Map Problem Was Solved. Princeton University Press, Princeton, NJ (2002).

Index

R. Rama, *Topics in Combinatorics and Graph Theory*,
https://doi.org/10.1007/978-3-031-74252-1

D

MIX
Papier aus verantwortungsvollen Quellen
Paper from responsible sources
FSC® C105338

If you have any concerns about our products,
you can contact us on
ProductSafety@springernature.com

In case Publisher is established outside the EU,
the EU authorized representative is:
Springer Nature Customer Service Center GmbH
Europaplatz 3, 69115 Heidelberg, Germany

Printed by Libri Plureos GmbH
in Hamburg, Germany